Springer Proceedings in Mathematics & Statistics

Volume 264

Springer Proceedings in Mathematics & Statistics

This book series features volumes composed of selected contributions from workshops and conferences in all areas of current research in mathematics and statistics, including operation research and optimization. In addition to an overall evaluation of the interest, scientific quality, and timeliness of each proposal at the hands of the publisher, individual contributions are all refereed to the high quality standards of leading journals in the field. Thus, this series provides the research community with well-edited, authoritative reports on developments in the most exciting areas of mathematical and statistical research today.

More information about this series at http://www.springer.com/series/10533

Zair Ibragimov · Norman Levenberg
Utkir Rozikov · Azimbay Sadullaev
Editors

Algebra, Complex Analysis, and Pluripotential Theory

2 USUZCAMP, Urgench, Uzbekistan, August 8–12, 2017

 Springer

Editors
Zair Ibragimov
Department of Mathematics
California State University
Fullerton, CA, USA

Norman Levenberg
Department of Mathematics
Indiana University
Bloomington, IN, USA

Utkir Rozikov
Institute of Mathematics
Uzbekistan Academy of Sciences
Tashkent, Uzbekistan

Azimbay Sadullaev
Department of Mathematics
National University of Uzbekistan
Tashkent, Uzbekistan

ISSN 2194-1009   ISSN 2194-1017   (electronic)
Springer Proceedings in Mathematics & Statistics
ISBN 978-3-030-13163-0   ISBN 978-3-030-01144-4   (eBook)
https://doi.org/10.1007/978-3-030-01144-4

Mathematics Subject Classification (2010): 17A32, 31B15, 32U30, 37K15, 41A17, 46L57, 46S10, 60J10

This Springer imprint is published by the registered company Springer Nature Switzerland AG
The registered company address is: Gewerbestrasse 11, 6330 Cham, Switzerland

Preface

The second USA-Uzbekistan Conference on Analysis and Mathematical Physics was held on August 8–12, 2017, at Urgench State University, Uzbekistan. The conference aimed to stimulate interactions among US mathematicians and their counterparts in Uzbekistan and other countries, and to serve as a catalyst for future collaborations. The main themes discussed were algebra and functional analysis; dynamical systems; mathematical physics and partial differential equations; probability theory and mathematical statistics; and pluripotential theory. A number of significant results were recently established in these areas; these were disseminated in our conference through the plenary talks. The invited talks in parallel sessions allowed for the presentation of a broad spectrum of further results. In addition, poster presentations afforded more junior mathematicians and students the opportunity to speak on their work.

This volume contains papers which were presented in the special sessions on algebra and functional analysis, complex analysis, and pluripotential theory. The research articles are devoted to topics such as slow convergence, spectral expansion, holomorphic extension, m-subharmonic functions; pseudo-Galilean groups; involutive algebras; log-integrable measurable functions; Gibbs measures; harmonic and analytic functions; local automorphisms; Lie algebras; and Leibniz algebras. Some extensive survey papers related to topics of the volume are also included.

Readers of the volume are anticipated to be graduate students and research mathematicians interested in functional analysis, complex analysis, operator algebras, and non-associative algebras.

Fullerton, USA   Zair Ibragimov
Bloomington, USA   Norman Levenberg
Tashkent, Uzbekistan   Utkir Rozikov
Tashkent, Uzbekistan   Azimbay Sadullaev

v

Acknowledgements

We thank all contributing authors and referees for their efforts. We also thank Springer for the opportunity to publish this volume.

Contents

Removable Singular Sets
of m-Subharmonic Functions

B. I. Abdullaev, S. A. Imomkulov and R. A. Sharipov

Abstract In this article we consider the removable singularities of $m - sh$ functions. We prove a few theorems on removable sets in terms of capacities and Hausdorff measure.

Keywords Subharmonic function · m-subharmonic function · Hausdorff measure · Polar sets · Capacity

1 Introduction

In this work, we study removable singular sets for the certain classes of subharmonic and $m-$subharmonic functions in the domain D of complex space $\mathbb{C}^n$.

Definition 1 The function $u(z) \in L^1_{loc}(D)$ given in the domain $D \subset \mathbb{C}^n$ is called $m - sh$ function in D (subharmonic function on m-dimensional complex planes, $1 \leq m \leq n$), if

1. it is upper semi-continuous in D, that is

$$\overline{\lim_{z \to z^0}} u(z) = \lim_{\varepsilon \to 0} \sup_{B(z^0, \varepsilon)} u(z) \leq u(z^0);$$

2. for any m-dimensional complex plane $\Pi \subset \mathbb{C}^n$ the restriction $u|_\Pi$ is subharmonic (sh) function on $\Pi \cap D$.

B. I. Abdullaev · R. A. Sharipov (✉)
Urgench State University, Urgench, Uzbekistan
e-mail: sharipovr80@mail.ru

B. I. Abdullaev
e-mail: abakhrom1968@mail.ru

S. A. Imomkulov
Navoi State Pedagogical Institute, Navoi, Uzbekistan
e-mail: sevdiyor_i@mail.ru

© Springer Nature Switzerland AG 2018
Z. Ibragimov et al. (eds.), *Algebra, Complex Analysis, and Pluripotential Theory*,
Springer Proceedings in Mathematics & Statistics 264,
https://doi.org/10.1007/978-3-030-01144-4_1

1

Note that the *upper semi-continuous function u(z) on D is m − sh function if and only if the current*

$$dd^c u \wedge \left(dd^c \, |z|^2\right)^{m-1} \geq 0 \text{ on } D, \text{ i.e.}$$

$$dd^c u \wedge \left(dd^c \, |z|^2\right)^{m-1} (\omega) = \int u \left(dd^c \, |z|^2\right)^{m-1} \wedge dd^c \omega \geq 0, \qquad (1)$$

for any $\omega \in F^{(n-m,n-m)}$, $\omega \geq 0$. Here the space $F^{(n-m,n-m)}$ is finite and C^∞- smooth on D differential forms of bidegree $(n-m, n-m)$.

The class of such functions are considered in works of Z. Khusanov [13, 14], R. Harvey and B. Lausson [10], M. Verbitsky [27], D. Joyce [12] and others. In works [1, 2] first author has proved a series of potential properties of $m-$ subharmonic functions.

The class of $m - sh$ functions is wider than the class of plurisubharmonic functions, but it strictly contains in the class of subharmonic functions. In addition, the class of $1 - sh$ functions coincides with the class of subharmonic functions and the class of $n - sh$ functions coincides with the class of subharmonic functions.

2 Preliminary Results on Removable Singularities

Definition 2 A closed subset E of a domain D is called *removable singular set* for some class of functions $\Im$, if for each function $u(x) \in \Im (D \backslash E)$ there exists a function $\tilde{u}(x) \in \Im (D)$ such that $\tilde{u}(x) = u(x)$ for any $x \in D \backslash E$.

For example, the polar set $E \subset D$ is removable singular set for functions, $m-$ subharmonic in $D \backslash E$ and locally bounded from above in D.

For the class of subharmonic functions (case $m = n$) and plurisubharmonic (case $m = 1$), there are sufficiently complete results on removable singular sets. In terms of a Hausdorff measure, the condition for removing a singular set of harmonic functions of Hölder class Lip_α was studied in the works of L. Carleson [5] (for $0 < \alpha \leq 1$) and E.P. Dolzhenko [7, 8] (for $0 < \alpha \leq 2$), R. Harvey and J. Polking [9], A. Dauzhanov [6]. Singular sets of subharmonic functions from the Lip_α class were studied in the work of V.L. Shapiro [25] (case $(0 < \alpha < 1)$) and in the work of A. Sadullaev and Zn. Yarmetov [21] (case $(1 \leq \alpha \leq 2)$).

Theorem 1 *Let E be a closed set in a domain $D \subset \mathbb{R}^n$ with the Hausdorff measure*

$$H_{n-2+\alpha} = 0, \quad 0 < \alpha \leq 2,$$

then any subharmonic function in $D \backslash E$ from the class $Lip_\alpha(D)$ subharmonically extends into the domain D.

In the works of A. Sadullaev and B. Abdullaev (see [20, 22, 23]) a number of theorems have been proved about removable singular sets for the bounded from above class $m - sh\,(D\backslash E)$ and for the class $Lip_\alpha\,(D) \cap m - sh\,(D\backslash E)$ of functions. Since the methods of proving these results are directly related to the problems studied here, we provide these results.

Theorem 2 (A. Sadullaev, B. Abdullaev [20]) *If a closed in D set E is polar, i.e. its newtonian capacity*

$$C(E) = 0,$$

then any function $u \in m - sh\,(D\backslash E)$ bounded from above is $m-$ subharmonically extends into the domain D, i.e. there is a function $w \in m - sh(D)$ such that $w|_{D\backslash E} \equiv u$.

In the proof of Theorem 2 used upper estimation for the capacities of $E \cap \Pi$, dim $\Pi = k$, which were obtained in the work of A. Sadullaev [24] (see also [15]): let E be a closed polar set in the space $\mathbb{R}^n(x) \times \mathbb{R}^k(y)$, $n \geq 1, k \geq 2$. Then for almost all (with respect to Lebesgue measure) $x^0 \in \mathbb{R}^n(x)$ the intersection $E \cap \{x = x^0\}$ is a polar set in space $\mathbb{R}^k(y)$. Note that instead of the family of parallel planes, one can consider other families of planes, for example, a pencil of planes passing through some fixed plane.

Since $m - sh(D) \subset sh(D)$, then the function $u(z)$, considered in the theorem is subharmonically extends into D (see, for example [11]). This extended function again we denote by $u(z)$. For almost all m–dimensional complex planes $\Pi \subset \mathbb{C}^n$ the intersection $E \cap \Pi$ is a polar set and for such planes the restriction $u|_\Pi$ subharmonically continues to $D \cap \Pi$. Since the composite set of these planes is dense in the space of all m-dimensional complex planes, then for all such planes $\Pi \subset \mathbb{C}^n$, restrictions $u|_\Pi$ are subharmonic functions. Consequently, it follows that $u(z) \in m - sh(D)$.

Theorem 3 (A. Sadullaev, B. Abdullaev [22]) *If the Hausdorff measure*

$$H_{2n-2+\alpha}(E) = 0,$$

where $0 < \alpha \leq 2$, then any function $u \in Lip_\alpha(D) \cap m - sh(D\backslash E)$ continues $m-$ subharmonically into the domain D, that is, $\exists w \in m - sh(D) : w|_{D\backslash E} \equiv u$.

In the proof of this theorem, authors used the following theorem of W. Schiffman [26] (see also [15, 24]): if $E \subset \mathbb{R}^n(x) \times \mathbb{R}^k(y)$, $n \geq 1, k \geq 1$, $H_{n+\alpha}(E) = 0, 0 \leq \alpha \leq k$, then for almost all $x^0 \in \mathbb{R}^n$ (x) the intersection $E \cap \Pi_{x^0}$, where $\Pi_{x^0} = \{x = x^0\}$, has a zero α–Hausdorff measure, $H_\alpha(E \cap \Pi_{x^0}) = 0$ in $\mathbb{R}^k$ (y).

Applying this theorem to $E \subset \mathbb{C}^n = \mathbb{C}^{(n-m)+m} \approx \mathbb{R}^{2(n-m)+2m}$, we obtain that for almost all $z' = (z_1, z_2, \ldots, z_{n-m}) \in \mathbb{C}^{n-m}$ the intersection $E \cap \Pi_{z'}$ has the zero $(2m - 2 + \alpha)$ Hausdorff measure. (The case $m = n$ is trivial. In this case $\Pi_{z'} \approx \mathbb{C}^n$ and $E \cap \Pi_{z'} = E$.)

From the definition of $m - sh$ functions, the restriction $u|_\Pi$ is subharmonic in $(D\backslash E) \cap \Pi_{z'}$ and belongs to the class $Lip_\alpha(D \cap \Pi_{z'})$. Hence, for the planes $\Pi_{z'}$ for which $H_{2m-2+\alpha}(E \cap \Pi_{z'}) = 0$ the restriction $u|_\Pi$ is subharmonic in $D \cap \Pi_{z'}$, because of the Theorem 1 stated above, since $\dim_\mathbb{C} \Pi_{z'} = m$.

Since, the set of such planes has full measure then, they are dense everywhere. From the fact that the function u belongs to the class Lip_α and is continuous in the domain D, it follows that the restriction $u|_\Pi$ is subharmonic in $D \cap \Pi_{z'}$ for all planes $\Pi_{z'}$. Considering unitary transformations of the space $\mathbb{C}^n$, we get that the restriction $u|_\Pi \in sh(D \cap \Pi)$ for any $\Pi \subset \mathbb{C}^n$, where dim $\Pi = m$. This means that $u \in m - sh(D)$.

3 $(n - s, q)$-Capacity

Below, we study removable singular sets of $m-$subharmonic functions that have certain smoothness. In order to formulate the main results, we need to introduce a quantity, $C_{n-s,q}$- capacity, $1 < q \leq \frac{n}{s}$. Consider in space $\mathbb{R}^n$ the kernel $K_s(x) = 1/|x|^{n-s}$, when $0 < s < n$ and the kernel $K_s(x) = \ln|x|$, when $s = n$. For a positive Borel measure μ, as usual, we define the Riesz potential:

$$U_{n-s}^\mu(x) = \int K_s(x - y) \, d\mu(y).$$

A capacitive quantity arbitrary compact set $E \subset \mathbb{R}^n$, is defined as follows

Definition 3 (*see* [18])

$$C_{n-s,q}(E) = \sup \mu(E), \quad 1 < q < +\infty,$$

where the upper bound is taken all over positive Borel measures concentrated on the set E and satisfying the condition

$$\left\| U_{n-s}^\mu(x) \right\|_p = \left[\int \left| U_{n-s}^\mu(x) \right|^p dx \right]^{\frac{1}{p}} \leq 1, \qquad \frac{1}{p} + \frac{1}{q} = 1. \qquad (2)$$

For $p > \frac{n}{n-s}$, that is $qs < n$ the integral

$$\int\limits_{|x| \geq 1} \left| U_{n-s}^\mu(x) \right|^p dx$$

exists. However, when $p = \frac{n}{n-s}$, i.e. $qs = n$, in the definition of the capacity $C_{n-s,q}$, which we mentioned above there is inconvenience that for such s a potential $U_{n-s}^\mu(x)$ may not belong to the class $L_p(\mathbb{R}^n)$ and the integral in (2), is not defined in general, because of the behavior of the kernel $K_s(x)$ in a neighborhood of infinity. In this case, we may assume that $E \subset B(0, 1)$ and define the capacity only for such sets:

$$C_{n-s,q}(E) = \sup \mu(E), \quad 1 < q < +\infty,$$

where, now the upper bound is taken over all positive Borel measures concentrated on the set E and satisfying condition

$$\left\| U_{n-s}^{\mu}(x) \right\|_p = \left[\int_{B(0,1)} \left| U_{n-s}^{\mu}(x) \right|^p dx \right]^{\frac{1}{p}} \leq 1, \quad p = \frac{n}{n-s}.$$

We provide a following equivalent definition of $(n-s, q)$-capacity (see [16, 17]).

Definition 4 Let E be a compact set in $\mathbb{R}^n$, and let

$$\Phi(E) = \left\{ \varphi \in C_0^{\infty}(\mathbb{R}^n) : \ \varphi(x) > 1 \ \text{ for any } \ x \in E \right\}.$$

We call $(n-s, q)$- capacity of the set E, a number

$$\gamma_{n-s,q}(E) = \inf \left\{ \|\varphi\|_{q,s} : \ \varphi \in \Phi(E) \right\}.$$

Here

$$\|\varphi\|_{q,s} = \left(\int_{\mathbb{R}^n} |\nabla_x^s \varphi(x)|^q dx \right)^{\frac{1}{q}} =$$

$$= \left\{ \int_{\mathbb{R}^n} \left[\sum_{\alpha_1+\alpha_2+\cdots+\alpha_n=s} \frac{s!}{\alpha_1!\alpha_2!\ldots\alpha_n!} \left(\frac{\partial^s \varphi}{\partial x_1^{\alpha_1} \partial x_2^{\alpha_2} \ldots \partial x_n^{\alpha_n}} \right)^2 \right]^{\frac{q}{2}} dx \right\}^{\frac{1}{q}}.$$

We note that there are constants $0 < A_1 < A_2$ depending only on n, s and q, the following estimation holds:

$$A_1 C_{n-s,q}(E) \leq \gamma_{n-s,q}(E) \leq A_2 C_{n-s,q}(E), \quad 1 < q \leq \frac{n}{s}. \tag{3}$$

$C_{n-s,q}$-capacity has the following metric properties (see [16]):
(a) if $qs < n, 0 < \alpha < n - qs$ and $H_{\alpha}(E) = 0$, then $C_{n-s,q}(E) = 0$;
(b) if $qs < n, n - qs < \alpha$ and $H_{\alpha}(E) > 0$, then $C_{n-s,q}(E) > 0$;
(c) if $qs = n, \varphi(r) = |\ln r|^{1-q}, q > 1$ and $H_{\varphi}(E) < \infty$, then $C_{n-s,q}(E) = 0$;
(d) if $qs = n, \alpha > 0$ and $H_{\alpha}(E) > 0$, then $C_{n-s,q}(E) > 0$.
It follows that the dimension the set of zero $C_{n-s,q}$-capacity is not greater than $n - qs$.

In the classical case $n > 2, 0 < s \leq m$, P. Matthila [15] (see also [24]) obtained the following result: *let E be a compact set in $\mathbb{R}^{n+m} = \mathbb{R}_x^n \times \mathbb{R}_y^m$. If the newtonian capacity $C_{n+m-s}(E) = 0, 0 < s \leq m$, defined by the Riesz kernel $K(x) = |x|^{s-m-n}$ is zero, then for almost all $a \in \mathbb{R}_x^n$ intersection $E \cap \{x = a\} \subset \mathbb{R}_y^m$ has a newtonian* *capacity zero in the plane* $\{x = a\} \in \mathbb{R}_y^m$: $C_{m-s}(E \cap \{x = a\}) = 0$ (*if* $m = s$, *then* C_0-*capacity in* $\mathbb{R}_y^m$ *by definition is logarithmic capacity*).

We provide more general theorem.

Theorem 4 *Let* $E \subset \mathbb{R}^{n+m}$ *be a compact set and let*

$$C_{n+m-s,q}(E) = 0, \quad \left(0 < s < m, 1 < q \le \frac{m}{s}\right).$$

Then, for almost all $a \in \mathbb{R}_x^n$, *the intersection* $E \cap \{x = a\} \subset \mathbb{R}_y^m$ *has zero* $(m - s, q)$-*capacity, i.e*

$$C_{m-s,q}(E \cap \{x = a\}) = 0.$$

The proof of Theorem 4 is based on the definition 4 of $(n - s, q)$-capacity.

Proof Let $E \subset \mathbb{R}^{n+m} = \mathbb{R}_x^n \times \mathbb{R}_y^m$ is a compact set and

$$\Phi(E) = \left\{\varphi \in C_0^\infty(\mathbb{R}^{n+m}) : \varphi(x, y) > 1, \ (x, y) \in E\right\}.$$

According to the Whitney's theorem (see [28]), for any plane $\Pi_a = \{x = a\}$, such that $E \cap \Pi_a \ne \emptyset$, the set of restrictions $\varphi|_{\Pi_a} = \{\psi(y) = \varphi(a, y), \varphi \in \Phi(E)\}$ of functions $\varphi(x, y) \in \Phi(E)$ coincides with the set of functions

$$\Phi(E \cap \Pi_a) = \left\{\psi(y) \in C_0^\infty(\mathbb{R}^m) : \psi(y) > 1, \ y \in E \cap \Pi_a\right\}.$$

Obviously,

$$\left(\int\limits_{\mathbb{R}^n} \left(\int\limits_{\mathbb{R}^m} |\nabla_y^s \varphi(x, y)|^q \, dy\right) dx\right)^{\frac{1}{q}} \le \left(\int\limits_{\mathbb{R}^n} \int\limits_{\mathbb{R}^m} |\nabla_{x,y}^s \varphi(x, y)|^q \, dx dy\right)^{\frac{1}{q}}$$

and

$$\left(\int\limits_{\mathbb{R}^n} \inf_\varphi \left(\int\limits_{\mathbb{R}^m} |\nabla_y^s \varphi(x, y)|^q \, dy\right) dx\right)^{\frac{1}{q}} \le \inf_\varphi \left(\int\limits_{\mathbb{R}^n} \left(\int\limits_{\mathbb{R}^m} |\nabla_y^s \varphi(x, y)|^q \, dy\right) dx\right)^{\frac{1}{q}} \le$$

$$\le \inf_\varphi \left(\int\limits_{\mathbb{R}^n} \int\limits_{\mathbb{R}^m} |\nabla_{x,y}^s \varphi(x, y)|^q \, dx dy\right)^{\frac{1}{q}}.$$

Consequently, and by the Definition 4, we get an inequality

$$\left(\int\limits_{\mathbb{R}^n} [\gamma_{m-s,q}(E \cap \Pi_x)]^q \, dx\right)^{\frac{1}{q}} \leq \gamma_{n+m-s,q}(E).$$

According to the conditions of the theorem and the estimation (3) we have

$$\int\limits_{\mathbb{R}^n} [\gamma_{m-s,q}(E \cap \Pi_x)]^q \, dx = 0.$$

Consequently, for almost all $x \in \mathbb{R}^n$ the value $\gamma_{m-s,q}(E \cap \Pi_x) = 0$ and, hence

$$C_{m-s,q}(E \cap \Pi_x) = 0.$$

The proof of the Theorem 4 is complete. $\qquad\qquad\qquad\qquad\qquad\qquad\square$

4 Main Results

The main results of this paper are Theorems 5 and 6 which we provide below.

Theorem 5 *Let F be a compact subset of the domain $D \subset \mathbb{C}^n$, $n \geq 2$, if a following equality holds*

$$C_{2n-2,q}(F) = 0,$$

where

$$q = \frac{p}{p-1} \ \text{and} \ m > 1, \ \frac{2m}{2m-2} \leq p < +\infty,$$

then it is removable for a functions $u \in m - sh\,(D \backslash F) \cap L_{p,loc}(D)$.

Theorem 6 *Let F be a compact subset of the domain $D \subset \mathbb{C}^n$, $n \geq 2$, if a following equality holds*

$$C_{2n-1,q}(F) = 0,$$

where

$$q = \frac{p}{p-1} \ \text{and} \ m \geq 1, \ \frac{2m}{2m-1} \leq p < +\infty,$$

then it is removable for a functions $u \in m - sh\,(D \backslash F) \cap L^1_{p,loc}(D)$.

In the proofs of these theorems we will use the Theorems 4, 7 and 8 which will be provided below on the properties of removable singular sets of subharmonic $(n - sh)$ functions in the class of L_p and L^1_p. Here $L^k_p(D)(k$ is a fixed entire number) denotes the class of functions having all derivatives up to k, furthermore k-order derivatives belonging to $L_p(D)$.

Theorem 7 (see [3]) *A closed set $E \subset D \subset \mathbb{R}^n$, $n > 2$, is removable for a class $L_{p,loc}(D) \cap sh(D \backslash E)$, $\frac{n}{n-2} \le p < \infty$, if and only if the capacity*

$$C_{n-2,q}(E) = 0, \quad q = \frac{p}{p-1}.$$

Sketch of proof. (1) First we assume that there exists a compact set E, which is removable for all subharmonic functions in $D \backslash E$ of the class L_p, $\frac{n}{n-2} \le p < +\infty$ and $C_{n-2,q}(E) > 0$, $q = \frac{p}{p-1}$. Then, by the definition of the capacity $C_{n-2,q}$, there exists a positive Borel measure, such that $\text{supp}\mu \subset E$, $\mu(E) > 0$ and the potential

$$u(x) = \int \frac{d\mu(y)}{|x-y|^{n-2}}$$

belongs to the class L_p. The function $u(x)$ is harmonic, and consequently subharmonic outside of E. By the assumption, it extends subharmonically into E. However, it is not subharmonic in D, since $-u(x)$ is a subharmonic function. We came to the contradiction, i.e., $C_{n-2,q}(E) = 0$.

(2) Now suppose that $C_{n-2,q}(E) = 0$ and $u(x)$ is subharmonic function in $D \backslash E$, and belong to $L_p(D)$, $\frac{1}{p} + \frac{1}{q} = 1$.

Further we use the following statement (see [19]): *let E be a compact subset of the space $\mathbb{R}^n$, and $D \supset E$ is a neighborhood of the set E. Then for a fixed integer k the following statements are equivalent:*

(1) $C_{k,q}(E) = 0$;

(2) *The set of test-functions $C_0^\infty(D \backslash E)$ is dense in the set of test-functions $C_0^\infty(D)$ with respect to the norm L_p^k.*

If now $\psi(x)$ is a positive test-function $\text{supp}\,\psi \Subset D$, then according to this statement there is a sequence of positive basic functions $\varphi_j(x)$, $\text{supp}\,\varphi_j \Subset D \backslash E$ and converges to $\psi(x)$ with respect to the norm L_p^k. Then

$$\left| \int u(x) \Delta(\psi - \varphi_j) dx \right| \le C \|u\|_{L_p} \|\psi - \varphi_j\|_{L_q^2},$$

where $C = const$. From this inequality it follows that

$$\int_D u \Delta \psi dx = \lim_{j \to \infty} \int_{D \backslash E} u \Delta \varphi_j dx \ge 0,$$

for any positive basic test-function $\psi(x) \in C_0^\infty(D)$. Therefore, the function $u(x)$ subharmonically extends in whole D. *The Theorem 7 is proved.* $\square$

Theorem 8 *A closed set $E \subset D \subset \mathbb{R}^n$, $n \ge 2$, is removable for a class $L_{p,loc}^1(D) \cap sh(D \backslash E)$, $\frac{n}{n-1} \le p < +\infty$, if and only if*

$$C_{n-1,q}(E) = 0, \quad q = \frac{p}{p-1}.$$

The proof of Theorem 8 is analogous to the proof of Theorem 7, and it is also carried out mainly due to Theorem 4.

Proof of Theorem 5. Since the case $m = n$ is considered in Theorems 7 and 8, we can assume that $m < n$. The m-subharmonic function is subharmonic, so the function $u(z)$ extends subharmonically into the domain D. The extended function we again denote by $u(z)$. Applying Theorem 4 for $F \subset \mathbb{C}^n = \mathbb{C}^{(n-m)+m} \approx \mathbb{R}^{2(n-m)+2m}$ we state the fact, that for almost all $z' = (z_1, \ldots, z_{n-m})$ the intersection $F \cap \Pi_{z'}$ has capacity

$$C_{2m-2,q}(F \cap \Pi_{z'}) = 0.$$

Hence, according to Theorem 7, we obtain that for almost all $\Pi_{z'}$ planes restriction $u(z)|_{\Pi_{z'}}$ is subharmonic in $D \cap \Pi_{z'}$.

This implies that the restriction $u(z)|_{\Pi_{z'}}$ is subharmonic in $D \cap \Pi_{z'}$ for all planes $\Pi_{z'}$. Using a unitary change of variables $\mathbb{C}^n$, we get that $u(z)|_{\Pi}$ is subharmonic in $D \cap \Pi$ for all planes Π. *Theorem 5 is proved.* $\square$

The proof of Theorem 6 is analogous to the proof of Theorem 5.

Remark. The class L_p^2 is characterized by the fact that functions from the class L_p^2 have derivatives up to the second order, with the second derivatives belonging to L_p. Therefore, for $p \geq 1$ a closed set $F \subset D, mes\, F = 0$, appears removable singular set for $m - sh(D \backslash F) \cap L_p^2$. If $u \in m - sh(D \backslash F) \cap C^2(D)$, then the set $F \subset D$, which is nowhere dense, is removable for u. Therefore, the cases $m - sh(D \backslash F) \cap L_p^2$ and $m - sh(D \backslash F) \cap C^2(D)$ are trivial and we shall not focus on them in this paper.

5 Appendix

For the completeness of the materials of subject, we present one theorem on removable singular sets of sub-solutions of elliptic operators from the class L_p^k (see [4]): let $G \subset \mathbb{R}^n$ be a domain in space $\mathbb{R}^n$ and $F'(G)$ be the space of distributions in the domain G and

$$P(D) = \sum_{|\alpha| \leq m} a_\alpha D^\alpha,$$

$$D^\alpha = \frac{\partial^{|\alpha|}}{\partial x_1^{\alpha_1} \partial x_2^{\alpha_2} \ldots \partial x_n^{\alpha_n}}, \quad \alpha = (\alpha_1, \alpha_2, \ldots, \alpha_n), \quad |\alpha| = |\alpha_1| + |\alpha_2| + \cdots + |\alpha_n|,$$

be elliptic differential operator of order $m \geq 2$ with constant coefficients a_α.

Definition 5 An upper semi-continuous function defined in a domain $G \subset \mathbb{R}^n$ is called sub-solution of the elliptic operator $P(D)$ in the domain G if $P(D)u(x) \geq 0$, in terms of distributions, that is,

$$(P(D)u, \varphi) = (u, P(D)\varphi) = \int_G u(x) \cdot P(D)\varphi(x)\, dx \geq 0$$

for all positive test-functions $\varphi \in F(G)$.

Theorem 9 *A compact set E in a domain $G \subset \mathbb{R}^n$, $n > 2$ is removable for subsolutions $u(x)$ from a class $L_p^k(G)$, of an elliptic operator $P(D)$ (of order m), $\frac{n}{n-m+k} \leq p < +\infty$, $0 < m - k < n$, if and only if*

$$C_{m-k,q}(E) = 0, \quad q = \frac{p}{p-1}.$$

Corollary 1 *A compact set E in a domain $G \subset \mathbb{R}^n$, $n > 2$ is removable for subsolutions $u(x)$ of an elliptic equation $P(D)u(x) = 0$ (order m) in $G \backslash E$ from a class $L_p^k(G)$, $\frac{n}{n-m+k} \leq p < +\infty$, $0 < m - k < n$, if and only if $C_{m-k,q}(E) = 0$, $q = \frac{p}{p-1}$.*

References

1. Abdullayev, B.I.: $\mathscr{P}$-measure in the class of $m - wsh$ functions. J. Sib. Fed. Univ. Math. Phys. **N.7**(1), 3–9 (2014)
2. Abdullayev, B.I.: Subharmonic functions on complex Hyperplanes of $\mathbb{C}^n$. J. Sib. Fed. Univ. Math. Phys.-Krasn. **N6**(4), 409–416 (2013)
3. Abdullaev, B.I., Imomkulov, S.A.: Removable singularities of subharmonic functions in the level L_p and L_p^1. Uzb. Math. J. **N.4**, 10–14 (1997)
4. Abdullaev, B.I., Yarmetov, Zh.R.: On singular sets of subsolutions of elliptic operators. Bull. KrasGAU **N9**, 74–80 (2006)
5. Carleson, L.: Selected Problems on Exceptional Sets. Van Nostrand, Toronto (1967)
6. Dauzhanov, A.Sh.: On the smoothness of generalized solutions of elliptic equations (Russian). Uzb. Math. J. **N2**, 9–19 (2006)
7. Dolzhenko, E.P.: On the singularities of continuous harmonic functions. Izv. Akad. Nauk. SSSR Ser. Math. **28**(6), 1251–1270 (1964)
8. Dolzhenko, E.P.: On the representation of continuous harmonic functions in the form of potentials. Izv. Akad. Nauk SSSR Ser. Math. **28**(5), 1113–1130 (1964)
9. Harve, R., Polking, J.C.: A notion of capacity which characterizes removable singularites. Trans. Am. Math. Soc. **169**, 183–195 (1968)
10. Harvey, F.R., Lawson Jr., H.B.: $p-$ convexity, $p-$ plurisubharmonicity and the Levi problem. Indiana Univ. Math. J. **62**(N1), 149–169 (2013)
11. Hayman, W.K., Kennedy, P.B.: Subharmonic Functions. Vol 1. London Mathematical Society Monographs, vol. 9. Academic Press, London (1976)
12. Joyce, D.: Riemannian Holonomy Groups and Calibrated Geometry. Oxford Graduate Texts in Mathematics, vol. 12. OUP, Oxford (2007)
13. Khusanov, Z.Kh.: Capacity properties of q-subharmonic functions. I, (Russian). Izv. Akad. Nauk UzSSR Ser. Fiz.-Mat. Nauk **97**(1), 41–45 (1990)
14. Khusanov, Z.Kh.: Capacity properties of q-subharmonic functions. II, (Russian). Izv. Akad. Nauk UzSSR Ser. Fiz.-Mat. Nauk **94**(5), 28–33 (1990)
15. Mattila, P.: Integral geometric properties of capacities. Trans. Am. Math. Soc. **266**, 539–554 (1981)
16. Maz'ya, V.G., Havin, V.P.: Non-linear potential theory. Russ. Math. Surv. **27**(6), 71–148 (1972)

17. Maz'ya, V.G.: On (p, l)-capacity, inbedding theorems, and the spectrum of a selfadjoint elliptic operator. Math. USSR-Izv. **7**(2), 357–387 (1973)
18. Mel'nikov, M.S., Sinanyan, S.O.: Questions of the theory of approximation of functions of a complex variable. J. Sov. Math. **5**(5), 688–752 (1976)
19. Maz'ya V.G.: Classes of sets and measures connected with imbedding theorems (Russian). In: Embedding Theorems and Their Applications. Proceedings of the symposium on imbedding theorems, pp. 142–159. M. Nauka (1970)
20. Sadullaev, A.S., Yarmetov, Zh.R.: Removable singularities of subharmonic functions of class Lip_α. Sb. Math. **186**(1), 133–150 (1995)
21. Sadullaev, A.S., Abdullaev, B.: A removable singularity of the bounded above $m - wsh$ functions, (Russian). Dok. Akad Nauk **N5**, 12–14 (2015)
22. Sadullaev, A.S., Abdullaev, B.: Removable singularity $m - wsh$ functions of the class Lip_α, (Russian). Bull. Natl. Univ. Uzb. **N1**, 4–6 (2015)
23. Sadullaev, A.S., Abdullaev, B.I., Sharipov, R.A.: A removable singularity of the bounded above $m - sh$ functions, (Russian). Uzbek Math. J. **N3**, 118–124 (2016)
24. Sadullaev, A.S.: Rational approximation and pluripolar sets. Math. USSR-Sb. **47**(1), 91–113 (1984)
25. Shapiro, V.L.: Subharmonic functions and Hausdorff measure. J. Differ. Equ. **27**(1), 28–45 (1978)
26. Shiffman, B.: On the removal of singularities of analytic sets. Mich. Math. J. **15**, 111–120 (1968)
27. Verbitsky, M.: Plurisubharmonic functions in calibrated geometry and q-convexity. Math. Z. **264**(4), 939–957 (2010)
28. Whitney, H.: Analytic extensions of differentiable functions defined in closed set. Trans. Am. Math. Soc. **36**, 63–89 (1934)

Extensions of Bernstein's Lethargy Theorem

Asuman Güven Aksoy

Abstract In this paper, we examine the aptly-named "Lethargy Theorem" of Bernstein and survey its recent extensions. We show that one of these extensions shrinks the interval for best approximation by half while the other gives a surprising connection to the space of bounded linear operators between two Banach spaces.

Keywords Best approximation · Bernstein's Lethargy theorem · Approximation numbers

Mathematics Subject Classification (2000) 41A25 · 41A50 · 46B20

1 Introduction

The formal beginnings of approximation theory date back to 1885, with Weierstrass' celebrated approximation theorem [31]. The discovery that every continuous function defined on a closed interval $[a, b]$ can be uniformly approximated as closely as desired by a polynomial function immediately prompted many new questions. One such question concerned approximating functions with polynomials of limited degree. That is, if we limit ourselves to polynomials of degree at most n, what can be said of the best approximation? As it turns out, there is no unified answer to this question. In fact, S. N. Bernstein [11] in 1938 showed that there exist functions whose best approximation converges arbitrarily slowly as the degree of the polynomial rises. In this paper, we take up this aptly-named "Lethargy Theorem" of Bernstein and present two extensions. For $f \in C([0, 1])$, the sequence of the best approximation errors is defined as:

$$\rho(f, P_n) = \inf\{||f - p|| : \ p \in P_n\} \tag{1}$$

A. G. Aksoy (✉)
Department of Mathematical Sciences, Claremont McKenna College,
Claremont, CA 91711, USA
e-mail: aaksoy@cmc.edu

© Springer Nature Switzerland AG 2018   13
Z. Ibragimov et al. (eds.), *Algebra, Complex Analysis, and Pluripotential Theory*,
Springer Proceedings in Mathematics & Statistics 264,
https://doi.org/10.1007/978-3-030-01144-4_2

where P_n denotes the space of all polynomials of degree $\leq n$. Clearly,

$$\rho(f, P_1) \geq \rho(f, P_2) \geq \cdots$$

and $\{\rho(f, P_n)\}$ form a non-increasing sequence of numbers. Bernstein [11] proved that if $\{d_n\}_{n \geq 1}$ is a non-increasing null sequence (i.e. $\lim_{n \to \infty} d_n = 0$) of positive numbers, then there exists a function $f \in C[0, 1]$ such that

$$\rho(f, P_n) = d_n, \text{ for all } n \geq 1.$$

This remarkable result is called Bernstein's Lethargy Theorem (BLT) and is used in the constructive theory of functions [37], and it has been applied to the theory of quasi analytic functions in several complex variables [33, 34]. Also see [18] and references therein for an application of BLT to the study Gonchar quasianalytic functions of several variables.

Bernstein's proof is based on a compactness argument and only works when the subspaces are finite dimensional. Note that the density of polynomials in $C[0, 1]$ (the Weierstrass Approximation Theorem) implies that

$$\lim_{n \to \infty} \rho(f, P_n) = 0.$$

Weierstrass Approximation Theorem gives no information about the speed of convergence for $\rho(f, P_n)$, but Bernstein's Lethargy Theorem does. Bernstein Lethargy Theorem has been extended, replacing $C[0, 1]$ by an arbitrary Banach space X and replacing P_n by arbitrary closed subspaces $\{Y_n\}$ of X where $Y_1 \subset Y_2 \subset \cdots$ by Shapiro [36]. Using Baire's category theorem and Riesz's lemma, he proved that for each null sequence $\{d_n\}$ of non-negative numbers, there exists a vector $x \in X$ such that

$$\rho(x, Y_n) \neq O(d_n), \text{ as } n \to \infty.$$

That is, there is no $M > 0$ such that

$$\rho(x, Y_n) \leq M d_n, \text{ for all } n \geq 1.$$

Note that Shapiro's result is not restricted to finite dimensional subspaces Y_n. This result was later strengthened by Tyuriemskih [39] . He showed that the sequence of errors of the best approximation from x to Y_n, $\{\rho(x, Y_n)\}$ may converge to zero at an arbitrary slow rate. More precisely, for any expanding sequence $\{Y_n\}$ of subspaces of X and for any null sequence $\{d_n\}$ of positive numbers, he constructed an element $x \in X$ such that

$$\lim_{n \to \infty} \rho(x, Y_n) = 0, \text{ and } \rho(x, Y_n) \geq d_n \text{ for all } n \geq 1.$$

However, it is also possible that the errors of the best approximation $\{\rho(x, Y_n)\}$ may converge to zero arbitrarily fast; for results of this type see [9].

We refer the reader to [14] for an application of Tyuriemskih's Theorem to convergence of sequences of bounded linear operators.

We also refer to [3, 5, 9, 10, 23, 24] for other versions of Bernstein's Lethargy Theorem and to [4, 6, 26, 32, 41] for Bernstein's Lethargy Theorem for Fréchet spaces.

Given an arbitrary Banach space X, a strictly increasing sequence $\{Y_n\}$ of subspaces of X and a non-increasing null sequence $\{d_n\}$ of non-negative numbers, one can ask the question whether there exists $x \in X$ such that $\rho(x, Y_n) = d_n$ for each n? For a long time, no sequence $\{d_n\}$ of this type was known for which such an element x exists for *all* possible Banach spaces X. The only known spaces X in which the answer is always "yes" are the Hilbert spaces (see [7, 8, 40]). For a general (separable) Banach space X, a solution x is known to exist whenever all Y_n are finite-dimensional (see [38]). Moreover, it is known that if X has the above property, then it is reflexive (see [40]).

2 Extensions of BLT to Banach Spaces

The following lemma is a Bernstein's Lethargy type of a result restricted to *finite number* of subspaces, and for the proof of this lemma we refer the reader to Timan's book [38].

Lemma 1 *Let $(X, \| \cdot \|)$ be a normed linear space, $Y_1 \subset Y_2 \subset \ldots \subset Y_n \subset X$ be a finite system of strictly nested subspaces, $d_1 > d_2 > \cdots > d_n \geq 0$ and $z \in X\backslash Y_n$. Then, there is an element $x \in X$ for which $\rho(x, Y_k) = d_k$ $(k = 1, \ldots, n)$, $\|x\| \leq d_1 + 1$, and $x - \lambda z \in Y_n$ for some $\lambda > 0$.*

Borodin [12] proved the above Lemma 1 by taking $(X, \| \cdot \|)$ to be a Banach space. Returning to the question posed before, namely given an arbitrary Banach space X, a strictly increasing sequence $\{Y_n\}$ of subspaces of X and a non-increasing null sequence $\{d_n\}$ of non-negative numbers, one can ask the question whether there exists $x \in X$ such that $\rho(x, Y_n) = d_n$ for each n? For a long time no sequence $\{d_n\}$ of this type was known for which such an element x exists for *all* possible Banach spaces X. Borodin in [12] uses the above lemma for Banach space to establish the existence of such an element in case of rapidly decreasing sequences; more precisely, in 2006 he proves the following theorem:

Theorem 1 (Borodin [12]) *Let X be an arbitrary Banach space (with finite or infinite dimension), $Y_1 \subset Y_2 \subset \cdots$ be an arbitrary countable system of strictly nested subspaces in X, and fix a numerical sequence $\{d_n\}_{n \geq 1}$ satisfying: there exists a natural number $n_0 \geq 1$ such that*

$$d_n > \sum_{k=n+1}^{\infty} d_k \text{ for all } n \geq n_0 \text{ at which } d_n > 0. \tag{2}$$

Then there is an element $x \in X$ such that

$$\rho(x, Y_n) = d_n, \text{ for all } n \geq 1. \tag{3}$$

The condition (2) on the sequence $\{d_n\}$ is the key to the derivation of (3) in Theorem 1. Based on this result, Konyagin [22] in 2013 takes a further step to show that, for a general non-increasing null sequence $\{d_n\}$, the deviation of $x \in X$ from each subspace Y_n can range in some interval depending on d_n.

Theorem 2 (Konyagin [22]) *Let X be a real Banach space, $Y_1 \subset Y_2 \subset \cdots$ be a sequence of strictly nested closed linear subspaces of X, and $d_1 \geq d_2 \geq \cdots$ be a non-increasing sequence converging to zero, then there exists an element $x \in X$ such that the distance $\rho(x, Y_n)$ satisfies the inequalities*

$$d_n \leq \rho(x, Y_n) \leq 8d_n, \text{ for } n \geq 1. \tag{4}$$

Note that the condition (2) is satisfied when $d_n = (2 + \varepsilon)^{-n}$ for $\varepsilon > 0$ arbitrarily small, however it is not satisfied when $d_n = 2^{-n}$. Of course there are two natural questions to ask:

Question 1   Is the condition (2) necessary for the results in Theorem 1 to hold, or does Theorem 1 still hold for the sequence $d_n = 2^{-n}, n \geq 1$?

Question 2   Under the same conditions given in Theorem 2, can the lower and upper bounds of $\rho(x, Y_n)$ in (4) be improved?

Both of these questions are answered and two improvements on a theorem of S. N. Bernstein for Banach spaces are presented in [2]. A positive answer to Question 1 is obtained in [2] by showing that Theorem 1 can be extended by weakening the strict inequality in (2) to a non-strict one:

$$d_n \geq \sum_{k=n+1}^{\infty} d_k, \text{ for every } n \geq n_0. \tag{5}$$

Clearly, the condition (5) is weaker than (2), but unlike the condition (2), (5) is satisfied by the sequences $\{d_n\}_{n \geq 1}$ verifying $d_n = \sum_{k=n+1}^{\infty} d_k$ for all $n \geq n_0$. For a typical example of such a sequence one can take $\{d_n\} = \{2^{-n}\}$. We have also shown that if X is an arbitrary infinite-dimensional Banach space, $\{Y_n\}$ is a sequence of strictly nested subspaces of X, and if $\{d_n\}$ is a non-increasing sequence of non-negative numbers tending to 0, then for any $c \in (0, 1]$ we can find $x_c \in X$, such that the distance $\rho(x_c, Y_n)$ from x_c to Y_n satisfies

$$cd_n \leq \rho(x_c, Y_n) \leq 4cd_n \text{ for all } n \in \mathbb{N}.$$

We prove the above inequality by first improving [12] result for Banach spaces by weakening his condition on the sequence $\{d_n\}$. The weakened condition on d_n requires a refinement of Borodin's construction to extract an element in X, whose distances from the nested subspaces are precisely the given values d_n.

Now, we are ready to state the following theorem [2] which improves the theorem of Borodin [12].

Theorem 3 (Aksoy, Peng [2]) *Let X be an arbitrary infinite-dimensional Banach space, $\{Y_n\}_{n \geq 1}$ be an arbitrary system of strictly nested subspaces with the property $\overline{Y}_n \subset Y_{n+1}$ for all $n \geq 1$, and let the non-negative numbers $\{d_n\}_{n \geq 1}$ satisfy the following property: there is an integer $n_0 \geq 1$ such that*

$$d_n \geq \sum_{k=n+1}^{\infty} d_k, \text{ for every } n \geq n_0.$$

Then, there exists an element $x \in X$ such that $\rho(x, Y_n) = d_n$ for all $n \geq 1$.

Proof of this theorem depends on some technical lemmas, details can be found in [2]. Next we state an improvement of the Konyagin's result.

Theorem 4 (Aksoy, Peng [2]) *Let X be an infinite-dimensional Banach space, $\{Y_n\}$ be a system of strictly nested subspaces of X satisfying the condition $\overline{Y}_n \subset Y_{n+1}$ for all $n \geq 1$. Let $\{d_n\}_{n \geq 1}$ be a non-increasing null sequence of strictly positive numbers. Assume that there exists an extension $\{(\tilde{d}_n, \tilde{Y}_n)\}_{n \geq 1} \supseteq \{(d_n, Y_n)\}_{n \geq 1}$ satisfying: $\{\tilde{d}_n\}_{n \geq 1}$ is a non-increasing null sequence of strictly positive values, $\tilde{Y}_n \subset \tilde{Y}_{n+1}$ for $n \geq 1$; and there is an integer $i_0 \geq 1$ and a constant $K > 0$ such that*

$$\{K2^{-n}\}_{n \geq i_0} \subseteq \{\tilde{d}_n\}_{n \geq 1}.$$

Then, for any $c \in (0, 1]$, there exists an element $x_c \in X$ (depending on c) such that

$$cd_n \leq \rho(x_c, Y_n) \leq 4cd_n, \text{ for } n \geq 1. \tag{6}$$

Proof We first show (6) holds for $c = 1$. By assumption, there is a subsequence $\{n_i\}_{i \geq i_0}$ of $\mathbb{N}$ such that
$$\tilde{d}_{n_i} = K2^{-i}, \text{ for } i \geq i_0.$$

Since the sequence $\{\tilde{d}_n\}_{n=1,2,\ldots,n_{i_0}-1} \cup \{\tilde{d}_{n_i}\}_{i \geq i_0}$ satisfies the condition (5) and $\tilde{Y}_n \subset \tilde{Y}_{n+1}$ for all $n \geq 1$, then we can apply Theorem 3 to get $x \in X$ so that

$$\rho(x, \tilde{Y}_n) = \tilde{d}_n, \text{ for } n = 1, \ldots, n_{i_0} - 1, \text{ and } \rho(x, \tilde{Y}_{n_i}) = \tilde{d}_{n_i}, \text{ for all } i \geq i_0. \tag{7}$$

Therefore for any integer $n \geq 1$, we can consider the following cases:

Case 1   if $n \leq n_{i_0} - 1$ or $n = n_i$ for some $i \geq i_0$, then it follows from (7) that

$$\rho(x, \tilde{Y}_n) = \tilde{d}_n;$$

Case 2 if $n_i < n < n_{i+1}$ for some $i \geq i_0$, then the fact that $\{\tilde{d}_n\}$ is non-increasing and $\tilde{Y}_{n_i} \subset \tilde{Y}_n \subset \tilde{Y}_{n_{i+1}}$ leads to

$$\rho(x, \tilde{Y}_n) \in \left(\rho(x, \tilde{Y}_{n_{i+1}}), \rho(x, \tilde{Y}_{n_i}) \right) = \left(K2^{-(i+1)}, K2^{-i} \right)$$

and

$$\tilde{d}_n \in \left[K2^{-i}, K2^{-i+1} \right].$$

It follows that

$$\frac{\rho(x, \tilde{Y}_n)}{\tilde{d}_n} \in \left(\frac{K2^{-i-1}}{K2^{-i+1}}, \frac{K2^{-i}}{K2^{-i}} \right) = \left(\frac{1}{4}, 1 \right).$$

Putting the above cases together yields

$$\frac{1}{4}\tilde{d}_n \leq \rho(x, \tilde{Y}_n) \leq \tilde{d}_n \text{ for all } n \geq 1.$$

For $c \in (0, 1]$, taking $x_c = 4cx$ in the above inequalities, we obtain

$$c\tilde{d}_n \leq \rho(x_c, \tilde{Y}_n) \leq 4c\tilde{d}_n, \text{ for all } n \geq 1.$$

Remembering that $\{(d_n, Y_n)\}_{n\geq1} \subseteq \{(\tilde{d}_n, \tilde{Y}_n)\}_{n\geq1}$, we then necessarily have

$$cd_n \leq \rho(x_c, Y_n) \leq 4cd_n, \text{ for all } n \geq 1.$$

Hence Theorem 4 is proved.

Remark 1 Taking $c = \dfrac{1}{4}$ in Theorem 4, we obtain existence of $x \in X$ for which

$$\frac{1}{4} \leq \frac{\rho(x, Y_n)}{d_n} \leq 1, \text{ for all } n \geq 1.$$

The interval length $\dfrac{3}{4}$ makes $\left[\dfrac{1}{4}, 1\right]$ the "narrowest" estimating interval of $\dfrac{\rho(x, Y_n)}{d_n}$ that Theorem 4 could provide.

The subspace condition given in Theorem 4 states that the nested sequence $\{Y_n\}$ has "enough gaps" so that the sequence

$$\{(d'_n, Y'_n)\}_{n\geq1} = \{(d_n, Y_n)\}_{n\geq1} \cup \{(K2^{-i}, \tilde{Y}_{n_i})\}_{i\geq i_0}$$

satisfies $d'_n \geq d'_{n+1} \to 0$ and $\overline{Y'_n} \subset Y'_{n+1}$ for all $n \geq 1$. For another subspace condition see the results in [1].

Observe that in Konyagin's paper [22] it is assumed that $\{Y_n\}$ are closed and strictly increasing. In Borodin's paper [12], this is not specified, but from the proof of Theorem 1 it is clear that his proof works only under the assumption that $\overline{Y}_n$ is strictly included in Y_{n+1}. The subspace condition $\overline{Y}_n \subset Y_{n+1}$ does not come at the expense of our assumption to weaken the condition on the sequence d_n. This is a natural condition. To clarify the reason why almost all Lethargy theorems have this condition on the subspaces, we give the following example.

Example 1 Let $X = L^\infty[0, 1]$ and consider $C[0, 1] \subset L^\infty[0, 1]$ and define the subspaces of X as follows:

1. $Y_1 =$ space of all polynomials;
2. $Y_{n+1} = \mathrm{span}[Y_n \cup \{f_n\}]$ where $f_n \in C[0, 1] \setminus Y_n$, for $n \geq 1$.

Observe that by the Weierstrass Theorem we have $\overline{Y}_n = C[0, 1]$ for all $n \geq 1$. Take any $f \in L^\infty[0, 1]$ and consider the following cases:

(a) If $f \in C[0, 1]$, then

$$\rho(f, Y_n) = \rho(f, C[0, 1]) = 0 \text{ for all } n \geq 1.$$

(b) If $f \in L^\infty[0, 1] \setminus C[0, 1]$, then

$$\rho(f, Y_n) = \rho(f, C[0, 1]) = d > 0 \text{ (independent of } n).$$

Note that in the above, we have used the fact that $\rho(f, Y_n) = \rho(f, \overline{Y}_n)$. Hence in this case BLT does not hold.

3 Extension of BLT to Fréchet Spaces

Fréchet spaces are locally convex spaces that are complete with respect to a translation invariant metric, and they are generalization of Banach spaces which are normed linear spaces, complete with respect to the metric induced by the norm. However, there are metric spaces which are not normed spaces. This can be easily seen by considering the space s the set of all sequence $x = (x_n)$ and defining $d(x, y) = \sum_{i=1}^\infty 2^{-i} \dfrac{|x_i - y_i|}{(1 + |x_i - y_i|)}$ as a metric on s. If we let $\lambda \in \mathbb{R}$ then

$$d(\lambda x, \lambda y) = \sum_{i=1}^\infty \frac{1}{2^i} \frac{|\lambda||x_i - y_i|}{(1 + |\lambda||x_1 - y_i|)} \neq |\lambda|d(x, y) \quad \text{(no homogeneity)}.$$

Definition 1 $(X, \| \cdot \|_F)$ is called a *Fréchet space*, if it is a metric linear space which is complete with respect to its F-norm $\|.\|_F$ giving the topology. An F-norm $\|.\|_F$ satisfies the following conditions [35]:

1. $||x||_F = 0$ if and only if $x = 0$,
2. $||\alpha x||_F = ||x||_F$ for all real or complex α with $|\alpha| = 1$ (no homogeneity),
3. $||x + y||_F \leq ||x||_F + ||y||_F$,

Many Fréchet spaces X can also be constructed using a countable family of semi-norms $||x||_k$ where X is a complete space with respect to this family of semi-norms. For example a translation invariant complete metric inducing the topology on X can be defined as

$$d(x, y) = \sum_{k=0}^{\infty} 2^{-k} \frac{||x - y||_k}{1 + ||x - y||_k} \quad \text{for } x, y \in X.$$

Clearly, every Banach space is a Fréchet space, and the other well known example of a Fréchet space is the vector space $C^{\infty}[0, 1]$ of all infinitely differentiable functions $f : [0, 1] \to \mathbb{R}$ where the semi-norms are

$$||f||_k = \sup\{|f^{(k)}(x)| : x \in [0, 1]\}.$$

For more information about Fréchet spaces the reader is referred to [20, 35].

Recently in [4] a version of Bernstein Lethargy Theorem (BLT) was given for Fréchet spaces. More precisely, let X be an infinite-dimensional Fréchet space and let $\mathcal{V} = \{V_n\}$ be a nested sequence of subspaces of X such that $\overline{V_n} \subseteq V_{n+1}$ for any $n \in \mathbb{N}$. Let d_n be a decreasing sequence of positive numbers tending to 0. Under an additional natural condition on $\sup\{\text{dist}(x, V_n)\}$, we proved that there exists $x \in X$ and $n_o \in \mathbb{N}$ such that

$$\frac{d_n}{3} \leq \rho(x, V_n) \leq 3d_n,$$

for any $n \geq n_o$. By using the above theorem, it is also possible to obtain an extension of both Shapiro's [36] and Tyuremskikh's [39] theorems for Fréchet spaces as well.

Notation. Let $(X, ||.||_F)$ be a Fréchet space and assume that $\mathcal{V} = \{V_n\}$ is a nested sequence of linear subspaces of X satisfying $\overline{V_n} \subset V_{n+1}$. Let $d_{n,\mathcal{V}}$ denote the deviation of V_n from V_{n+1} defined as:

$$d_{n,\mathcal{V}} = \sup\{\rho(v, V_n) : v \in V_{n+1}\} \tag{8}$$

and throughout this paper we assume:

$$d_{\mathcal{V}} = \inf\{d_{n,\mathcal{V}} : n \in \mathbb{N}\} > 0. \tag{9}$$

The necessity of this assumption is illustrated in the following example.

Example 2 Let $X = \{(x_n) : x_n \in \mathbb{R} \text{ for any } n \in \mathbb{N}\}$ equipped with the F-norm $||x||_F = \sum_{j=1}^{\infty} \frac{|x_j|}{2^j(1 + |x_j|)}$, where $x = (x_1, \ldots, x_j, \ldots)$. Let

$$V_n = \{x \in X : x_k = 0 \text{ for } k \geq n + 1\}.$$

It is easy to see that for any $x \in X$

$$\rho(x, V_n) = \sum_{j=n+1}^{\infty} \frac{|x_j|}{2^j(1 + |x_j|)} \leq \frac{1}{2^n}.$$

Let $d_n = \dfrac{2}{n}$ and observe that for any $x \in X$, $\rho(x, V_n) \leq \dfrac{1}{2^n} < \dfrac{2}{n}$. Also observe that $d_{n,\mathscr{V}} = \dfrac{1}{2^{n+1}}$ which implies that $d_{\mathscr{V}} = 0$.

Thus in the case when $d_{\mathscr{V}} = 0$, we cannot even hope to prove Shapiro's theorem. Above example also shows that it is impossible to prove the Tyuriemskih Theorem or Konyagin's type result in Fréchet spaces without additional assumptions, because they are stronger statements than Shapiro's theorem. Note that if X is a Banach space, then the condition $d_{\mathscr{V}} = \inf\{d_{n,\mathscr{V}} : n \in \mathbb{N}\} > 0$ is satisfied automatically. It can be seen easily that $d_{n,\mathscr{V}} = +\infty$ for Banach spaces. because

$$\rho(tx, V_n) = t\rho(x, V_n)$$

and the supremum taking over all $v \in V_{n+1}$ and V_n is strictly included in V_{n+1}. The next, example illustrates that there is a natural way to build Fréchet spaces where $d_{n,\mathscr{V}} = 1$.

Example 3 Let $(X, \|.\|)$ be a Banach space. Define in X an F-norm $\| \| \|_F$ by: $\|x\|_F = \|x\|/(1 + \|x\|)$. Then $d_{n,\mathscr{V}} = 1$ for any $n \in \mathbb{N}$ independently of $\mathscr{V}$. Because the mapping

$$t \mapsto \frac{t}{1 + t}$$

is increasing for $t > -1$ and

$$\rho_F(tx, V_n) = \frac{\rho(tx, V_n)}{1 + \rho(tx, V_n)} \to 1$$

as $t \to \infty$.

Theorem 5 (Aksoy, Lewicki, [4]) *Let X be a Fréchet space and and assume that $\mathscr{V} = \{V_n\}$ is a nested sequence of linear subspaces of X satisfying $\overline{V_n} \subseteq V_{n+1}$, where the closure is taken with respect to $\| \cdot \|_F$. Let $d_{n,\mathscr{V}}$ be defined as above and $\{e_n\}$ be a decreasing sequence of positive numbers satisfying*

$$\sum_{j=n}^{\infty} 2^{j-n}(\delta_j + e_j) < \min\{d_{n,\mathscr{V}}, e_{n-1}\}$$

with a fixed sequence of positive numbers δ_j. Then, there exists $x \in X$ such that

$$\rho(x, V_n) = e_n \quad \text{for any } \ n \in \mathbb{N}.$$

Remark 2 Note that the condition given in the above theorem extends Borodin's condition. In the case when X is a Banach space, we have $d_{n,\mathscr{V}} = +\infty$, and the inequality

$$\sum_{j=n}^{\infty} 2^{j-n} e_j < \min\{d_{n,\mathscr{V}}, e_{n-1}\}$$

reduces to

$$e_{n-1} > \sum_{j=n}^{\infty} 2^{j-n} e_j$$

compared with the condition $e_{n-1} > \sum_{j=n}^{\infty} e_j$ given in Borodin's theorem.

The idea of the proof of the Theorem 5 above lies in the following claims:

(a) $F_n = \{v \in V_{n+1} : \rho(v, V_j) = e_j \text{ for } j = 1, \ldots, n\} \neq \emptyset.$
(b) F_n consists of elements of the form described as $w_n = \sum_{j=1}^{n} q_{j,n}$, where

$$q_{j,n} = t_{j,n} v_j - z_{j,n},$$

with $t_{j,n} \in [0, 1]$, $z_{j,n} \in Z_j$ (where Z_j a finite subset of V_j) for $j = 1, \ldots, n$ and v_j are given by the equation

$$\sum_{j=n}^{\infty} 2^{j-n} (e_j + \delta_j) = \rho(v_n, V_n) \geq \|v_n\| - \delta_n.$$

Moreover,

$$\|q_{j,n}\| < \sum_{l=j}^{\infty} 2^{l-j} (e_l + \delta_l).$$

Using a diagonal argument, select a subsequence $\{n_k\}$ with $\|q_{j,n_k} - q_j\| \to 0$ and $q_j \in V_{j+1}$. Fix $j_0 \in \mathbb{N}$. Then, for $k \geq k_0$ and $n_k \geq j_0$, to construct $x \in X$ with $\rho(x, V_{j_0}) = e_{j_0}$, set $s_k = \sum_{j=1}^{k} q_j$, show $\{s_k\}$ is a Cauchy sequence and converges to $x = \sum_{j=1}^{\infty} q_j$ and then show that $\|w_{n_k} - x\| \to 0$. Since $w_{n_k} \in F_{n_k}$, we have:

$$\rho(w_{n_k}, V_{j_0}) = e_{j_0} \quad \text{for } \ k \geq k_0$$

hence

$$\rho(x, V_{j_0}) = \lim_k \rho(w_{n_k}, V_{j_0}) = e_{j_0}.$$

For the details of the proof we refer the reader to [4].

Theorem 6 (Aksoy, Lewicki [4]) *Let X be a infinite-dimensional Fréchet space and let $\mathscr{V} = \{V_n\}$ be a nested sequence of subspaces of X such that $\overline{V_n} \subseteq V_{n+1}$ for every $n \in \mathbb{N}$. Let e_n be a decreasing sequence of positive numbers tending to 0. Assume that*

$$d_{\mathscr{V}} = \inf\{d_{n,\mathscr{V}} : n \in \mathbb{N}\} > 0. \tag{10}$$

Then there exists $n_o \in \mathbb{N}$ and $x \in X$ such that for any $n \geq n_o$

$$\frac{e_n}{3} \leq \rho(x, V_n) \leq 3e_n. \tag{11}$$

Theorem 7 (Shapiro's Theorem for Fréchet spaces) *Let the assumptions of Theorem 6 be satisfied. Then, there exists $x \in X$ such that $\rho(x, V_n) \neq O(e_n)$.*

Proof By Theorem 6 applied to the sequence $\{\sqrt{e_n}\}$ there exists $x \in X$ and $n_o \in \mathbb{N}$ such that

$$\frac{\sqrt{e_n}}{3} \leq \rho(x, V_n) \leq 3\sqrt{e_n}$$

for $n \geq n_o$. Since $e_n \to 0$, it is obvious that $\rho(x, V_n) \neq O(e_n)$.

Theorem 8 (Tyuremskikh's Theorem for Fréchet spaces) *Let the assumptions of Theorem 6 be satisfied. Then, there exists $x \in X$ and $n_o \in \mathbb{N}$ such that $\rho(x, V_n) \geq e_n$ for $n \geq n_o$.*

Proof By Theorem 6 applied to the sequence $\{3\sqrt{e_n}\}$ there exists $x \in X$ and $n_o \in \mathbb{N}$ such that

$$\sqrt{e_n} \leq \rho(x, V_n) \leq 9\sqrt{e_n}$$

for $n \geq n_o$. Since $e_n \to 0$, it is obvious that $\rho(x, V_n) \geq \sqrt{e_n} \geq e_n$ for $n \geq n_o$.

4 Bernstein Pairs

Suppose X and Y are infinite dimensional Banach spaces. Let $\mathscr{L}(X, Y)$ denote the normed space of all bounded linear maps from X to Y. In this section we investigate the existence of an operator $T \in \mathscr{L}(X, Y)$ whose sequence of approximation numbers $\{a_n(T)\}$ behaves like the prescribed sequence $\{d_n\}$ given in the Bernstein's Lethargy Theorem.

Definition 2 Let X and Y be Banach spaces. For every operator $T \in \mathscr{L}(X, Y)$ the n-th approximation number of T is defined as:

$$a_n(T) = \inf\{\|T - S\| : \quad S \in \mathscr{L}(X, Y) \quad \text{rank } S < n\}.$$

Clearly,
$$a_n(T) = \rho(T, \mathscr{F}_n)$$

where $\mathscr{F}_n$ is the space of all bounded linear maps from X into Y with rank at most n, and
$$\|T\| = a_1(T) \geq a_2(T) \geq \cdots \geq 0.$$

The connection with Bernstein's Lethargy theorem and the approximation numbers is based on the important relation between analytical entities for bounded linear maps such as eigenvalues and the geometrical quantities typified by approximation numbers. If we focus on compact liner maps, then there are known inequalities between the approximation numbers and the non-zero eigenvalues of such maps. For example, the well known inequality of Weyl [42], which states that if the approximation numbers form a sequence in ℓ^p for $0 < p < \infty$, then so do the eigenvalues. Even though Weyl's original result is for Hilbert spaces, there is a remarkable theorem of König [21], for the absolute value of the eigenvalues of a compact map in terms of approximation numbers. More precisely, Let X be a complex Banach space and let T be a compact operator on X. Assume that the nonzero eigenvalues of T are ordered in non-increasing modulus and counted according to their (finite) multiplicity, and denote them by $\lambda_n(T)$. König proves that

$$|\lambda_n(T)| = \lim_{m \to \infty} (a_n(T^m))^{\frac{1}{m}}.$$

It is known that the approximation numbers are the largest s-numbers; for the axiomatic theory of s-numbers we refer the reader to [29, 30].

If X and Y are Hilbert spaces, then the sequence of s-numbers $(s_n(T))$ or, in particular, the sequence of approximation numbers $(a_n(T))$ are the singular numbers of T. Moreover if T is a compact operator, then $a_n(T) = \lambda_n(T^*T)^{1/2}$ where

$$\lambda_1(T^*T) \geq \lambda_2(T^*T) \geq \cdots \geq 0$$

are the eigenvalues of T^*T ordered as above. We refer the reader to [13, 28] for general information about approximation and other s-numbers. We say T is *approximable* if $\lim_{n \to \infty} a_n(T) = 0$. Any approximable operator is compact, but the converse is not true due to the existence of Banach spaces without the approximation property [16].

Definition 3 Two Banach spaces X and Y are said to form a Bernstein pair (BP) if for any positive monotonic null sequence (d_n) there is an operator $T \in \mathscr{L}(X, Y)$ and a constant M depending only on T and (d_n) such that

$$d_n \leq a_n(T) \leq M d_n \qquad \text{for all} \quad n.$$

In this case we say that $(a_n(T))$ is equivalent to (d_n) and we write (X, Y) to denote Bernstein pair and in the case $M = 1$, then (X, Y) is called an exact Bernstein pair.

Note that by $a_n(T)$ we mean the nth approximation number of T defined above. Note that if $X = Y = H$ where H is a Hilbert space and if $\{d_n\}$ is a sequence decreasing to 0, then there exists *a compact operator T on H* such that

$$a_n(T) = d_n$$

(see [29]). Therefore (H, H) forms an exact Bernstein pair.

All of the above mentioned desirable properties of $a_n(T)$ for operators between Banach spaces prompts the following questions:

Questions: Suppose $\{d_n\}_{n \geq 1}$ is a non-increasing null sequence (i.e., $d_n \searrow 0^+$) of positive numbers. Does there exist $T \in \mathscr{L}(X, Y)$ such that the sequence $(a_n(T))$ behaves like the sequence $\{d_n\}$?
In other words, suppose $d_n \searrow 0^+$,

(a) Does there exist $T \in \mathscr{L}(X, Y)$ such that $a_n(T) \geq d_n$ for any n?
(b) Does there exist $T \in \mathscr{L}(X, Y)$ such that $a_n(T) = d_n$ for any n?
(c) Does there exist $T \in \mathscr{L}(X, Y)$ and a constant M such that

$$\frac{d_n}{M} \leq a_n(T) \leq M d_n$$

for any n?

First results of this kind appeared in [17] in the context of Banach spaces with Schauder basis.

An operator acting on a Banach space X is called an H-operator (see [25]) if its spectrum is real and its resolvent satisfies

$$\|(T - \lambda I)^{-1}\| \leq C |\text{Im}\lambda|^{-1} \quad \text{where} \quad \text{Im}\lambda \neq 0.$$

Here C is independent of the points of the resolvent. An operator on a Hilbert space is an H-operator with constant $C = 1$ if and only if it is a self-adjoint operator. Therefore the concept of an H-operator is the generalization in a Banach space the concept of a self-adjoint operator. In [25] some examples and a number of properties of H-operators are given.

Definition 4 We define nth Kolmogorov diameter of T, $d_n(T)$, by $d_n(T) = d_n$ $(T(U_X))$ where for a bounded subset A of X, the n-th Kolmogoroff diameter of A is given by

$$d_n(A) = \inf_L [\inf\{\varepsilon > 0 : A \subset \varepsilon U_X + L\}],$$

where the infimum is taken over all at most n-dimensional subspaces L of Y.

It is clear that $d_n(T)$ and $a_n(T)$ are monotone decreasing sequences and that

$$\lim_n a_n(T) = 0 \quad \text{if and only if} \quad T \in \mathscr{F}(X, Y)$$

and

$$\lim_n d_n(T) = 0 \quad \text{if and only if} \quad T \in \mathscr{K}(X, Y)$$

where $\mathscr{K}(X, Y)$ and $\mathscr{F}(X, Y)$ is the collection of all compact operators and approximable operators by operators of finite rank from X to Y. For more on Kolmogorov diameters and their relations to approximation numbers consult [29].

Theorem 9 (Marcus [25]) *If T is a compact H-operator on a Banach space X, then*

$$d_n(T) \leq a_n(t) \leq 2\sqrt{2} \; C|\lambda_n(T)| \leq 8C(C+1)d_n(T),$$

where C is the constant given in the definition of an H-operator and $d_n(T)$ is the n-th Kolmogoroff diameter of T.

Recall that a biorthogonal system (x_n, f_n) in X is sequence $\{x_n\} \subset X$ and $\{f_n\} \subset X^*$ such that $f_i(x_j) = \delta_{ij}$. A Schauder basis X is a biorthogonal system (x_n, f_n) such that for each $x \in X$, $x = \sum_{i=1}^{\infty} f_i(x)x_i$.

Using the above estimate of Marcus, Hutton, Morell and Retherford proved the following :

Theorem 10 (Hutton, Morrell and Retherford [17]) *Suppose $\{d_n\}_{n\geq 1}$ is a non-increasing null sequence of positive numbers and X is a Banach space with Schauder basis. Then there exists $T \in B(X)$ such that*

$$d_m \leq a_m(T) \leq K.d_m$$

where the constant K depends on X.

Proof Suppose (x_n) is a Schauder basis for X. If $\{\lambda_n\} \searrow 0$, then the operator

$$T = \sum_{n=1}^{\infty} \lambda_n f_n \otimes x_n$$

is a compact H-operator with the sequence $\{\lambda_n\}$ as eigenvalues. Here (x_n, f_n) is a biorthogonal system in X, then the result follows from Theorem 9.

These results are sharpened in [5] as follows:

Theorem 11 (Aksoy, Lewicki [5]) *Suppose* $\{d_n\}_{n\geq 1}$ *is a non-increasing null sequence of positive numbers. In each of the following cases there exists* $T \in B(X, Y)$ *such that*

$$a_m(T) = d_m \quad \text{for each} \quad m$$

(a) *X and Y have 1-unconditional basis.*
    (For example $X = Y = \ell_p$ for $1 \leq p < \infty$ and $X = Y = c_0$).
(b) *$X \in \{c_0, \ell_\infty\}$ and Y has 1-symmetric basis.*
(c) *$Y = \ell_1$, X has 1-symmetric basis.*

In [5], among other things, it was proved that if we assume (X, Y) is a Bernstein pair with respect to $\{a_n\}$ and suppose that a Banach space W contains an isomorphic and complementary copy of X, and a Banach space V contains an isomorphic copy of Y, then (W, V) is a Bernstein Pair with respect to $\{a_n\}$. This implies that there are some natural pairs of Banach spaces that form a Bernstein pair as shown in the following examples.

Example 4 For $1 < p < \infty$ and $1 \leq q < \infty$, the couple $(L_p[0, 1], L_q[0, 1])$ form a Bernstein pair.

The above example follows from the fact that (ℓ_2, ℓ_2) is a Bernstein pair with respect to $\{a_n\}$, [28], and the fact that for every p, $1 \leq p < \infty$, $L_p[0, 1]$ contains a subspace isomorphic to ℓ_2 and complemented in $L_p[0, 1]$ for $p > 1$ (see [43], p. 85). If one assumes that the Banach spaces X and Y have certain properties then one can generate Bernstein pairs as illustrated in the following "conditional" example.

Example 5 (a) If $(\ell_\infty, \ell_\infty)$ is BP, then (X, Y) is BP provided that both X and Y contain an isomorphic copy of ℓ_∞ (Phillips Theorem).
(b) If (c_0, c_0) is a BP, then (X, Y) is BP provided that both X and Y each contain an isomorphic copy of c_0 and X is separable (Sobczyk's Theorem).
(c) if (ℓ_1, ℓ_1) is a BP, then (X, Y) is BP provided that both X and Y each contain an isomorphic copy of ℓ_1 and X is a non reflexive subspace of $L_1[0, 1]$ (Pelczynski–Kadeč Theorem).

For the statements of Phillips, Sobczyk and Pelczynski–Kadeč theorems we refer the reader to [15]. The most recent result on the rate of decay of approximation numbers can be found in [27]. It is shown there that for X and Y, infinite dimensional Banach spaces, and $\{d_n\}_{n\geq 1}$ a non-increasing null sequence, there exists an approximable $T : X \rightarrow Y$ such that

$$\|T\| \leq 2d_1 \quad \text{and} \quad d_m/9 \leq a_m(T) \leq 3d_{[m/4]} \quad \text{for any } m,$$

where $[m/4]$ denotes the integral part of $m/4$. It is also worth mentioning that in [19] operators with prescribed eigenvalue sequences are constructed and in [30] a Shapiro's type of theorem was proved for approximation numbers, with quite elementary arguments.

References

1. Aksoy, A.G., Al-Ansari, M., Case, C., Peng, Q.: Subspace condition for Bernstein's lethargy theorem. Turk. J. Math. **41**(5), 1101–1107 (2017)
2. Aksoy, A.G., Peng, Q.: Constructing an element of a Banach space with given deviation from its nested subspaces. Khayyam J. Math. **4**(1), 59–76 (2018)
3. Aksoy, A.G., Almira, J.: On Shapiro's lethargy theorem and some applications. Jaén J. Approx. **6**(1), 87–116 (2014)
4. Aksoy, A.G., Lewicki, G.: Bernstein's lethargy theorem in Fréchet spaces. J. Approx. Theory **209**, 58–77 (2016)
5. Aksoy, A.G., Lewicki, G.: Diagonal operators, s-numbers and Bernstein pairs. Note Mat. **17**, 209–216 (1999)
6. Albinus, G.: Remarks on a theorem of S. N. Bernstein. Stud. Math. **38**, 227–234 (1970)
7. Almira, J.M., Luther, U.: Compactness and generalized approximation spaces. Numer. Funct. Anal. Optim. **23**(1–2), 1–38 (2002)
8. Almira, J.M., Luther, U.: Generalized approximation spaces and applications. Math. Nachr. **263**(264), 3–35 (2004)
9. Almira, J.M., del Toro, N.: Some remarks on negative results in approximation theory. In: Proceedings of the Fourth International Conference on Functional Analysis and Approximation Theory, vol. I (Potenza, 2000). Rend. Circ. Mat. Palermo (2) Suppl. No. 68, Part I, pp. 245–256 (2002)
10. Almira, J.M., Oikhberg, T.: Approximation schemes satisfying Shapiro's theorem. J. Approx. Theory **164**(5), 534–571 (2012)
11. Bernstein, S.N.: On the inverse problem in the theory of best approximation of continuous functions, Collected works (in Russian), Izd. Akad. Nauk, USSR, vol. II, pp. 292–294 (1954)
12. Borodin, P.A.: On the existence of an element with given deviations from an expanding system of subspaces. Math. Notes **80**(5), 621–630 (2006). (translated from Mat. Zametki **80**(5), 657–667)
13. Carl, B., Stephani, I.: Entropy, Compactness and the Approximation of Operators. Cambridge University Press, Cambridge (1990)
14. Deutsch, F., Hundal, H.: A generalization of Tyuriemskih's lethargy theorem and some applications. Numer. Func. Anal. Optim. **34**(9), 1033–1040 (2013)
15. Diestel, J.: Sequences and Series in Banach Spaces. Springer, New York (1984)
16. Enflo, P.: A counterexample to the approximation property. Acta Math. **13**, 308–317 (1973)
17. Hutton, C., Morrell, J.S., Retherford, J.R.: Diagonal operators, approximation numbers and Kolmogorow diameters. J. Approx. Theory **16**, 48–80 (1976)
18. Imomkulov, S.A., Ibragimov, Z.Sh: Uniqueness property for Gonchar quasianalytic functions of several variables. Topics in Several Complex Variables. Contemporary Mathematics, vol. 662, pp. 121–129. American Mathematical Society, Providence (2016)
19. Kaiser, R., Retherford, J.: Eigenvalue distribution of nuclear operators: a survey. Vector Measures and Integral Representation of Operators, pp. 245–287. Essen University Press, Essen (1983)
20. Kalton, N.J., Peck, N.T., Roberts, J.W.: An F-space Sampler. London Mathematical Society Lecture Notes Series. Cambridge University Press, Cambridge (1984)
21. König, H.: A formula for the eigenvalues of a compact operator. Stud. Math. **65**, 141–146 (1979)
22. Konyagin, S.V.: Deviation of elements of a Banach space from a system of subspaces. Proc. Steklov Inst. Math. **284**(1), 204–207 (2014)
23. Lewicki, G.: Bernstein's "lethargy" theorem in metrizable topological linear spaces. Monatsh. Math. **113**, 213–226 (1992)
24. Lewicki, G.: A theorem of Bernstein's type for linear projections, Univ. Iagel. Acta Math. **27**, 23–27 (1988)
25. Marcus, A.S.: Some criteria for the completeness of a system of root vectors of a linear operator in a Banach space. Trans. Am. Math. Soc. **85**, 325–349 (1969)

26. Micherda, B.: Bernstein's lethargy theorem in SF-spaces. Z. Anal. Anwendungen **22**(1), 3–16 (2003)
27. Oikhberg, T.: Rate of decay of s-numbers. J. Approx. Theory **163**, 311–327 (2011)
28. Pietch, A.: Eigenvalues and s-Numbers. Cambridge University Press, Cambridge (1987)
29. Pietch, A.: Operator Ideals. North Holland, Amsterdam (1980)
30. Pietch, A.: HIstory of Banach Spaces and Linear Operators. Boston, Birkhauser (2007)
31. Pinkus, A.: Weierstrass and approximation theory. J. Approx. Theory **107**(1), 1–66 (2000)
32. Pleśniak, W.: On a theorem of S. N. Bernstein in F-spaces. Zeszyty Naukowe Uniwersytetu Jagiellonskiego, Prace Mat. **20**, 7–16 (1979)
33. Pleśniak, W.: Quasianalytic functions in the sense of Bernstein. Diss. Math. **147**, 1–70 (1977)
34. Pleśniak, W.: Characterization of quasi-analytic functions of several variables by means of rational approximation. Ann. Pol. Math. **27**, 149–157 (1973)
35. Rolewicz, S.: Metric Linear Spaces. PWN, Warszawa (1982)
36. Shapiro, H.S.: Some negative theorems of approximation theory. Mich. Math. J. **11**, 211–217 (1964)
37. Singer, I.: Best Approximation in Normed Linear Spaces by Elements of Linear Subspaces. Springer, Berlin (1970)
38. Timan, A.F.: Theory of Approximation of Functions of a Real Variable. Dover publications, New York (1994)
39. Tyuriemskih, I.S.: On one problem of S. N. Bernstein. In: Scientific Proceedings of Kaliningrad State Pedagogical Institute, vol. 52, pp. 123–129 (1967)
40. Tyuriemskih, I.S.: The B-property of hilbert spaces. Uch. Zap. Kalinin. Gos. Pedagog. Inst. **39**, 53–64 (1964)
41. Vasil'ev, A.I.: The inverse problem in the theory of best approximation in F-spaces, (In Russian). Dokl. Ross. Akad. Nauk. **365**(5), 583–585 (1999)
42. Weyl, H.: Inequalities between two kinds of eigenvalues of a linear trasformation. Proc. Natl. Acad. Sci. USA **35**, 408–411 (1949)
43. Wojtaszczyk, P.: Banach Spaces for Analysts. Studies in Advanced Mathematics, vol. 25. Cambridge University Press, Cambridge (1991)

Local Automorphisms on Finite-Dimensional Lie and Leibniz Algebras

Shavkat Ayupov and Karimbergen Kudaybergenov

Abstract We prove that a linear mapping on the algebra $\mathfrak{sl}_n$ of all trace zero complex matrices is a local automorphism if and only if it is an automorphism or an anti-automorphism. We also show that a linear mapping on a simple Leibniz algebra of the form $\mathfrak{sl}_n \dotplus \mathscr{I}$ is a local automorphism if and only if it is an automorphism. We give examples of finite-dimensional nilpotent Lie algebras $\mathscr{L}$ with $\dim \mathscr{L} \geq 3$ which admit local automorphisms which are not automorphisms.

Keywords Simple Lie algebra · Simple Leibniz algebra · Nilpotent Lie algebra · Automorphism · Local automorphism

1 Introduction

In last decades a series of papers have been devoted to study of mappings which are close to automorphism and derivation of associative algebras (especially of operator algebras and C*-algebras). Namely, the problems of describing so-called local automorphisms (respectively, local derivations) and 2-local automorphisms (respectively, 2-local derivations) have been considered. Later similar problems were extended for non associative algebras, in particular, for the case of Lie algebras.

Linear preserver problems (LPP) represent one of the most active research areas in matrix theory. According to the linear character of matrix theory, preserver problems mean here the characterizations of all linear transformations on a given linear space of matrices that leave certain functions, subsets, relations, etc. invariant (see for

S. Ayupov
V.I.Romanovskiy Institute of Mathematics, Uzbekistan Academy of Sciences,
81, Mirzo Ulughbek street, 100170 Tashkent, Uzbekistan
e-mail: sh_ayupov@mail.ru

K. Kudaybergenov (✉)
Ch. Abdirov 1, Department of Mathematics, Karakalpak State University,
Nukus 230113, Uzbekistan
e-mail: karim2006@mail.ru

© Springer Nature Switzerland AG 2018   31
Z. Ibragimov et al. (eds.), *Algebra, Complex Analysis, and Pluripotential Theory*,
Springer Proceedings in Mathematics & Statistics 264,
https://doi.org/10.1007/978-3-030-01144-4_3

example [17]). In this paper we present some applications of LPP to the study of local automorphisms of finite dimensional Lie and Leibniz algebras.

Let $\mathscr{A}$ be an associative algebra. Recall that a linear mapping Φ of $\mathscr{A}$ into itself is called a local automorphism (respectively, a local derivation) if for every $x \in \mathscr{A}$ there exists an automorphism (respectively, a derivation) Φ_x of $\mathscr{A}$, depending on x, such that $\Phi_x(x) = \Phi(x)$. These notions were introduced and investigated independently by Kadison [14] and Larson and Sourour [16]. Later, in 1997, P. Šemrl [19] introduced the concepts of 2-local automorphisms and 2-local derivations. A map $\Phi : \mathscr{A} \to \mathscr{A}$ (not linear in general) is called a 2-local automorphism (respectively, a 2-local derivation) if for every $x, y \in \mathscr{A}$, there exists an automorphism (respectively, a derivation) $\Phi_{x,y} : \mathscr{A} \to \mathscr{A}$ (depending on x, y) such that $\Phi_{x,y}(x) = \Phi(x)$, $\Phi_{x,y}(y) = \Phi(y)$. In [19], P. Šemrl described 2-local derivations and 2-local automorphisms on the algebra $B(H)$ of all bounded linear operators on the infinite-dimensional separable Hilbert space H by proving that every 2-local automorphism (respectively, 2-local derivation) on $B(H)$ is an automorphism (respectively, a derivation). A similar result for finite-dimensional case appeared later in [15]. Further, in [1], a new techniques was introduced to prove the same result for an arbitrary Hilbert space H (no separability is assumed.)

Afterwards the above considerations gave arise to similar questions in von Neumann algebras framework. First positive results have been obtained in [2] and [3] for finite and semi-finite von Neumann algebras respectively, by showing that all 2-local derivations on these algebras are derivations. Finally, in [4], the same result was obtained for purely infinite von Neumann algebras. This completed the solution of the above problem for arbitrary von Neumann algebras.

It is natural to study the corresponding analogues of these problems for automorphisms or derivations of non-associative algebras.

Let $\mathscr{L}$ be a Lie algebra. A derivation D of $\mathscr{L}$ is a linear map $D : \mathscr{L} \to \mathscr{L}$ which satisfies the condition $D([x, y]) = [D(x), y] + [x, D(y)]$ for all $x, y \in \mathscr{L}$. An automorphism (respectively, an anti-automorphism) Φ of $\mathscr{L}$ is an invertible linear map $\Phi : \mathscr{L} \to \mathscr{L}$ which satisfies the condition $\Phi([x, y]) = [\Phi(x), \Phi(y)]$ (respectively, $\Phi([x, y]) = [\Phi(y), \Phi(x)]$) for all $x, y \in \mathscr{L}$. The set of all automorphisms of a Lie algebra $\mathscr{L}$ is denoted by $\mathrm{Aut}\mathscr{L}$.

The notions of a local derivation (respectively, a local automorphism) and a 2-local derivation (respectively, a 2-local automorphism) for Lie algebras are defined as above, similar to the associative case. Every derivation (respectively, automorphism) of a Lie algebra $\mathscr{L}$ is a local derivation (respectively, local automorphism) and a 2-local derivation (respectively, 2-local automorphism). For a given Lie algebra $\mathscr{L}$, the main problem concerning these notions is to prove that they automatically become a derivation (respectively, an automorphism) or to give examples of local and 2-local derivations or automorphisms of $\mathscr{L}$, which are not derivations or automorphisms, respectively. For a finite-dimensional semi-simple Lie algebra $\mathscr{L}$ over an algebraically closed field of characteristic zero, the derivations and automorphisms of $\mathscr{L}$ are completely described in [12].

Recently in [6] we have proved that every local derivation on semi-simple Lie algebras is a derivation and gave examples of nilpotent finite-dimensional Lie algebras with local derivations which are not derivations.

Earlier in [5] the authors have proved that every 2-local derivation on a semi-simple Lie algebra $\mathscr{L}$ is a derivation, and showed that each finite-dimension nilpotent Lie algebra, with dimension larger than two, admits a 2-local derivation which is not a derivation.

In [11], Chen and Wang initiated study of 2-local automorphisms of finite-dimensional Lie algebras. They prove that if $\mathscr{L}$ is a simple Lie algebra of type A_l $(l \geq 1)$, D_l $(l \geq 4)$, or E_k $(k = 6, 7, 8)$ over an algebraically closed field of characteristic zero, then every 2-local automorphism of $\mathscr{L}$, is an automorphism. Finally, in [7] Ayupov and Kudaybergenov generalized this result of [11] and proved that every 2-local automorphism of a finite-dimensional semi-simple Lie algebra over an algebraically closed field of characteristic zero is an automorphism. Moreover, they show also that every nilpotent Lie algebra with finite dimension larger than two admits 2-local automorphisms which are not automorphisms. It should be noted that similar problems for local automorphism of finite-dimensional Lie algebras still remain open.

In the present paper we prove that a linear mapping on the algebra of all trace zero complex matrices $\mathfrak{sl}_n$ is a local automorphism if and only if it is an automorphism or an anti-automorphism. We also show that a linear mapping on a simple Leibniz algebra of the form $\mathfrak{sl}_n \dot{+} \mathscr{I}$ is a local automorphism if and only if it is an automorphism. We also give examples of finite-dimensional nilpotent Lie algebras $\mathscr{L}$ with dim $\mathscr{L} \geq 3$ which admit local automorphisms which are not automorphisms.

2 Local Automorphisms on $\mathfrak{sl}_n$

In this section we study local automorphisms of the simple Lie algebra $\mathfrak{sl}_n$ of all trace zero complex $n \times n$-matrices.

Firstly we shall present the following two Theorems.

Theorem 1 [16, Theorem 2.2]. *If $M_n(\mathbb{C})$ is the algebra of all complex $n \times n$-matrices, then $\Delta : M_n(\mathbb{C}) \to M_n(\mathbb{C})$ is a local automorphism iff is an automorphism or an anti-automorphism, i.e., either $\Delta(x) = axa^{-1}$ or $\Delta(x) = ax^t a^{-1}$ for a fixed a and for all $x \in M_n(\mathbb{C})$.*

Recall that here x^t denotes the transpose of the matrix x.

We say that a matrix x is square-zero if $x^2 = 0$. A mapping $\Phi : \mathfrak{sl}_n \to \mathfrak{sl}_n$ preserves square-zero matrices if $x \in \mathfrak{sl}_n$ and $x^2 = 0$ imply $\Phi(x)^2 = 0$.

It is clear that the linear mappings defined as follows

$$\Phi(x) = \lambda axa^{-1}, \ x \in \mathfrak{sl}_n \tag{1}$$

and

$$\Phi(x) = \lambda a x^t a^{-1}, \ x \in \mathfrak{sl}_n. \tag{2}$$

preserve square-zero matrices, where a is an invertible matrix from $M_n(\mathbb{C})$ and $\lambda \in \mathbb{C}$.

The following result about linear mappings preserving square-zero due to P. Šemrl (see [18]).

Theorem 2 (see [18, Corollary 2]). *Assume that $\Phi : \mathfrak{sl}_n \to \mathfrak{sl}_n$ is a bijective linear mapping preserving square-zero matrices. Then Φ is either of the form (1) or of the form (2).*

It is well-known [13] that for any automorphism Φ on $\mathfrak{sl}_n$ there exists an invertible matrix $a \in M_n(\mathbb{C})$ such that either $\Phi(x) = axa^{-1}$ for all $x \in \mathfrak{sl}_n$, or $\Phi(x) = -ax^t a^{-1}$ for all $x \in \mathfrak{sl}_n$. Recall that automorphisms of the form $\Phi(x) = axa^{-1}$ ($x \in M_n(\mathbb{C})$) is said to be inner.

Theorem 3 *A linear mapping $\Delta : \mathfrak{sl}_n \to \mathfrak{sl}_n$ is a local automorphism if and only if Δ is either an automorphism or an anti-automorphism, i.e., it has either the form*

(i) $\Delta(x) = \pm axa^{-1}, \ x \in \mathfrak{sl}_n$ *or*
(ii) $\Delta(x) = \pm ax^t a^{-1}, \ x \in \mathfrak{sl}_n.$

Proof Let Δ be a local automorphism on $\mathfrak{sl}_n$. Let us show that Δ is a bijective linear map preserving square-zero matrices.

Let $x \in \mathfrak{sl}_n$ be a non zero element. By the definition there exists an automorphism Φ^x on $\mathfrak{sl}_n$ such that $\Delta(x) = \Phi^x(x)$. Since Φ^x is an automorphism, it follows that $\Phi^x(x) \neq 0$. Thus $\Delta(x) \neq 0$, and therefore the kernel of Δ is trivial. Hence Δ is bijective.

Let now $x \in \mathfrak{sl}_n$ be a square-zero matrix, that is $x^2 = 0$. By the definition of local automorphism there exists an invertible matrix $a_x \in M_n(\mathbb{C})$ such that $\Delta(x) = a_x x a_x^{-1}$ or $\Delta(x) = -a_x x^t a_x^{-1}$. We have

$$\Delta(x)^2 = \left(a_x x a_x^{-1}\right)^2 = a_x x^2 a_x^{-1} = 0$$

or

$$\Delta(x)^2 = \left(-a_x x^t a_x^{-1}\right)^2 = a_x (x^2)^t a_x^{-1} = 0.$$

So, Δ is a bijective linear map preserving square-zero matrices. By Theorem 2 there exist an invertible matrix $a \in M_n(\mathbb{C})$ and a nonzero scalar λ such that either

$$\Delta(x) = \lambda axa^{-1}, \ \forall x \in \mathfrak{sl}_n \ or \ \Delta(x) = \lambda ax^t a^{-1}, \ \forall x \in \mathfrak{sl}_n. \tag{3}$$

Take the diagonal matrix $y = \begin{pmatrix} 1 & 0 & 0 & \cdots & 0 \\ 0 & -1 & 0 & \cdots & 0 \\ 0 & 0 & 0 & \cdots & 0 \\ & \cdot & \cdot & \cdot & \cdot \cdot \cdot \\ 0 & 0 & 0 & \cdots & 0 \end{pmatrix}$ in $\mathfrak{sl}_n$. By the definition of local

automorphism there exists an invertible matrix $a_y \in M_n(\mathbb{C})$ such that

$$\Delta(y) = a_y y a_y^{-1} \quad \text{or} \quad \Delta(y) = -a_y y^t a_y^{-1}. \tag{4}$$

Let $p_x(t) = \det(t\mathbf{1} - x)$ be the characteristic polynomial of the matrix x, where $\mathbf{1}$ is the unit matrix in $M_n(\mathbb{C})$.

We shall find the characteristic polynomial of the matrix $\Delta(y)$, using the equalities (3) and (4). Taking into account that $y = y^t$ and comparing the equalities (3) and (4), we obtain that $\Delta(y) = \lambda a y a^{-1} = \pm a_y y a_y^{-1}$. Then

$$p_{\Delta(y)}(t) = p_{\lambda a y a^{-1}}(t) = \det(t\mathbf{1} - \lambda a y a^{-1}) = \det(a(t\mathbf{1} - \lambda y)a^{-1}) =$$
$$= \det(t\mathbf{1} - \lambda y) = (t - \lambda)(t + \lambda)t^{n-2}$$

and

$$p_{\Delta(y)}(t) = p_{\pm a_y y a_y^{-1}}(t) = \det(t\mathbf{1} \mp a_y y a_y^{-1}) = \det(a_y(t\mathbf{1} \mp y)a_y^{-1}) =$$
$$= \det(t\mathbf{1} \mp y) = (t + 1)(t - 1)t^{n-2}$$

Thus $(t - \lambda)(t + \lambda)t^{n-2} = (t - 1)(t + 1)t^{n-2}$, and therefore $\lambda = \pm 1$. So, $\Delta(x) = \pm a x a^{-1}$ or $\Delta(x) = \pm a x^t a^{-1}$ for all $x \in \mathfrak{sl}_n$.

Now we shall show that every anti-automorphism on $\mathfrak{sl}_n$ is a local automorphism. Let us first show that the mappings defined by

$$\Delta(x) = x^t, \ x \in \mathfrak{sl}_n \tag{5}$$

and

$$\Delta(x) = -x, \ x \in \mathfrak{sl}_n \tag{6}$$

are local automorphisms.

Let $x \in \mathfrak{sl}_n$ be an arbitrary matrix. By Theorem 1, the mapping on $M_n(\mathbb{C})$ defined as $z \to z^t$ is a local associative automorphism. Therefore there exists an invertible matrix $a_x \in M_n(\mathbb{C})$ such that $x^t = a_x x a_x^{-1}$. Hence

$$x^t = a_x x a_x^{-1} \text{and} - x = -a_x^t x^t (a_x^t)^{-1}.$$

This means that the mappings defined as (5) and (6) are local automorphisms on $\mathfrak{sl}_n$. Finally, since every anti-automorphism on $\mathfrak{sl}_n$ is a superposition of an anti-automorphism of the form (5) or (6) and an inner automorphism, it follows that every anti-automorphism on $\mathfrak{sl}_n$ is a local automorphism. The proof is complete.

3 Local Automorphisms of Algebras $\mathfrak{sl}_n \dotplus \mathscr{I}$

In this section we study local automorphisms of simple Leibniz algebras of the form $\mathfrak{sl}_n \dotplus \mathscr{I}$.

An algebra $(\mathscr{L}, [\cdot, \cdot])$ over a field $\mathbb{F}$ is called a Leibniz algebra if it satisfies the property

$$x, [y, z]] = [[x, y], z] - [[x, z], y] \text{ for all } x, y \in \mathscr{L},$$

which is called Leibniz identity.

For a Leibniz algebra $\mathscr{L}$, a subspace generated by squares of its elements $\mathscr{I} = span\{[x, x] : x \in \mathscr{L}\}$ due to Leibniz identity becomes an ideal, and the quotient $\mathscr{G}_{\mathscr{L}} = \mathscr{L}/\mathscr{I}$ is a Lie algebra called liezation of $\mathscr{L}$. Moreover, $[\mathscr{L}, \mathscr{I}] = 0$. In general, $[\mathscr{I}, \mathscr{L}] \neq 0$. Since we are interested in Leibniz algebras which are not Lie algebras, we will always assume that $\mathscr{I} \neq 0$.

A Leibniz algebra $\mathscr{L}$ is called *simple* if its liezation is a simple Lie algebra and the ideal $\mathscr{I}$ is a simple ideal. Equivalently, $\mathscr{L}$ is simple iff $\mathscr{I}$ is the only non-trivial ideal of $\mathscr{L}$.

Let $\mathscr{G}$ be a Lie algebra and $\mathscr{V}$ a (right) $\mathscr{G}$-module. Endow the vector space $\mathscr{L} = \mathscr{G} \oplus \mathscr{V}$ with the bracket product as follows:

$$[(g_1, v_1), (g_2, v_2)] := ([g_1, g_2], v_1.g_2),$$

where $v.g$ (sometimes denoted as $[v, g]$) is the action of an element g of $\mathscr{G}$ on $v \in \mathscr{V}$. Then $\mathscr{L}$ is a Leibniz algebra, denoted as $\mathscr{G} \ltimes V$.

The following Theorem is the main result of this section.

Theorem 4 *Let $\mathfrak{sl}_n \dotplus \mathscr{I}$ be a simple Leibniz algebra with $\mathscr{I} \neq \{0\}$. Then a linear mapping $\Delta : \mathfrak{sl}_n \dotplus \mathscr{I} \to \mathfrak{sl}_n \dotplus \mathscr{I}$ is a local automorphism if and only if Δ is an automorphism.*

First we give necessary notations concerning the algebra $\mathfrak{sl}_n$. Let $\{e_{ij} : 1 \leq i, j \leq n\}$ be the system of matrix units in $M_n(\mathbb{C})$. A subalgebra

$$\mathfrak{h} = \left\{ h = \sum_{i=1}^{n} a_i e_{ii} : \sum_{i=1}^{n} a_i = 0 \right\}$$

is a Cartan subalgebra of $\mathfrak{sl}_n$. For any $i, j \in \{1, \ldots, n\}$ we have

$$[h, e_{ij}] = (a_i - a_j)e_{ij}.$$

Let $\mathfrak{h}^*$ be the space of all linear functionals on $\mathfrak{h}$. Denote by ε_i, $1 \leq i \leq n$, the elements of $\mathfrak{h}^*$ defined by

$$\langle \varepsilon_i, h \rangle = a_i.$$

The root system R consists the elements of the form $\varepsilon_i - \varepsilon_j$, $i \neq j$, and $\mathbb{C}e_{ij}$ are the corresponding root subspaces, moreover, $R^+ = \{\alpha = \varepsilon_i - \varepsilon_j : i < j\}$ is the set of all positive roots. Denote

$$\alpha_i = \varepsilon_i - \varepsilon_{i+1}, \ i = 1, \ldots, n-1.$$

From the relation

$$\pm(\varepsilon_i - \varepsilon_j) = \pm \sum_{k=i}^{j-1} \alpha_k, \ i < j$$

we see that the set Π formed by the elements α_i, $i = 1, \ldots, n-1$, is a base of R (see [12]).

By [10, Lemma 2.2], there exists an element $h_0 \in \mathfrak{h}$ such that $\alpha(h_0) \neq \beta(h_0)$ for every $\alpha, \beta \in R$, $\alpha \neq \beta$. Such elements h_0 are called *strongly regular* elements of $\mathfrak{sl}_n$. Again by [10, Lemma 2.2], every strongly regular element h_0 is a *regular semi-simple element*, i.e.

$$\{x \in \mathfrak{sl}_n : [h_0, x] = 0\} = \mathfrak{h}.$$

Lemma 1 *Let Φ be an automorphism on $\mathfrak{sl}_n$ and let h_0 be a strongly regular element in $\mathfrak{h}$ such that $\Phi(h_0) = -h_0$. Then there exists an invertible diagonal matrix $a \in M_n(\mathbb{C})$ such that $\Phi(x) = -ax^t a^{-1}$.*

Proof Let Θ be an automorphism on $\mathfrak{sl}_n$ defined by $\Theta(x) = -x^t$, $x \in \mathfrak{sl}_n$. Since $(\Theta \circ \Phi)(h_0) = h_0$, by [7, Lemma 2.2], the automorphism $\Theta \circ \Phi$ leaves every element of $\mathfrak{h}$ fixed. Further by [21, P. 109, Lemma 2], there exists an invertible diagonal matrix $b \in M_n(\mathbb{C})$ such that $(\Theta \circ \Phi)(x) = bxb^{-1}$. Thus $\Phi(x) = -b^t x^t (b^t)^{-1}$ for all $x \in \mathfrak{sl}_n$. The proof is complete.

Let $\mathscr{S}$ be a simple Lie algebra. For any $\mathscr{S}$-module $\mathscr{I}$ and any automorphism σ of $\mathscr{S}$, we define the new $\mathscr{S}$-module structure $\mathscr{I}^\sigma$ on $\mathscr{I}$, given by the action

$$v \cdot x = v\sigma(x), \ \forall v \in \mathscr{I}, x \in \mathscr{S}.$$

We know from [12] that $\mathscr{I} \simeq \mathscr{I}^\sigma$ if σ is an inner automorphism of $\mathscr{S}$. If σ is not an inner automorphism of $\mathscr{S}$, we generally do not have $\mathscr{I} \simeq \mathscr{I}^\sigma$.

Let $\mathscr{L} = \mathscr{S} \dot{+} \mathscr{I}$ be a simple complex Leibniz algebra. In [9] it was proved that an automorphism σ of $\mathscr{S}$ can be extended to an automorphism φ of $\mathscr{L}$ if and only if $\mathscr{I} \simeq \mathscr{I}^\sigma$ as $\mathscr{S}$-modules. In [9] we also have presented an example which shows the existence of automorphism of $\mathscr{S}$ which can not be extended to the whole algebra $\mathscr{L}$.

Let $\Phi : \mathscr{L} \to \mathscr{L}$ be an automorphism. Then Φ can be represented as (see [8])

$$\Phi = \begin{pmatrix} \Phi_{\mathscr{S}} & 0 \\ \Phi_{\mathscr{S},\mathscr{I}} & \Phi_{\mathscr{I}} \end{pmatrix}, \tag{7}$$

where $\Phi_{\mathscr{S}}$ is an automorphism of $\mathscr{S}$, $\Phi_{\mathscr{S},\mathscr{I}}$ is a $\mathscr{S}$-module homomorphism from $\mathscr{S}$ to $\mathscr{I}^{\Phi_{\mathscr{S}}}$ and $\Phi_{\mathscr{I}}$ is a $\mathscr{S}$-module isomorphism from $\mathscr{I}$ onto $\mathscr{I}^{\Phi_{\mathscr{S}}}$, i.e.,

$$\left[\Phi_{\mathscr{S},\mathscr{I}}(x), \Phi_{\mathscr{S}}(y) \right] = \Phi_{\mathscr{S},\mathscr{I}}([x, y])$$

and

$$\left[\Phi_{\mathscr{I}}(i), \Phi_{\mathscr{S}}(y) \right] = \Phi_{\mathscr{I}}([i, y])$$

for all $x, y \in \mathscr{S}$ and $i \in \mathscr{I}$. Note that $\Phi_{\mathscr{S},\mathscr{I}} = \omega\theta \circ \Phi_{\mathscr{S}}$, $(\omega \in \mathbb{C})$ where θ is a $\mathscr{S}$-module isomorphism from $\mathscr{S}$ onto $\mathscr{I}^{\Phi_{\mathscr{S}}}$ and for dim $\mathscr{S} \neq$ dim $\mathscr{I}$, we have $\Phi_{\mathscr{S},\mathscr{I}} = 0$.

Thus every local automorphism Δ on $\mathscr{L} = \mathscr{S}\dot{+}\mathscr{I}$ is also represented as

$$\Delta = \begin{pmatrix} \Delta_{\mathscr{S}} & 0 \\ \Delta_{\mathscr{S},\mathscr{I}} & \Delta_{\mathscr{I}} \end{pmatrix}, \tag{8}$$

where $\Delta_{\mathscr{S}}$ is a local automorphism of $\mathscr{S}$.

Below we shall use the following properties of local automorphisms. Let Δ be a local automorphism of $\mathscr{S}\dot{+}\mathscr{I}$ and let $x = x_{\mathscr{S}} + x_{\mathscr{I}} \in \mathscr{S} + \mathscr{I}$. Take an automorphism Φ^x such that $\Delta(x) = \Phi^x(x)$. Then (7) and (8) imply that $\Phi^x_{\mathscr{S}}(x_{\mathscr{S}}) = \Delta_{\mathscr{S}}(x_{\mathscr{S}})$ and $\Phi^x_{\mathscr{S},\mathscr{I}}(x_{\mathscr{S}}) + \Phi^x_{\mathscr{I}}(x_{\mathscr{I}}) = \Delta_{\mathscr{S},\mathscr{I}}(x_{\mathscr{S}}) + \Delta_{\mathscr{I}}(x_{\mathscr{I}})$.

Let $\mathscr{S}\dot{+}\mathscr{I}$ be a simple complex Leibniz algebra. From the representation theory of semisimple Lie algebras [12] we have that a Cartan subalgebra $\mathscr{H}$ of the simple Lie algebra $\mathscr{S}$ acts diagonalizable on $\mathscr{S}$-module $\mathscr{I}$:

$$\mathscr{I} = \bigoplus_{\beta \in \Gamma} \mathscr{I}_{\beta},$$

where

$$\mathscr{I}_{\beta} = \{i \in \mathscr{I} : [i, h] = \beta(h)i, \ \forall h \in \mathscr{H}\},$$
$$\Gamma = \{\beta \in \mathscr{H}^* : \mathscr{I}_{\beta} \neq \{0\}\}$$

and $\mathscr{H}^*$ is the space of all linear functionals on $\mathscr{H}$. Elements of Γ are called weights of $\mathscr{I}$.

Lemma 2 *Let Φ be an automorphism on $\mathfrak{sl}_n\dot{+}\mathscr{I}$ and let h_0 be a strongly regular element in $\mathfrak{h}$ such that $\Phi_{\mathfrak{sl}_n}(h_0) = h_0$. Then there exists $\lambda_\beta \in \mathbb{C}$ such that $\Phi(y_\beta) = \lambda_\beta u_\beta$, where β is a highest weight of $\mathscr{I}$.*

Proof Since $\Phi_{\mathfrak{sl}_n}(h_0) = h_0$, by [7, Lemma 2.2] for every root $\alpha \in R$ there exists non zero $c_\alpha \in \mathbb{C}$ such that $\Phi_{\mathfrak{sl}_n}(e_\alpha) = c_\alpha e_\alpha$. Then

$$[\Phi(y_\beta), e_\alpha] = c_\alpha^{-1}[\Phi(y_\beta), \Phi(e_\alpha)] = c_\alpha^{-1}\Phi([y_\beta, e_\alpha]) = c_\alpha^{-1}\Phi(0) = 0$$

for each positive root α. Since the highest weight subspace $\mathscr{I}_\beta$ is one dimensional, it follows that $\Phi(y_\beta) = \lambda_\beta y_\beta$. The proof is complete.

From now on $\mathfrak{sl}_n \dot{+} \mathscr{I}$ is a simple Leibniz algebra with $\mathscr{I} \neq \{0\}$.

Lemma 3 *Let* $\Delta : \mathfrak{sl}_n \dot{+} \mathscr{I} \to \mathfrak{sl}_n \dot{+} \mathscr{I}$ *be a linear mapping such that* $\Delta_{\mathfrak{sl}_n}(x) = x^t$ *for all* $x \in \mathfrak{sl}_n$. *Then* Δ *is not a local automorphism.*

Proof Suppose that Δ is a local automorphism. Take an element $x = h_0 + y_\beta \in \mathfrak{sl}_n \dot{+} \mathscr{I}$, where $h_0 \in \mathfrak{h}$ is a strongly regular element and y_β is a highest weight vector of $\mathscr{I}$, i.e., $[y_\beta, e_\alpha] = 0$ for every positive root α. Take an automorphism Φ^{h_0} on $\mathfrak{sl}_n \dot{+} \mathscr{I}$ such that $\Delta(h_0) = \Phi^{h_0}(h_0)$. Then

$$h_0 + \Delta_{\mathfrak{sl}_n, \mathscr{I}}(h_0) = \Delta_{\mathfrak{sl}_n}(h_0) + \Delta_{\mathfrak{sl}_n, \mathscr{I}}(h) = \Delta(h_0) =$$
$$= \Phi^{h_0}(h_0) = \Phi^{h_0}_{\mathfrak{sl}_n}(h_0) + \Phi^{h_0}_{\mathfrak{sl}_n, \mathscr{I}}(h_0).$$

Thus

$$\Phi^{h_0}_{\mathfrak{sl}_n}(h_0) = h_0,$$

and therefore

$$\Phi^{h_0}_{\mathfrak{sl}_n, \mathscr{I}}(h_0) = \omega_1 \theta(\Phi^{h_0}_{\mathfrak{sl}_n}(h_0)) = \omega_1 \theta(h_0) = y_0^{(1)} \in \mathscr{I}_0.$$

Further

$$\Delta(h_0 + y_\beta) = \Delta(h_0) + \Delta(y_\beta) = \Phi^{h_0}(h_0) + \Delta(y_\beta) =$$
$$= \Phi^{h_0}_{\mathfrak{sl}_n}(h_0) + \Phi^{h_0}_{\mathfrak{sl}_n, \mathscr{I}}(h_0) + \Delta(y_\beta) = h_0 + y_0^{(1)} + \Delta(y_\beta).$$

Take an automorphism Φ^x on $\mathfrak{sl}_n \dot{+} \mathscr{I}$ such that $\Delta(x) = \Phi^x(x)$. Then

$$\Delta(h_0 + y_\beta) = \Phi^x(h_0 + y_\beta) = \Phi^x_{\mathfrak{sl}_n}(h_0) + \Phi^x_{\mathfrak{sl}_n, \mathscr{I}}(h_0) + \Phi^x_{\mathscr{I}}(y_\beta).$$

Comparing the last two equalities, we obtain that $\Phi^x_{\mathfrak{sl}_n}(h_0) = h_0$, and

$$\Phi^x_{\mathfrak{sl}_n, \mathscr{I}}(h_0) = \omega_2 \theta(\Phi^x_{\mathfrak{sl}_n}(h_0)) = \omega_2 \theta(h_0) = y_0^{(2)} \in \mathscr{I}_0.$$

By Lemma 2 it follows that $\Phi^x_{\mathscr{I}}(y_\beta) = \lambda_\beta y_\beta$. Hence

$$\Delta(y_\beta) = \lambda_\beta y_\beta + y_0,$$

where $y_0 = y_0^{(2)} - y_0^{(1)} \in \mathscr{I}_0$.

Now let us take an element $z = e_\alpha + y_\beta$, where β is the highest weight of $\mathscr{I}$ and $\alpha \in R$ is a positive root. Taking into account that $[y_\beta, e_\alpha] = 0$, we have

$$[\Delta(e_\alpha + y_\beta), \Delta(e_\alpha + y_\beta)] = [\Phi^z(e_\alpha + y_\beta), \Phi^z(e_\alpha + y_\beta)] = \Phi^z([e_\alpha + y_\beta, e_\alpha + y_\beta]) =$$
$$= \Phi^z([y_\beta, e_\alpha]) = 0.$$

On the other hand,

$$[\Delta_{\mathscr{S},\mathscr{I}}(e_\alpha), e_{-\alpha}] = [\Delta_{\mathscr{S},\mathscr{I}}(e_\alpha), \Delta_{\mathscr{S}}(e_\alpha)] =$$
$$= [\Delta_{\mathscr{S}}(e_\alpha) + \Delta_{\mathscr{S},\mathscr{I}}(e_\alpha), \Delta_{\mathscr{S}}(e_\alpha) + \Delta_{\mathscr{S},\mathscr{I}}(e_\alpha)] =$$
$$= [\Delta(e_\alpha), \Delta(e_\alpha)] = [\Phi^{e_\alpha}(e_\alpha), \Phi^{e_\alpha}(e_\alpha)] = \Phi^{e_\alpha}([e_\alpha, e_\alpha]) = 0$$

and

$$[\Delta(e_\alpha + y_\beta), \Delta(e_\alpha + y_\beta)] = [\Delta(e_\alpha) + \Delta(y_\beta), \Delta(e_\alpha) + \Delta(y_\beta)] =$$
$$= [\Delta_{\mathscr{S}}(e_\alpha) + \Delta_{\mathscr{S},\mathscr{I}}(e_\alpha) + \Delta(y_\beta),$$
$$\Delta_{\mathscr{S}}(e_\alpha) + \Delta_{\mathscr{S},\mathscr{I}}(e_\alpha) + \Delta(y_\beta)] =$$
$$= [e_{-\alpha} + \Delta_{\mathscr{S},\mathscr{I}}(e_\alpha) + \lambda_\beta y_\beta + y_0,$$
$$e_{-\alpha} + \Delta_{\mathscr{S},\mathscr{I}}(e_\alpha) + \lambda_\beta y_\beta + y_0] =$$
$$= [\lambda_\beta y_\beta, e_{-\alpha}] + [y_0, e_{-\alpha}] \in \mathscr{I}_{\beta-\alpha} + \mathscr{I}_{-\alpha},$$

Thus $[\lambda_\beta y_\beta, e_{-\alpha}] = 0$ for all positive roots α, and therefore $[\lambda_\beta y_\beta, e_\gamma] = 0$ for all roots $\gamma \in R$. This means that $[\lambda_\beta y_\beta, \mathfrak{sl}_n] = 0$. Since $\mathscr{I}$ is an irreducible $\mathfrak{sl}_n$-module, it follows that $\lambda_\beta = 0$. So $\Phi^x_{\mathscr{I}}(y_\beta) = \lambda_\beta y_\beta = 0$, which contradicts that $\Phi^x_{\mathscr{I}}$ is invertible, because $\Phi^x_{\mathscr{I}}$ is an autmorphism. The proof is complete.

Lemma 4 *Let* $\Delta : \mathfrak{sl}_n \dot{+} \mathscr{I} \to \mathfrak{sl}_n \dot{+} \mathscr{I}$ *be a linear mapping such that* $\Delta_{\mathfrak{sl}_n}(x) = -x$ *for all* $\mathfrak{sl}_n$. *Then* Δ *is not a local automorphism.*

Proof Assume Δ is a local automorphism. Take an element $x = h_0 + y_\beta \in \mathfrak{sl}_n \dot{+} \mathscr{I}$, where $h_0 \in \mathfrak{h}$ is a strongly regular element and y_β is a highest weight vector of $\mathscr{I}$, i.e., $[y_\beta, e_\alpha] = 0$ for all positive root α. We have

$$\Delta(h_0 + y_\beta) = \Delta(h_0) + \Delta(y_\beta) = \Phi^h(h_0) + \Delta(y_\beta) =$$
$$= -h_0 + y_0^{(1)} + \Delta(y_\beta).$$

Take an automorphism Φ^x on $\mathfrak{sl}_n \dot{+} \mathscr{I}$ such that $\Delta(x) = \Phi^x(x)$. Then

$$\Delta(h_0 + y_\beta) = \Phi^x(h_0 + y_\beta) = \Phi^x_{\mathfrak{sl}_n}(h_0) + \Phi^x_{\mathfrak{sl}_n, \mathscr{I}}(h_0) + \Phi^x_{\mathscr{I}}(y_\beta).$$

Comparing the last two equalities, we obtain that

$$\Phi^x_{\mathfrak{sl}_n}(h_0) = -h_0.$$

Thus $\Phi^x_{\mathfrak{sl}_n, \mathscr{I}}(h_0) = y_0^{(2)} \in \mathscr{I}_0$ and $\Phi^x_{\mathscr{I}}(y_\beta) = \lambda_\beta y_{-\beta}$. Hence

$$\Delta(y_\beta) = \lambda_\beta y_{-\beta} + y_0.$$

Let $x = e_\alpha + y_\beta$, where β be the highest weight of $\mathscr{I}$ and α be a positive root of $\mathfrak{sl}_n$. As in the proof of the previous Lemma, we obtain that

$$
\begin{aligned}
\left[\Delta(e_\alpha + y_\beta), \Delta(e_\alpha + y_\beta)\right] &= \left[\Delta(e_\alpha) + \Delta(y_\beta), \Delta(e_\alpha) + \Delta(y_\beta)\right] = \\
&= [-e_\alpha + \Delta_{\mathfrak{sl}_n, \mathscr{I}}(e_\alpha) + \lambda_\beta y_{-\beta} + y_0, \\
&\quad\quad -e_\alpha + \Delta_{\mathfrak{sl}_n, \mathscr{I}}(e_\alpha) + \lambda_\beta y_{-\beta} + y_0] = \\
&= -\lambda_\beta \left[y_{-\beta}, e_\alpha\right] + [y_0, e_{-\alpha}]
\end{aligned}
$$

and

$$
\begin{aligned}
\left[\Delta(e_\alpha + y_\beta), \Delta(e_\alpha + y_\beta)\right] &= \left[\Phi^x(e_\alpha + y_\beta), \Phi^x(e_\alpha + y_\beta)\right] = \Phi^x\left(\left[e_\alpha + y_\beta, e_\alpha + y_\beta\right]\right) = \\
&= \Phi^x([y_\beta, e_\alpha]) = 0,
\end{aligned}
$$

because $\left[y_\beta, e_\alpha\right] = 0$. Thus $\lambda_\beta \left[y_{-\beta}, e_\alpha\right] = 0$ for all α. Hence $\lambda_\beta = 0$ and therefore $\Delta(y_\beta) = 0$, which contradicts the inversibility of Δ. The proof is complete.

The following Lemma is a particular case of [8, Theorem 6.9].

Lemma 5 *Let* $\Delta : \mathfrak{sl}_n \dot{+} \mathscr{I} \to \mathfrak{sl}_n \dot{+} \mathscr{I}$ *be a local automorphism such that* $\Delta_{\mathfrak{sl}_n}(x) = x$ *for all* $x \in \mathfrak{sl}_n$. *Then* Δ *is an automorphism.*

Denote by $\tau : \mathfrak{sl}_n \to \mathfrak{sl}_n$ the automorphism of $\mathfrak{sl}_n$ defined by

$$\tau(x) = -x^t, \quad x \in \mathfrak{sl}_n.$$

Lemma 6 *Let* $\Delta : \mathfrak{sl}_n \dot{+} \mathscr{I} \to \mathfrak{sl}_n \dot{+} \mathscr{I}$ *be a local automorphism such that* $\Delta_{\mathfrak{sl}_n} = \tau$. *Then* $\mathscr{I} \cong \mathscr{I}^\tau$ *and* Δ *is an automorphism.*

Proof Let $\Delta : \mathfrak{sl}_n \dot{+} \mathscr{I} \to \mathfrak{sl}_n \dot{+} \mathscr{I}$ be a local automorphism such that $\Delta_{\mathfrak{sl}_n}(x) = -x^t$ for all $x \in \mathfrak{sl}_n$. Let us show that $\mathscr{I} \cong \mathscr{I}^\tau$.

Take an automorphism Φ^{h_0} of $\mathfrak{sl}_n \dot{+} \mathscr{I}$ such that $\Phi^{h_0}_{\mathfrak{sl}_n}(h_0) = \Delta_{\mathfrak{sl}_n}(h_0) = -h_0$, where $h_0 \in \mathfrak{h}$ is a strongly regular element. By Lemma 1 it follows that $\Phi^{h_0}_{\mathfrak{sl}_n}(x) = -ax^t a^{-1}$ for all $x \in \mathfrak{sl}_n$, where a is an invertible diagonal matrix. Then $\widetilde{\Phi}_a^{-1} \circ \Phi^{h_0}$ is an automorphism of $\mathfrak{sl}_n \dot{+} \mathscr{I}$ such that $\left(\widetilde{\Phi}_a^{-1} \circ \Phi^{h_0}\right)_{\mathfrak{sl}_n}(x) = -x^t$ for all $x \in \mathfrak{sl}_n$, where $\widetilde{\Phi}_a$ is an extension onto $\mathfrak{sl}_n \dot{+} \mathscr{I}$ of the inner automorphism of $\mathfrak{sl}_n$, generated by the element a. Thus the restriction $\widetilde{\Phi}_a^{-1} \circ \Phi^{h_0}|_{\mathscr{I}}$ is a $\mathfrak{sl}_n$-module isomorphism from $\mathscr{I}$ onto $\mathscr{I}^\tau$, that is $\mathscr{I} \cong \mathscr{I}^\tau$. Further $\Delta \circ \widetilde{\Phi}_a^{-1} \circ \Phi^{h_0}$ is a local automorphism such that $\left(\Delta \circ \widetilde{\Phi}_a^{-1} \circ \Phi^{h_0}\right)_{\mathfrak{sl}_n}$ is an identical map on $\mathfrak{sl}_n$. Lemma 5 implies that Δ is an automorphism. The proof is complete.

Proof of Theorem 4. It is suffices to show that every local automorphism on $\mathfrak{sl}_n \dot{+} \mathscr{I}$ is an automorphism.

Let Δ be a local automorphism on $\mathfrak{sl}_n \dot{+} \mathscr{I}$. By Theorem 3, the restriction $\Delta_{\mathfrak{sl}_n}$ is either an automorphism or an anti-automorphism, i.e., there exists an invertible matrix a such that $\Delta_{\mathfrak{sl}_n}(x) = \pm axa^{-1}$ or $\Delta_{\mathfrak{sl}_n}(x) = \pm ax^t a^{-1}$ for all $x \in \mathfrak{sl}_n$. Thus $\left(\widetilde{\Phi}_a^{-1} \circ \Delta\right)_{\mathfrak{sl}_n}(x) = \pm x$ or $\left(\widetilde{\Phi}_a^{-1} \circ \Delta\right)_{\mathfrak{sl}_n}(x) = \pm x^t$, where $\widetilde{\Phi}_a$ is an extension onto $\mathfrak{sl}_n \dotplus \mathscr{I}$ of the inner automorphism of $\mathfrak{sl}_n$, generated by the element a. By Lemmata 3 and 4, it follows that $\left(\widetilde{\Phi}_a^{-1} \circ \Delta\right)_{\mathfrak{sl}_n}$ can not be an anti-automorphism. So it is an automorphism, i.e., $\left(\widetilde{\Phi}_a^{-1} \circ \Delta\right)_{\mathfrak{sl}_n}(x) = x$ or $\left(\widetilde{\Phi}_a^{-1} \circ \Delta\right)_{\mathfrak{sl}_n}(x) = -x^t$. Further Lemmata 5 and 6 imply that $\widetilde{\Phi}_a^{-1} \circ \Delta$ is an automorphism and therefore Δ is also an automorphism. The proof is complete.

4 Local Automorphisms of Filiform Lie Algebras

In this section we consider a special class of nilpotent Lie algebras, so-called filiform Lie algebras, and show that they admit local automorphisms which are not automorphisms.

A Lie algebra $\mathscr{L}$ is called *nilpotent* if $\mathscr{L}^k = \{0\}$ for some $k \in \mathbb{N}$, where $\mathscr{L}^0 = \mathscr{L}$, $\mathscr{L}^k = [\mathscr{L}^{k-1}, \mathscr{L}]$, $k \geq 1$.

In particular, a nilpotent Lie algebra $\mathscr{L}$ is called *filiform* if $\dim \mathscr{L}^k = n - k - 1$ for $1 \leq k \leq n - 1$.

Theorem 5 *Let $\mathscr{L}$ be a finite-dimensional filiform Lie algebra with $\dim \mathscr{L} \geq 3$. Then $\mathscr{L}$ admits a local automorphism which is not an automorphism.*

It is known [20] that there exists a basis $\{e_1, e_2, \ldots, e_n\}$ of $\mathscr{L}$ such that

$$[e_1, e_i] = e_{i+1} \tag{9}$$

for all $i \in \overline{2, n-1}$.

Note that a filiform Lie algebra $\mathscr{L}$ besides (9) may have also other non-trivial commutators.

From (9) it follows that $\{e_{k+2}, \ldots, e_n\}$ is a basis in $\mathscr{L}^k$ for all $1 \leq k \leq n-2$ and $e_n \in Z(\mathscr{L})$. Since $[\mathscr{L}^1, \mathscr{L}^{n-3}] \subseteq \mathscr{L}^{n-1} = \{0\}$, it follows that

$$[e_i, e_{n-1}] = 0 \tag{10}$$

for all $i = 3, \ldots, n$.

Define a linear operator Φ on $\mathscr{L}$ by

$$\Phi(x) = x + \alpha x_2 e_{n-1} + x_3 e_n, \quad x = \sum_{k=1}^{n} x_k e_k \in \mathscr{L}, \tag{11}$$

where $\alpha \in \mathbb{C}$.

Lemma 7 *A linear operator Φ on $\mathscr{L}$ defined by (11) is an automorphism if and only if $\alpha = 1$.*

Proof Let $x = \sum_{k=1}^{n} x_k e_k, \; y = \sum_{k=1}^{n} y_k e_k \in \mathscr{L}$. Then $[x, y] = (x_1 y_2 - x_2 y_1) e_3 + z$, where $z \in \mathscr{L}^2$. Thus

$$\Phi([x, y]) = [x, y] + (x_1 y_2 - x_2 y_1) e_n.$$

Taking into account (10), we obtain that

$$[\Phi(x), \Phi(y)] = [x + \alpha x_2 e_{n-1} + x_3 e_n . y + \alpha y_2 e_{n-1} + y_3 e_n] =$$
$$= [x, y] + \alpha(x_1 y_2 - x_2 y_1) e_n.$$

Comparing the last two equalities we obtain that the map Φ is an automorphism if and only if $\alpha = 1$. The proof is complete.

Consider the linear operator Δ defined by (11) with $\alpha = 0$.

Lemma 8 *The linear operator Δ is a local automorphism which is not an automorphism.*

Proof By Lemma 7, Δ is not an automorphism.

Let us show that Δ is a local automorphism. Let Ψ_β be a linear operator on $\mathscr{L}$ defined by

$$\Psi_\beta(u) = u + \beta u_2 e_n, \; u = \sum_{k=1}^{n} u_k e_k \in \mathscr{L}.$$

It is clear that Ψ_β is bijective. Since $\Psi_\beta|_{[\mathscr{L}, \mathscr{L}]} \equiv \mathrm{id}_{[\mathscr{L}, \mathscr{L}]}$ and $e_n \in Z(\mathscr{L})$, it follows that

$$\Psi_\beta([x, y]) = [x, y] = [\Psi_\beta(x), \Psi_\beta(y)]$$

for all $x, y \in \mathscr{L}$. So, Ψ_β is an automorphism.

Finally, for any $x = \sum_{k=1}^{n} x_k e_k$ let us show an automorphism that coincides with Δ at the point x. Denote by Φ the automorphism defined by (11) with $\alpha = 1$.

Case 1. $x_2 = 0$. Then

$$\Phi(x) = x + x_3 e_n = \Delta(x).$$

Case 2. $x_2 \neq 0$. Set $\beta = \frac{x_3}{x_2}$. Then

$$\Psi_\beta(x) = x + \beta x_2 e_n = x + x_3 e_n = \Delta(x).$$

The proof is complete.

Acknowledgements The motivation to study the problems considered in this paper came out from discussions made with Professor E.Zelmanov during the Second USA–Uzbekistan Conference held at the Urgench State University on August, 2017, which the authors gratefully acknowledge. The authors are indebted to Professor Mauro Costantini from the University of Padova for valuable comments to the initial version of the present paper.

References

1. Ayupov, Sh.A, Kudaybergenov, K.K.: 2-local derivations and automorphisms on $B(H)$. J. Math. Anal. Appl. **395**, 15–18 (2012)
2. Ayupov, Sh.A., Kudaybergenov, K.K., Nurjanov, B.O., Alauadinov, A.K.: Local and 2-local derivations on noncommutative Arens algebras. Math. Slovaca **64**, 423–432 (2014)
3. Ayupov, Sh.A., Arzikulov, F.N.: 2-local derivations on semi-finite von Neumann algebras. Glasgow Math. J. **56**, 9–12 (2014)
4. Ayupov, Sh.A., Kudaybergenov, K.K.: 2-local derivations on von Neumann algebras. Positivity **19**, 445–455 (2015)
5. Ayupov, Sh.A., Kudaybergenov, K.K., Rakhimov, I.S.: 2-Local derivations on finite-dimensional Lie algebras. Linear Algebr. Appl. **474**, 1–11 (2015)
6. Ayupov, Sh.A., Kudaybergenov, K.K.: Local derivations on finite dimensional Lie algebras. Linear Algebr. Appl. **493**, 381–398 (2016)
7. Ayupov, Sh.A., Kudaybergenov, K.K.: 2-Local automorphisms on finite-dimensional Lie algebras. Linear Algebr. Appl. **507**, 121–131 (2016)
8. Ayupov, Sh.A., Kudaybergenov, K.K., Omirov, A.B.: Local and 2-local derivations and automorphisms on simple Leibniz algebras (2017). arXiv:1703.10506
9. Ayupov, Sh.A., Kudaybergenov, K.K., Omirov, A.B., Zhao, K.: Semisimple Leibniz algebras and their derivations and automorphisms (2017). arXiv:1708.08082
10. Chen, Z., Wang, D.: Nonlinear maps satisfying derivability on standard parabolic subalgebras of finite-dimensional simple Lie algebras. Linear Multilinear Algebr. **59**, 261–270 (2011)
11. Chen, Z., Wang, D.: 2-Local automorphisms of finite-dimensional simple Lie algebras. Linear Algebr. Appl. **486**, 335–344 (2015)
12. Humphreys, J.E.: Introduction to Lie Algebras and Representation Theory. Springer, New York (1972)
13. Jacobson, N.: Lie Algebras. Dover Publications. Inc., New York (1979)
14. Kadison, R.V.: Local derivations. J. Algebr. **130**, 494–509 (1990)
15. Kim, S.O., Kim, J.S.: Local automorphisms and derivations on M_n. Proc. Am. Math. Soc. **132**, 1389–1392 (2004)
16. Larson, D.R., Sourour, A.R.: Local derivations and local automorphisms of $B(X)$. Proc. Sympos. Pure Math. **51**, 187–194 (1990)
17. Molnar, L.: Selected Preserver Problems on Algebraic Structures of Linear Operators and on Function Spaces. Springer, Berlin (2007)
18. Šemrl, P.: Linear mappings preserving square-zero matrices. Bull. Aust. Math. Soc. **48**, 365–370 (1993)
19. Šemrl, P.: Local automorphisms and derivations on $B(H)$. Proc. Am. Math. Soc. **125**, 2677–2680 (1997)
20. Vergne, M.: Réductibilité de la varireté des algébres de Lie nilpotentes. C. R. Acad. Sci. Paris. **263**, 4–6 (1966)
21. Wan, Z.: Lie Algebras. Pergamon Press Ltd, Oxford (1975)

Integration of the Toda-Type Chain with a Special Self-consistent Source

B. A. Babajanov

Abstract In this paper, it is shown that the solutions of the Toda-type chain with a special self-consistent source can be found by the inverse scattering method for the discrete Sturm–Liuville operator with moving eigenvalues.

Keywords Toda-type chain · Self-consistent source · Inverse scattering method · Discrete Sturm-Liouville operator · One-soliton solution

1 Introduction

The Toda lattice [1, 2] is a simple model for a nonlinear one-dimensional crystal that describes the motion of a chain of particles with exponential interactions of the nearest neighbors. The equation of motion for such a system is given by

$$\frac{d^2 u_n}{dt^2} = \exp\left(u_{n-1} - u_n\right) - \exp\left(u_n - u_{n+1}\right), n \in Z,$$

where $u_n(t)$ the coordinate of the nth atom in a lattice. Using Flaschka's variables [3]

$$a_n = \frac{1}{2} \exp\left(\frac{u_n - u_{n+1}}{2}\right), b_n = \frac{1}{2}\dot{a}_n,$$

the Toda equation can be rewritten in the form

$$\begin{cases} \dot{a}_n = a_n(b_n - b_{n+1}), \\ \dot{b}_n = 2(a_{n-1}^2 - a_n^2), \quad n \in Z. \end{cases}$$

Soliton solutions of the Toda lattice are obtained in the works [4, 5]. Soliton equations with self-consistent sources have received much attention in the recent research

B. A. Babajanov (✉)
Urgench State Universiy, Urgench, Uzbekistan
e-mail: a.murod@mail.ru

© Springer Nature Switzerland AG 2018
Z. Ibragimov et al. (eds.), *Algebra, Complex Analysis, and Pluripotential Theory*,
Springer Proceedings in Mathematics & Statistics 264,
https://doi.org/10.1007/978-3-030-01144-4_4

literature. They have important applications in plasma physics, hydrodynamics, solid state physics, etc. [6–12]. For example, the KdV equation, which is included an integral type self-consistent source, was considered in [9]. By this type equation the interaction of long and short capillary-gravity waves can be described [10]. Other important soliton equations with self-consistent source are the nonlinear Schrodinger equation which describes the nonlinear interaction of an ion acoustic wave in the two component homogeneous plasma with the electrostatic high frequency wave [11]. Different techniques have been used to construct their solutions, such as inverse scattering [4, 7, 8, 13, 14], Darboux transformation [15–18] or Hirota bi-linear methods [19–21].

In the work [22] it is shown that Toda lattice with special self-consistent source can be integrated by Inverse Scattering Method for the discrete Sturm–Liuville operator.

In this work, we explore the Toda-type chain with a special self-consistent source. The purpose of this paper is to derive representations for the solutions of the Toda-type chain with a special self-consistent source in the framework of the inverse scattering method for the discrete Sturm–Liuville operator. An effective method of integration of the Toda-type chain with a special self-consistent source is presented.

The considering system, similarly to [23, 24], can be used in some models of special types of electric transmission line.

The paper is organized as follows. In Sect. 2, we present the formulation of the problem which is being considered. We give some basic information about the scattering theory for the discrete Sturm–Liouville operator in Sect. 3. Section 4 is devoted to deducing the evolution of the scattering data corresponding to the problem in question.

2 The Formulation of Problem

We consider the system of equations with a special self-consistent source

$$
\begin{cases}
\frac{da_n}{dt} = a_n(a_{n+1}^2 - a_{n-1}^2) + a_n(b_{n+1}^2 - b_n^2) + a_n \sum_{i=1}^{N}(f_{n+1}^i g_{n+1}^i - f_n^i g_n^i), \\
\frac{db_n}{dt} = 2a_n^2(b_{n+1} + b_n) - 2a_{n-1}^2(b_n + b_{n-1}) + a_n \sum_{i=1}^{N}(f_n^i g_{n+1}^i + f_{n+1}^i g_n^i) - \\
\quad -a_{n-1} \sum_{i=1}^{N}(f_n^i g_{n-1}^i + f_{n-1}^i g_n^i), \\
a_{n-1} f_{n-1}^k + b_n f_n^k + a_n f_{n+1}^k = \lambda_k f_n^k, \\
a_{n-1} g_{n-1}^k + b_n g_n^k + a_n g_{n+1}^k = \lambda_k g_n^k, \quad k = 1, 2, \ldots, N, \quad n \in Z,
\end{cases}
\tag{1}
$$

and the initial condition

$$
a_n(0) = a_n^0, \quad b_n(0) = b_n^0, \quad n \in Z.
\tag{2}
$$

The initial data $\{a_n^0\}_{-\infty}^{\infty}$, $\{b_n^0\}_{-\infty}^{\infty}$ have the following properties:

1. $a_n^0 > 0$, $Im b_n^0 = 0$, $n \in Z$,
2. $\sum_{n=-\infty}^{\infty} |n| \left(\left| a_n^0 - \frac{1}{2} \right| + |b_n^0| \right) < \infty$,

3. the discrete Sturm–Liouville equation

$$a_{n-1}^0 y_{n-1} + b_n^0 y_n + a_n^0 y_{n+1} = \lambda y_n, \quad n \in Z$$

has exactly N eigenvalues $\lambda_1(0), \lambda_2(0), \ldots, \lambda_N(0)$ outside the segment $[-1; 1]$. In system (1), $\{a_n(t)\}_{-\infty}^{\infty}$, $\{b_n(t)\}_{-\infty}^{\infty}$, $\{f_n^k(t)\}_{-\infty}^{\infty}$, $\{g_n^k(t)\}_{-\infty}^{\infty}$, $k = 1, 2, \ldots, N$ are the unknown functions, and $\{f_n^k(t)\}_{-\infty}^{\infty}$ is the eigenvector of the operator

$$(L(t)y)_n \equiv a_{n-1} y_{n-1} + b_n y_n + a_n y_{n+1},$$

associated with the eigenvalue $\lambda_k = \frac{z_k + z_k^{-1}}{2}$, $k = 1, 2, \ldots, N$. The function $\{g_n^k(t)\}_{-\infty}^{\infty}$ that solves the equation

$$a_{n-1} g_{n-1}^k + b_n g_n^k + a_n g_{n+1}^k = \lambda_k g_n^k, \quad k = 1, 2, \ldots, N, \quad n \in Z.$$

and the eigenvector $\{f_n^k(t)\}_{-\infty}^{\infty}$ are linearly independent. We assume that

$$W\left\{f_n^k, g_n^k\right\} \equiv a_n\left(f_n^k g_{n+1}^k - f_{n+1}^k g_n^k\right) = \omega_k(t), \quad k = 1, 2, \ldots, N, \qquad (3)$$

where $\omega_k(t)$ are initially specified real continuous functions of t, satisfying the condition

$$\left| \frac{\lambda_k(0)}{2} + \int_0^t \omega_k(t)dt \right| > \frac{1}{2}, \quad k = 1, 2, \ldots, N, \qquad (4)$$

for all nonnegative t. We seek the solution of problem (1)–(4) in the class of functions $\{a_n(t)\}_{-\infty}^{\infty}$ and $\{b_n(t)\}_{-\infty}^{\infty}$ belonging to $C^1[0, \infty)$ and satisfying the conditions

$$a_n(t) > 0, \quad Imb_n = 0, \quad n \in Z, \quad \sum_{n=-\infty}^{\infty} |n| \left(\left| a_n(t) - \frac{1}{2} \right| + |b_n(t)| \right) < \infty.$$
$$(5)$$

Our main purpose is to obtain a representation for the solutions $\{a_n(t)\}_{-\infty}^{\infty}$ $b_n(t)_{-\infty}^{\infty}$, $\{f_n^k(t)\}_{-\infty}^{\infty}$ $g_n^k(t)_{-\infty}^{\infty}$, $k = 1, 2, \ldots, N$ of the problem (1)–(5) in the framework of inverse scattering method for the operator $L(t)$.

3 Basic Facts About the Scattering Problem

In this section, we momentarily omit the dependence of the functions $\{a_n(t)\}_{-\infty}^{\infty}$, $\{b_n(t)\}_{-\infty}^{\infty}$ on variable t.

We start with the following discrete Sturm–Liouville equation

$$(Ly)_n \equiv a_{n-1} y_{n-1} + b_n y_n + a_n y_{n+1} = \lambda y_n, \quad n \in Z, \qquad (6)$$

with spectral parameter $\lambda = \frac{z+z^{-1}}{2}$. We suppose that the sequences $\{a_n\}_{-\infty}^{\infty}$, $\{b_n\}_{-\infty}^{\infty}$ satisfy the conditions

$$a_n > 0, \quad Imb_n = 0, \quad n \in Z, \quad \sum_{n=-\infty}^{\infty} |n| \left(\left| a_n - \frac{1}{2} \right| + |b_n| \right) < \infty. \quad (7)$$

In this section we give basic information about the theory of direct and inverse scattering problem for the operator L. This theory was developed in the work [25].

If condition (7) is valid, then Eq. (6) has Jost solutions with the asymptotics:

$$\begin{aligned} \varphi_n(z) &= z^n + o(1) \quad as \quad n \to \infty, \quad |z| = 1, \\ \psi_n(z) &= z^{-n} + o(1) \quad as \quad n \to -\infty, \quad |z| = 1. \end{aligned} \quad (8)$$

Under the condition (7) it is known that such solutions exist, moreover they are identified by asymptotics (8) unique and analytically expended into the circle $|z| < 1$.

The function $\varphi_n(z)$ admits the following representation

$$\varphi_n(z) = \sum_{n'=n}^{\infty} K(n, n') z^{n'}, \quad (9)$$

where the coefficients $K(n, n')$ is independent on z, and is related to a_n and b_n by formulas

$$a_n = \frac{1}{2} \frac{K(n+1, n+1)}{K(n, n)}, \quad b_n = \frac{1}{2} \left(\frac{K(n, n+1)}{K(n, n)} - \frac{K(n-1, n)}{K(n-1, n-1)} \right). \quad (10)$$

For $|z| = 1$ the pairs $\{\varphi_n(z), \varphi_n(z^{-1})\}$ and $\{\psi_n(z), \psi_n(z^{-1})\}$ are the pairs of linearly independent solutions of (6), therefore

$$\psi_n(z) = \alpha(z)\varphi_n(z^{-1}) + \beta(z)\varphi_n(z) \quad (11)$$

where

$$\alpha(z) = \frac{2}{z - z^{-1}} W\{\psi_n(z), \varphi_n(z)\}, \quad (12)$$

and $W\{\psi_n(z), \varphi_n(z)\} \equiv a_n(\psi_n(z)\varphi_{n+1}(z) - \psi_{n+1}(z)\varphi_n(z))$. The scattering function is given by the formula $R(z) = \frac{\beta(z)}{\alpha(z)}$. The function $\alpha(z)$ can be analytically continued into the circle $|z| < 1$ and there it has finitely many zeros $z_1, z_2, \ldots, z_N$. The points $\lambda_k = \frac{z_k + z_k^{-1}}{2}$, $k = 1, 2, \ldots, N$ correspond to eigenvalues of the operator L. From (12) it follows that

$$\psi_n^k = \beta_k \varphi_n^k, \quad k = 1, 2, \ldots, N, \quad (13)$$

where $\psi_n^k \equiv \psi_n(z_k)$.

Let $\left\{ \xi_n^k \right\}_{n=-\infty}^{\infty}$ be the eigenvector corresponding to eigenvalues $\lambda_k = \frac{z_k + z_k^{-1}}{2}$ that is normalized by following condition

$$\sum_{n=-\infty}^{\infty} (\xi_n^k)^2 = 1.$$

It is obvious that $\xi_n^k = C_k \varphi_n^k$, $k = 1, 2, \ldots, N$. From known equality

$$\dot\alpha(z_k) = -\frac{1}{z_k} \sum_{j=-\infty}^{\infty} \varphi_j(z_k)\psi_j(z_k),$$

we get

$$C_k^2 = -\frac{\beta_k}{z_k \dot\alpha(z_k)}, \quad k = 1, 2, \ldots, N. \tag{14}$$

The set of quantities $\{R(z), z_1, z_2, \ldots, z_N, C_1, C_2, \ldots, C_N\}$ is called the scattering data for Eq. (6). The coefficients $K(n, n')$ given in representation (9) satisfy the equation of Gelfand–Levitan–Marchenko type

$$\chi(n, m) + F(n + m) + \sum_{n'=n+1}^{\infty} \chi(n, n')F(n' + m) = 0, \quad m > n,$$

$$(K(n, n))^{-2} = 1 + F(2n) + \sum_{n'=n+1}^{\infty} \chi(n, n')F(n' + n),$$

where

$$F(n) = \frac{1}{2\pi i} \oint_{|z|=1} R(z)z^{n-1}dz + \sum_{k=1}^{N} C_k^2 z_k^n, \quad \chi(n, m) = \frac{K(n, m)}{K(n, n)}.$$

Now $\{a_n\}_{-\infty}^{\infty}$ and $\{b_n\}_{-\infty}^{\infty}$ can be expressed via the scattering data by the formulas (10). It is worthy to remark that the vectors

$$h_n^k = \frac{d}{dz}(\psi_n(z) - \beta_k \varphi_n(z))\Big|_{z = z_k},$$

are solutions of the equations $Ly = \lambda_k y$, $k = 1, 2, \ldots, N$. From the Eq. (12), as $|z| < 1$ we deduce that

$$\varphi_n(z) \to \alpha(z) z^n, \quad n \to -\infty$$

therefore

$h_n^k \to -\beta_k \dot{\alpha}(z_k) z_k^n$ as $n \to -\infty$, $k = 1, 2, \ldots, N$, where $\dot{\alpha}(z_k) = \left. \frac{d\alpha(z)}{dz} \right|_{z=z_k}$.

From asymptotes (8) and (15) we get

$$W\{h_n^k, \psi_n^k\} = \frac{\beta_k \dot{\alpha}(z_k)(z_k - z_k^{-1})}{2}. \tag{15}$$

It is easy to see that the following statement is true.

Lemma 1 *If $\{x_n(\lambda)\}_{-\infty}^{\infty}$ and $\{y_n(\mu)\}_{-\infty}^{\infty}$ are solutions of equations $Lx = \lambda x$ and $Ly = \mu y$. Then the identity*

$$(\mu - \lambda)x_n(\lambda)y_n(\mu) = W\{x_n(\lambda), y_n(\mu)\} - W\{x_{n-1}(\lambda), y_{n-1}(\mu)\}, \quad n \in Z,$$

holds.

The lemma is proved by direct verification.

4 Evolution of the Scattering Data

Let $\left\{ F_n^k(t) \right\}_{-\infty}^{\infty}$ be the normalized eigenvector of the operator $L(t)$, associated with the eigenvalue $\lambda_k(t)$, $k = 1, 2, \ldots, N$, i.e.

$$a_{n-1}F_{n-1}^k + b_n F_n^k + a_n F_{n+1}^k = \lambda_k F_n^k, n \in Z.$$

After differentiating these relations with respect to t and then multiplying them by F_n^k, we sum over n from $-\infty$ to ∞. Using Eq. (1) and recalling that the operator $L(t)$ is self-adjoint, we obtain

$$\frac{d\lambda_k}{dt} = \sum_{i=1}^{N} \sum_{n=-\infty}^{\infty} (a_{n-1}(f_n^i g_n^i - f_{n-1}^i g_{n-1}^i)F_{n-1}^k F_n^k + a_n(f_n^i g_{n+1}^i + f_{n+1}^i g_n^i)(F_n^k)^2 -$$

$$-a_{n-1}(f_n^i g_{n-1}^i + f_{n-1}^i g_n^i)(F_n^k)^2 + a_n(f_{n+1}^i g_{n+1}^i - f_n^i g_n^i)F_{n+1}^k F_n^k).$$

Last equality can be written as

$$\frac{d\lambda_k}{dt} = \sum_{i=1}^{N} \sum_{n=-\infty}^{\infty} \left[2a_n(f_{n+1}^i g_{n+1}^i - f_n^i g_n^i)F_{n+1}^k F_n^k + a_n(f_n^i g_{n+1}^i + f_{n+1}^i g_n^i)((F_n^k)^2 - (F_{n+1}^k)^2) \right].$$

Grouping the terms yields

$$\frac{d\lambda_k}{dt} = \sum_{i=1}^{N} \sum_{n=-\infty}^{\infty} \left[f_{n+1}^i F_{n+1}^k W\{F_n^k, g_n^i\} + g_{n+1}^i F_{n+1}^k W\{F_n^k, f_n^i\} + \right.$$

$$\left. + f_n^i F_n^k W\{F_n^k, g_n^i\} + g_n^i F_n^k W\{F_n^k, f_n^i\} \right]. \tag{16}$$

Lemma 2 *For $i \neq k$ the equality*

$$\sum_{n=-\infty}^{\infty} \left[W\{F_n^k, g_n^i\}(f_{n+1}^i F_{n+1}^k + f_n^i F_n^k) + W\{F_n^k, f_n^i\}(g_{n+1}^i F_{n+1}^k + g_n^i F_n^k) \right] = 0$$

is satisfied.

Proof For convenience, we introduce the notation

$$W_n = W\{F_n^k, g_n^i\}, V_n = W\{F_n^k, f_n^i\}.$$

According to Lemma 1 we get

$$\sum_{n=-\infty}^{\infty} \left[W_n(f_{n+1}^i F_{n+1}^k + f_n^i F_n^k) + V_n(g_{n+1}^i F_{n+1}^k + g_n^i F_n^k) \right] =$$

$$= \sum_{n=-\infty}^{\infty} \left[f_n^i F_n^k (W_n + W_{n-1}) + g_n^i F_n^k (V_n + V_{n-1}) \right] =$$

$$= \frac{1}{\lambda_i - \lambda_k} \sum_{n=-\infty}^{\infty} \left[(V_n - V_{n-1})(W_n + W_{n-1}) + (W_n - W_{n-1})(V_n + V_{n-1}) \right] =$$

$$= \frac{2}{\lambda_i - \lambda_k} \sum_{n=-\infty}^{\infty} (W_n V_n - W_{n-1} V_{n-1}) = 0.$$

which was to be proved. Since $W\{F_n^k, f_n^k\} = 0$, then due to (3) and Lemma 2 equality (16) takes form

$$\frac{d\lambda_k}{dt} = 2\omega_k(t), \quad k = 1, 2, \ldots, N. \tag{17}$$

Now we consider system of equations

$$(Ly)_n \equiv a_{n-1} y_{n-1} + b_n y_n + a_n y_{n+1} = \lambda y_n, \quad n \in Z, \tag{18}$$

$$p_{n+1}^k - p_n^k = f_{n+1}^k y_{n+1} + f_n^k y_n, \quad k = 1, 2, \ldots, N, \tag{19}$$

for unknown vector functions $\left\{ p_n^k(z, t) \right\}, \quad k = 1, 2, \ldots, N$. Using an arbitrary solution of this system, we define

$$S_n^0 = \frac{\partial y_n}{\partial t} + a_n a_{n+1} y_{n+2} + a_n(b_{n+1} + b_n) y_{n+1} - a_{n-1}(b_n + b_{n-1}) y_{n-1} -$$

$$- a_{n-1} a_{n-2} y_{n-2} + \sum_{m=1}^{N} g_n^m p_n^m, \tag{20}$$

$$S_n^k = a_n \left(f_{n+1}^k y_n - f_n^k y_{n+1} \right) + a_{n-1} \left(f_n^k y_{n-1} - f_{n-1}^k y_n \right) + (\lambda - \lambda_k) p_n^k, \quad k = 1, 2, \ldots, N. \tag{21}$$

It is easy to show that

$$S_n^k - S_{n-1}^k = 0, \quad k = 1, 2, \ldots, N, \quad n \in Z, \tag{22}$$

i.e., the quantities S_n^k are independent of n for $k = 1, 2, \ldots N$. In addition, it follows from (1), (18) and (20) that

$$(L - \lambda) S_n^0 = - \sum_{m=1}^{N} g_n^m S_n^m, \quad n \in Z. \tag{23}$$

Let $\varphi_n(z, t)$ and $\psi_n(z, t)$ be solutions of Jost equation (18) with asymptotic forms (8). Setting $y_n^+ \equiv \varphi_n(z)$ and $y_n^- \equiv \psi_n(z)$ in (19), we define

$$\begin{aligned} p_n^{k-}(z) &= f_n^k \psi_n(z) + 2 \sum_{j=-\infty}^{n-1} f_j^k \psi_j(z), \\ p_n^{k+}(z) &= -f_n^k \varphi_n(z) - 2 \sum_{j=n+1}^{\infty} f_j^k \varphi_j(z), \quad k = 1, 2, \ldots, N. \end{aligned} \tag{24}$$

For a finding S_n^{i+}, S_n^{i-}, $i = 0, 1, \ldots, N$ we can use quantities $y_n^+ \equiv \varphi_n(z)$, $y_n^- \equiv \psi_n(z)$, p_n^{1+}, p_n^{1-}, p_n^{2+}, p_n^{2-}, $\ldots$, p_n^{N+}, p_n^{N-} and formulas (20) and (21). Based on (19) and (21), we can state that

$$S_n^{i+} \underset{n \to \infty}{\to} 0, \quad S_n^{i-} \underset{n \to -\infty}{\to} 0, \quad i = 1, \ldots, N.$$

Therefore, from the equality (22) we get

$$S_n^{i+} = S_n^{i-} = 0, \quad i = 1, \ldots, N, \quad n \in Z. \tag{25}$$

Substituting (25) into (23) we obtain

$$(L - \lambda) S_n^{0+} = (L - \lambda) S_n^{0-} = 0. \tag{26}$$

As we know the sequences of functions $\{f_n^k(t)\}_{-\infty}^{\infty}$ is the eigenvector of operator $L(t)$ corresponding to eigenvalue $\lambda_k = \frac{z_k + z_k^{-1}}{2}$, so it follows that

$$f_n^k = c_k^- \psi_n(z_k, t) = c_k^+ \varphi_n(z_k, t), \tag{27}$$

i.e. $c_k^+ = \beta_k c_k^-$, $k = 1, 2, \ldots, N$. Due to (3) and (27), we have

$$g_n^k \underset{n \to \infty}{\sim} -\frac{2\omega_k}{c_k^+(z_k - z_k^{-1})} \cdot z_k^{-n}, \quad g_n^k \underset{n \to -\infty}{\sim} \frac{2\omega_k}{c_k^-(z_k - z_k^{-1})} \cdot z_k^n, \quad k = 1, 2, \ldots, N.$$

(28)

By using (19), (27) and asymptotic (8), we find

$$p_n^{k-} \underset{n \to -\infty}{\sim} \frac{c_k^-(1 + z_k z)}{1 - z_k z} \cdot (z_k z)^{-n}, \quad p_n^{k+} \underset{n \to \infty}{\sim} -\frac{c_k^+(1 + z_k z)}{1 - z_k z} \cdot (z_k z)^n, \quad k = 1, 2, , N.$$

As result of (5), (20) and (28), we can write

$$S_n^{0-} \underset{n \to -\infty}{\sim} \left(\frac{z^{-2} - z^2}{4} + 2 \sum_{m=1}^{N} \frac{(1 + z_m z)\,\omega_m}{(1 - z_m z)(z_m - z_m^{-1})} \right) \cdot z^{-n},$$

$$S_n^{0+} \underset{n \to \infty}{\sim} \left(\frac{z^2 - z^{-2}}{4} + 2 \sum_{m=1}^{N} \frac{(1 + z_m z)\,\omega_m}{(1 - z_m z)(z_m - z_m^{-1})} \right) \cdot z^n.$$

Hence, in accordance with Eq. (26)

$$S_n^{0-} = \left(\frac{z^{-2} - z^2}{4} + 2 \sum_{m=1}^{N} \frac{(1 + z_m z)\,\omega_m}{(1 - z_m z)\left(z_m - z_m^{-1}\right)} \right) \psi_n(z),$$

$$S_n^{0+} = \left(\frac{z^2 - z^{-2}}{4} + 2 \sum_{m=1}^{N} \frac{(1 + z_m z)\,\omega_m}{(1 - z_m z)\left(z_m - z_m^{-1}\right)} \right) \varphi_n(z). \qquad (29)$$

We define a quantity G_0 as

$$G_0 = S_n^{0-}(z) - \alpha(z) S_n^{0+}(z^{-1}) - \beta(z) S_n^{0+}(z).$$

According to (11) and (29) we have

$$G_0 = 2 \sum_{m=1}^{N} \left(\frac{1 + z_m z}{1 - z_m z} - \frac{1 + z_m z^{-1}}{1 - z_m z^{-1}} \right) \frac{\omega_m}{z_m - z_m^{-1}} \alpha(z)\varphi_n(z^{-1}) + \frac{z^{-2} - z^2}{2} \beta(z)\varphi_n(z).$$

(30)

On the other hand, it follows from formulas (11) and (20) that

$$S_n^{0-}(z) = \frac{\partial \alpha(z)}{\partial t}\varphi_n(z^{-1}) + \frac{\partial \beta(z)}{\partial t}\varphi_n(z) + \alpha(z) S_n^{0+}(z^{-1}) + \beta(z) S_n^{0+}(z) +$$

$$+ \sum_{m=1}^{N} g_n^m \left(p_n^{m-}(z) - \alpha(z) p_n^{m+}(z^{-1}) - \beta(z) p_n^{m+}(z) \right). \qquad (31)$$

Using (11), (24) and Lemma 1, we find

$$p_n^{m-}(z) - \alpha(z)p_n^{m+}(z^{-1}) - \beta(z)p_n^{m+}(z) = 2\sum_{j=-\infty}^{\infty} f_j^m \psi_j(z) = 0.$$

Therefore, by definition of G_0 and equality (31), we obtain

$$G_0 = \frac{\partial \alpha(z)}{\partial t}\varphi_n(z^{-1}) + \frac{\partial \beta(z)}{\partial t}\varphi_n(z). \tag{32}$$

Comparing (30) and (32), we get

$$\frac{\partial \alpha(z)}{\partial t} = 2\sum_{m=1}^{N}\left(\frac{1+z_m z}{1-z_m z} - \frac{1+z_m z^{-1}}{1-z_m z^{-1}}\right)\frac{\omega_m}{z_m - z_m^{-1}}\alpha(z);$$

$$\frac{\partial \beta(z)}{\partial t} = \frac{z^{-2}-z^2}{2}\beta(z), \quad |z| = 1.$$

Hence,

$$R(z,t) = R(z,0)\exp\theta(z,t), \tag{33}$$

where

$$\theta(z,t) = \frac{z^{-2}-z^2}{2}t - 2\sum_{m=1}^{N}\int_0^t\left(\frac{1+z_m(\tau)z}{1-z_m(\tau)z} - \frac{1+z_m(\tau)z^{-1}}{1-z_m(\tau)z^{-1}}\right)\frac{\omega_m(\tau)}{z_m(\tau) - z_m^{-1}(\tau)}d\tau.$$

We set

$$G_k = S_n^{0-}(z_k) - \beta_k S_n^{0+}(z_k), \quad k = 1, 2, \ldots, N. \tag{34}$$

Then, based on (13) and (31), we deduce

$$G_k = \frac{z_k^{-2} - z_k^2}{2}\beta_k\varphi_n(z_k), \quad k = 1, 2, \ldots, N. \tag{35}$$

Differentiating equality (13) with respect to t and recalling definition of vector h_n^k, we find

$$\frac{\partial \psi_n(z_k)}{\partial t} = \frac{d\beta_k}{dt}\varphi_n(z_k) + \beta_k\frac{\partial \varphi_n(z_k)}{\partial t} - \frac{dz_k}{dt}h_n^k.$$

In view of (17) this can be rewritten as

$$\frac{\partial \psi_n(z_k)}{\partial t} = \frac{d\beta_k}{dt}\varphi_n(z_k) + \beta_k\frac{\partial \varphi_n(z_k)}{\partial t} - \frac{4z_k^2\omega_k}{z_k^2 - 1}h_n^k, \quad k = 1, 2, \ldots, N. \tag{36}$$

Using (20) and (36), we have

$$S_n^{0-}(z_k) = \frac{\partial \beta_k}{\partial t} \varphi_n(z_k) + \beta_k S_n^{0+}(z_k) - \frac{4z_k^2 \omega_k}{z_k^2 - 1} h_n^k + \sum_{m=1}^{N} g_n^m \left(p_n^{m-}(z_k) - \beta_k p_n^{m+}(z_k) \right).$$

(37)

On the other hand, in accordance with formulas (13) and (24)

$$p_n^{m-}(z_k) - \beta_k p_n^{m+}(z_k) = 2\beta_k \sum_{j=-\infty}^{\infty} f_j^m \varphi_j(z_k).$$

Because eigenvectors associated with different eigenvalues are orthogonal according to Lemma 1, we find

$$\sum_{m=1}^{N} g_n^m \left(p_n^{m-}(z_k) - \beta_k p_n^{m+}(z_k) \right) = 2g_n^k \sum_{j=-\infty}^{\infty} f_j^k \psi_j(z_k), \quad k = 1, 2, \ldots, N.$$

(38)

By the definition of the vectors h_n^k, we have

$$h_n^k = \delta_k g_n^k + \gamma_k \psi_n(z_k).$$

(39)

Substituting equality (39) into (15), and taking account formulas (3) and (27), we obtain

$$\delta_k = -\frac{\beta_k c_k^- \left(z_k - z_k^{-1} \right) \dot{\alpha}(z_k)}{2\omega_k} = -\frac{c_k^+ \left(z_k - z_k^{-1} \right) \dot{\alpha}(z_k)}{2\omega_k}.$$

(40)

By virtue of formulas $\dot{\alpha}(z_k) = -\frac{1}{z_k} \sum_{j=-\infty}^{\infty} \varphi_j(z_k) \psi_j(z_k)$ and (38), equality (37) can be rewritten as

$$S_n^{0-}(z_k) - \beta_k S_n^{0+}(z_k) = \frac{\partial \beta_k}{\partial t} \varphi_n(z_k) - \frac{4z_k^2 \omega_k}{z_k^2 - 1} h_n^k - 2z_k c_k^+ \dot{\alpha}(z_k) g_n^k.$$

Hence, using formula (34), (39) and (40), we have

$$G_k = \left(\frac{\partial \beta_k}{\partial t} - \frac{4z_k^2 \gamma_k \omega_k}{z_k^2 - 1} \beta_k \right) \varphi_n(z_k).$$

(41)

Comparing (35) and (41), we get

$$\frac{d\beta_k}{dt} = \left(\frac{z_k^{-2} - z_k^2}{2} + \frac{4z_k^2 \gamma_k \omega_k}{z_k^2 - 1} \right) \beta_k, \quad k = 1, 2, \ldots, N.$$

The normalization constants $C_k(t)$ are now found from the relations

$$C_k^2(t) = -\frac{\beta_k(t)}{z_k(t)\dot{\alpha}(z_k(t), t)}, \quad k = 1, 2, \dots, N. \tag{42}$$

We have thus proved the following theorem.

Theorem 1 *If $a_n(t)$, $b_n(t)$, $f_n^k(t)$, $g_n^k(t)$, $k = 1, 2, \dots, N$, $n \in Z$ are solutions of the problem (1)–(5), then the scattering data of the operator*

$$(L(t)y)_n \equiv a_{n-1}y_{n-1} + b_n y_n + a_n y_{n+1}$$

are given by the Eqs. (17), (33) *and* (42).

The obtained relations completely specify the evolution of the scattering data for $L(t)$; together with (39), this allows using the inverse scattering method to find solutions of problem (1)–(5).

Acknowledgements This work was supported by the International Erasmus+ Program KA106-2, Keele University, UK.

References

1. Toda, M.: One-dimensional dual transformation. J. Phys. Soc. Jpn. **22**, 431–436 (1967)
2. Toda, M.: Wave propagation in anharmonic lattice. J. Phys. Soc. Jpn. **23**, 501–506 (1967)
3. Flaschka, H.: Toda lattice. I. Phys. Rev. **B9**, 1924–1925 (1974)
4. Manakov, S.V.: Complete integrability and stochastization of discrete dynamical systems. Zh. Eksp. Teor. Fiz. **67**, 543–555 (1974)
5. Khanmamedov, AgKh: The rapidly decreasing solution of the initial-boundary value problem for the Toda lattice. Ukr. Math. J. **57**, 1144–1152 (2005)
6. Mel'nikov, V.K.: A direct method for deriving a multisoliton solution for the problem of interaction of waves on the x, y plane. Commun. Math. Phys. **112**, 639–52 (1987)
7. Mel'nikov, V.K.: Integration method of the Korteweg-de Vries equation with a self-consistent source. Phys. Lett. A **133**, 493–6 (1988)
8. Mel'nikov, V.K.: Integration of the nonlinear Schrodinger equation with a self-consistent source. Commun. Math. Phys. **137**, 359–81 (1991)
9. Mel'nikov, V.K.: Integration of the Korteweg-de Vries equation with a source. Inverse Probl. **6**, 233–46 (1990)
10. Leon, J., Latifi, A.: Solution of an initial-boundary value problem for coupled nonlinear waves. J. Phys. A Math. Gen. **23**, 1385–403 (1990)
11. Claude, C., Latifi, A., Leon, J.: Nonliear resonant scattering and plasma instability: an integrable model. J. Math. Phys. **32**, 3321–3330 (1991)
12. Shchesnovich, V.S., Doktorov, E.V.: Modified Manakov system with self-consistent source. Phys. Lett. A **213** 23–31 (1996)
13. Cabada, A., Urazboev, G.U.: Integration of the Toda lattice with an integral-type source. Inverse Probl. **26**, 085004(12pp) (2010)
14. Lin, R.L., Zeng, Y.B., Ma, W.X.: Solving the KdV hierarchy with self-consistent sources by inverse scattering method. Phys. A **291**, 287–98 (2001)

15. Zeng, Y.B., Ma, W.X., Shao, Y.J.: Two binary Darboux transformations for the KdV hierarchy with self-consistent sources. J. Math. Phys. **42**, 2113–28 (2001)
16. Zeng, Y.B., Shao, Y.J., Ma, W.X.: Integral-type Darboux transformations for the mKdV hierarchy with self-consistent sources. Commun. Theor. Phys. (Beijing) **38**, 641–8 (2002)
17. Zeng, Y.B., Shao, Y.J., Xue, W.M.: Negaton and positon solutions of the soliton equation with self-consistent sources. J. Phys. A Math. Gen. **36**, 5035–43 (2003)
18. Xiao, T., Zeng, Y.B.: Generalized Darboux transformations for the KP equation with self-consistent sources. J. Phys. A Math. Gen. **37**, 7143–62 (2004)
19. Matsuno, Y.: Bilinear Backlund transformation for the KdV equation with a source. J. Phys. A Math. Gen. **24**, L273–7 (1991)
20. Deng, S.F., Chen, D.Y., Zhang, D.J.: The multisoliton solutions of the KP equation with self-consistent sources. J. Phys. Soc. Jpn. **72**, 2184–92 (2003)
21. Zhang, D.J., Chen, D.Y.: The N-soliton solutions of the sine-Gordon equation with self-consistent sources. Phys. A **321**, 467–81 (2003)
22. Urazboev, G.U.: Toda lattice with a special self-consistent source. Theor. Math. Phys. **154**, 305–315 (2008)
23. David, C., Niels, G.-J., Bishop, A.R., Findikoglu, A.T., Reago, D.: A perturbed Toda lattice model for low loss nonlinear transmission lines. Phys. D Nonlinear Phenom. **123**, 291–300 (1998)
24. Garnier, J., Kh, Abdullaev F.: Soliton dynamics in a random Toda chain Phys. Rev. E **67**, 026609–1 (2003)
25. Case, K., Kac, M.: A discrete version of the inverse scattering problem. J. Math. Phys. **14**, 594–603 (1973)

Ground States for Potts Model with a Countable Set of Spin Values on a Cayley Tree

G. I. Botirov and M. M. Rahmatullaev

Abstract We consider Potts model, with competing interactions and countable spin values $\Phi = \{0, 1, \ldots\}$ on a Cayley tree of order three. We study periodic ground states for this model.

Keywords Potts model · Configuration · Ground state · Weakly periodic ground state · Countable set of spin values

1 Introduction

Each Gibbs measure is associated with a single phase of physical system. As is known, the phase diagram of Gibbs measures for a Hamiltonian is close to the phase diagram of isolated (stable) ground states of this Hamiltonian. At low temperatures, a periodic ground state corresponds to a periodic Gibbs measures, see [1–9]. The problem naturally aries on description of periodic ground states. In [3, 4] for the Ising model with competing interactions, periodic and weakly periodic ground states were studied.

In [1] ground states were described and the Peierls condition for the Potts model is verified. Using a contour argument authors showed the existence of three different Gibbs measures associated with translation invariant ground states.

In [5], (1), [8] studying periodic and weakly periodic ground states for the Potts model with competing interactions on a Cayley tree. In the present paper, we consider Potts model, with competing interactions and a countable set of spin values $\Phi = \{0, 1, \ldots\}$ on a Cayley tree of order three. We study periodic ground states.

In [10] the 3-state Potts model with competing binary interactions (with couplings J and J_p) on a Bethe lattice of order two is considered. The set of ground states of

G. I. Botirov · M. M. Rahmatullaev (✉)
Institute of Mathematics, Tashkent, Uzbekistan
e-mail: mrahmatullaev@rambler.ru

G. I. Botirov
e-mail: botirovg@yandex.ru

© Springer Nature Switzerland AG 2018
Z. Ibragimov et al. (eds.), *Algebra, Complex Analysis, and Pluripotential Theory*,
Springer Proceedings in Mathematics & Statistics 264,
https://doi.org/10.1007/978-3-030-01144-4_5

the one-level model is completely described. The critical temperature of a phase transition is exactly found and the phase diagram is presented.

In [11] an exact phase diagram of the Potts model with next nearest neighbor interactions on the Cayley tree of order two is found.

The paper is organized as follows. In Sect. 2, we recall the main definitions and known facts. In Scct. 3, we study periodic ground states.

2 Main Definitions and Known Facts

Cayley tree. The Cayley tree (Bethe lattice) Γ^k of order $k \geq 1$ is an infinite tree, i.e., a graph without cycles, such that exactly $k + 1$ edges originate from each vertex (see [12]). Let $\Gamma^k = (V, L)$ where V is the set of vertices and L the set of edges. Two vertices x and y are called *nearest neighbors* if there exists an edge $l \in L$ connecting them and we denote $l = \langle x, y \rangle$.

On this tree, there is a natural distance to be denoted $d(x; y)$, being the number of nearest neighbor pairs of the minimal path between the vertices x and y (by path one means a collection of nearest neighbor pairs, two consecutive pairs sharing at least a given vertex).

For a fixed $x^0 \in V$, the root, let

$$W_n = \{x \in V : d(x, x^0) = n\}, \quad V_n = \{x \in V : d(x, x^0) \leq n\};$$

be respectively the sphere and the ball of radius n with center at x^0.

It is well-known that there exists a one-to-one correspondence between the set V of vertices of the Cayley tree of order $k \geq 1$ and the group G_k of the free products of $k + 1$ cyclic groups of second order with generators $a_1, a_2, \ldots, a_{k+1}$ (see [2, 13]).

3 Configuration Space and the Model

For each $x \in G_k$, let $S(x)$ denote the set of direct successors of x, i.e., if $x \in W_n$ then

$$S(x) = \{y \in W_{n+1} : d(x, y) = 1\}.$$

For each $x \in G_k$, let $S_1(x)$ denote the set of all neighbors of x, i.e. $S_1(x) = \{y \in G_k : \langle x, y \rangle \in L\}$. The set $S_1(x) \setminus S(x)$ is a singleton. Let $x_\downarrow$ denote the (unique) element of this set.

We consider the models in which the spin takes values in the set $\Phi = \{1, 2, \ldots\}$. A *configuration* σ on the set V is defined as a function $x \in V \rightarrow \sigma(x) \in \Phi$; the set of all configurations coincides with $\Omega = \Phi^V$.

Let G_k^* be a subgroup of index $r \geq 1$. Consider the set of right coset $G_k/G_k^* = \{H_1, \ldots, H_r\}$, where G_k^* is a subgroup.

Definition 1 A configuration $\sigma(x)$ is said to be G_k^* - *periodic* if $\sigma(x) = \sigma_i$ for all $x \in H_i$. A G_k-periodic configuration is said to be *translation invariant*.

The period of a periodic configuration is the index of the corresponding subgroup.

Definition 2 A configuration $\sigma(x)$ is said to be G_k^* - *weakly periodic* if $\sigma(x) = \sigma_{ij}$ for all $x \in H_i$ and $x_\downarrow \in H_j$.

The Hamiltonian of the Potts model with competing interactions has the form

$$H(\sigma) = J_1 \sum_{\substack{\langle x,y \rangle: \\ x,y \in V}} \delta_{\sigma(x)\sigma(y)} + J_2 \sum_{\substack{x,y \in V: \\ d(x,y)=2}} \delta_{\sigma(x)\sigma(y)}, \qquad (1)$$

where $J_1, J_2 \in \mathbf{R}$ and

$$\delta_{uv} = \begin{cases} 1, & u = v, \\ 0, & u \neq v. \end{cases}$$

4   Ground States

For pair of configurations σ and φ coinciding almost everywhere, i.e., everywhere except at a finite number of points, we consider the relative Hamiltonian $H(\sigma, \varphi)$ of the difference between the energies of the configurations σ and φ, i.e.,

$$H(\sigma, \varphi) = J_1 \sum_{\substack{\langle x,y \rangle, \\ x,y \in V}} (\delta_{\sigma(x)\sigma(y)} - \delta_{\varphi(x)\varphi(y)}) + J_2 \sum_{\substack{x,y \in V: \\ d(x,y)=2}} (\delta_{\sigma(x)\sigma(y)} - \delta_{\varphi(x)\varphi(y)}), \qquad (2)$$

where $J = (J_1, J_2) \in \mathbf{R}^2$ is an arbitrary fixed parameter.

Let M be the set of all unit balls with vertices in V. By the *restricted configuration* σ_b we mean the restriction of a configuration σ to a ball $b \in M$. The energy of a configuration σ_b on b is defined by the formula

$$U(\sigma_b) \equiv U(\sigma_b, J) = \frac{1}{2} J_1 \sum_{\substack{\langle x,y \rangle, \\ x,y \in b}} \delta_{\sigma(x)\sigma(y)} + J_2 \sum_{\substack{x,y \in b: \\ d(x,y)=2}} \delta_{\sigma(x)\sigma(y)}, \qquad (3)$$

where $J = (J_1, J_2) \in \mathbf{R}^2$.

The following assertion is known (see [2–8])

Lemma 1 *The relative Hamiltonian (2) has the form*

$$H(\sigma, \varphi) = \sum_{b \in M} (U(\sigma_b) - U(\varphi_b)).$$

Note that, in [5] in the case $k = 2$ and $\Phi = \{1, 2, 3\}$ all periodic (in particular translation-invariant) ground states for the Potts model (1) are given. In [9] the set

of weakly periodic ground states corresponding to index-two normal divisors of the
group representation of the Cayley tree is given. In [8] the sets of periodic and weakly
periodic ground states corresponding to normal subgroups of the group representation
of the Cayley tree of index 4 are described.

We consider the case $k = 3$. It is easy to see that $U(\sigma_b) \in \{U_1, U_2, \ldots, U_{12}\}$ for
any σ_b, where

$$U_1 = 2J_1 + 6J_6, \quad U_2 = \frac{3}{2}J_1 + 3J_2, \quad U_3 = J_1 + 2J_2, \quad U_4 = \frac{1}{2}J_1 + 3J_2,$$

$$U_5 = 6J_2, \quad U_6 = \frac{1}{2}J_1, \quad U_7 = 3J_2, \quad U_8 = J_2,$$

$$U_9 = J_1 + J_2, \quad U_{10} = \frac{1}{2}J_1 + J_2, \quad U_{11} = 2J_2, \quad U_{12} = 0.$$

Definition 3 A configuration φ is called a *ground state* of the relative Hamiltonian
H if $U(\varphi_b) = \min\{U_1, U_2, \ldots, U_{12}\}$ for any $b \in M$.

We set $C_i = \{\sigma_b : U(\sigma_b) = U_i\}$ and $U_i(J) = U(\sigma_b, J)$ if $\sigma_b \in C_i$, $i = 1, 2, \ldots,$
12.

If a ground state is a periodic (weakly periodic, translation invariant) configuration
then we call it a *periodic (weakly periodic, translation invariant) ground state*.

Let

$$A \subset \{1, 2, \ldots, k+1\}, \quad H_A = \{x \in G_k : \sum_{j \in A} w_j(x) \text{ is even}\},$$

$$G_k^{(2)} = \{x \in G_k : |x| \text{ is even}\}, \quad G_k^{(4)} = H_A \cap G_k^{(2)},$$

where $w_j(x)$ is the number of occurrences of a_j in x and $|x|$ is the length of x, i.e.
$|x| = \sum_{j=1}^{k+1} w_j(x)$. Notice that $G_k^{(4)}$ is a normal subgroup of index 4 of G_k.

Then we have

$$G_k^{(4)} = \{x \in G_k : |x| \text{ is even}, \sum_{j \in A} w_j(x) \text{ is even}\}.$$

If $A = \{1, 2, \ldots, k+1\}$ then the normal subgroup H_A coincides with the group $G_k^{(2)}$.
For any $i = 1, 2, \ldots, 12$ we put

$$A_i = \{J \in \mathbf{R}^2 : U_i = \min\{U_1, U_2, \ldots, U_{12}\}\}. \tag{4}$$

Quite cumbersome but not difficult calculations show that

$$A_1 = \{J \in \mathbf{R}^2 : J_1 \leq 0, J_2 \leq 0\} \cup \{J \in \mathbf{R}^2 : J_1 \leq -6J_2, J_2 \geq 0\},$$

$$A_2 = \{J \in \mathbf{R}^2 : J_1 \geq 0, \, -6J_2 \leq J_1 \leq -4J_2\},$$

$$A_3 = A_4 = A_{10} = \{J \in \mathbf{R}^2 : J_1 = 0, \, J_2 = 0\},$$

$$A_5 = \{J \in \mathbf{R}^2 : J_1 \geq 0, \, J_2 \leq 0\},$$

$$A_6 = \{J \in \mathbf{R}^2 : J_2 \geq 0, \, -2J_2 \leq J_1 \leq 0\},$$

$$A_7 = A_8 = A_{11} = \{J \in \mathbf{R}^2 : J_1 \geq 0, \, J_2 = 0\},$$

$$A_9 = \{J \in \mathbf{R}^2 : J_2 \leq 0, \, -4J_2 \leq J_1 \leq -2J_2\},$$

$$A_{12} = \{J \in \mathbf{R}^2 : J_1 \leq 0, \, J_2 \leq 0\}, \quad \text{and} \quad \mathbf{R}^2 = \bigcup_n A_n.$$

Theorem 1 *For any class C_i, $i = 1, 2, \ldots 12$, and any bounded configuration $\sigma_b \in C_i$, there exists a periodic configuration φ (on the Cayley tree) such that $\varphi_{b'} \in C_i$ for any $b' \in M$ and $\varphi_b = \sigma_b$.*

Proof For an arbitrary class C_i, $i = 1, 2, \ldots, 12$, and $\sigma_b \in C_i$, we construct the configuration φ as follows: without loss of generality, we can take the ball centered at $e \in G_3$ (where e is the unit element of G_3) for the ball b, i.e., $b = \{e, a_1, a_2, a_3, a_4\}$.

We consider several cases.

Case C_1. In this case, we have $\sigma(x) = i$, $i \in \Phi$, for any $x \in b$. The configuration φ hence coincides with the translation-invariant configuration, i.e. $\varphi^{(i)} = \{\varphi^{(i)}(x) = i\}$, where $i \in \Phi$.

Case C_2. This case is considered in [8]. Let $H_{\{i\}}$ be normal subgroup of index two, and $G_k/H_{\{i\}} = \{H_0^{(2)}, H_1^{(2)}\}$ is quotient group, for any $i \in \{1, 2, 3, 4\}$. For any $l, m \in \Phi$, $l \neq m$, considering the following $H_{\{i\}}$−periodic configuration:

$$\varphi_2^{(lm)}(x) = \begin{cases} l, & \text{if } x \in H_0^{(2)}, \\ m, & \text{if } x \in H_1^{(2)}. \end{cases}$$

In [8] it was proved, that $(\varphi_2^{(lm)})_{b'} \in C_2$ for any $b' \in M$.

Case C_3. Let $H_{\{i,j\}}$, $i, j \in \{1, 2, 3, 4\}$ and $i \neq j$ be a normal subgroup of index two, and $G_k/H_{\{i,j\}} = \{H_0^{(3)}, H_1^{(3)}\}$ be the quotient group. For any $l, m \in \Phi$, $l \neq m$, consider the following $H_{\{i,j\}}$− periodic configuration:

$$\varphi_3^{(lm)}(x) = \begin{cases} l, & \text{if } x \in H_0^{(3)}, \\ m, & \text{if } x \in H_1^{(3)}. \end{cases}$$

So we thus obtain a periodic configuration $\varphi_3^{(lm)}$ with period $p = 2$ (equal to the index of the subgroup); then by construction $(\varphi_3^{(lm)})_b = \sigma_b$. Now we shall prove that all restrictions $(\varphi_3^{(lm)})_{b'}$ for any $b' \in M$ of the configuration $\varphi_3^{(lm)}$ belong to C_3.

Let $q_j(x) = |S_1(x) \cap H_j^{(3)}|$, $j = 0, 1$; where $S_1(x) = \{y \in G_k : \langle x, y \rangle\}$, the set of all nearest neighbors of $x \in G_k$. Denote $Q(x) = (q_0(x), q_1(x))$. Clearly $q_0(x)$ (resp. $q_1(x)$) is the number of points y in $S_1(x)$ such that $\varphi_3^{(lm)}(y) = l$ (resp. $\varphi_3^{(lm)}(y) = m$). We note (see Chap. 1 of [2]) that for every $x \in G_k$ there is a permutation π_x of the coordinates of the vector $Q(e)$ (where e as before is the identity of G_k) such that

$$\pi_x Q(e) = Q(x).$$

Moreover $Q(x) = Q(e)$ if $x \in H_0^{(3)}$ and $Q(x) = (q_1(e), q_0(e))$ if $x \in H_1^{(3)}$. Thus for any $b' \in M$ we have (i) if $c_{b'} \in H_0^{(3)}$ (where $c_{b'}$ is the center of b') then $(\varphi_3^{(lm)})_{b'} \in C_3$, (ii) if $c_{b'} \in H_1^{(3)}$, then $(\varphi_3^{(lm)})_{b'} \in C_3$.

Case C_4. Let $H_{\{i,j,r\}}$ be a normal subgroup of index two, and $G_k/H_{\{i,j,r\}} = \{H_0^{(4)}, H_1^{(4)}\}$ is the quotient group, for any $i, j, r \in \{1, 2, 3, 4\}$ and $i \neq j, i \neq r, j \neq r$. For any $l, m \in \Phi, l \neq m$, considering the following $H_{\{i,j,r\}}$– periodic configuration:

$$\varphi_4^{(lm)}(x) = \begin{cases} l, & \text{if } x \in H_0^{(4)}, \\ m, & \text{if } x \in H_1^{(4)}. \end{cases}$$

We thus obtain a periodic configuration $\varphi_4^{(lm)}$ with period $p = 2$; it is clear that $(\varphi_4^{(lm)})_{b'} \in C_4$ for any $b' \in M$.

Case C_5. Let $G_k/G_k^{(2)} = \{H_0^{(5)}, H_1^{(5)}\}$ is quotient group. For any $l, m \in \Phi, l \neq m$, consider the following $G_k^{(2)}$– periodic configuration:

$$\varphi_5^{(lm)}(x) = \begin{cases} l, & \text{if } x \in H_0^{(5)}, \\ m, & \text{if } x \in H_1^{(5)}. \end{cases}$$

It is easy to see (see [8]) that for each $b' \in M$ we have $(\varphi_5^{(lm)})_{b'} \in C_5$.

Case C_6. Let $G_3^{(6)} = H_i \cap H_j \cap H_r$, for any $i, j, r \in \{1, 2, 3, 4\}, i \neq j, i \neq r$, $j \neq r$. We note (see [2]) that $G_3^{(6)}$ is a normal index-eight subgroup in G_3, and $G_3/G_3^{(6)} = \{H_0^{(6)}, H_1^{(6)}, \ldots, H_7^{(6)}\}$ is quotient group, where
$H_0^{(6)} = G_3^{(6)} = \{x \in G_3 : w_i(x) \text{ is even, } w_j(x) \text{ is even, } w_r(x) \text{ is even}\}$,
$H_1^{(6)} = \{x \in G_3 : w_i(x) \text{ is even, } w_j(x) \text{ is even, } w_r(x) \text{ is odd}\}$,
$H_2^{(6)} = \{x \in G_3 : w_i(x) \text{ is even, } w_j(x) \text{ is odd, } w_r(x) \text{ is even}\}$,
$H_3^{(6)} = \{x \in G_3 : w_i(x) \text{ is even, } w_j(x) \text{ is odd, } w_r(x) \text{ is odd}\}$,
$H_4^{(6)} = \{x \in G_3 : w_i(x) \text{ is odd, } w_j(x) \text{ is even, } w_r(x) \text{ is even}\}$,
$H_5^{(6)} = \{x \in G_3 : w_i(x) \text{ is odd, } w_j(x) \text{ is even, } w_r(x) \text{ is odd}\}$,
$H_6^{(6)} = \{x \in G_3 : w_i(x) \text{ is odd, } w_j(x) \text{ is odd, } w_r(x) \text{ is even}\}$,
$H_7^{(6)} = \{x \in G_3 : w_i(x) \text{ is odd, } w_j(x) \text{ is odd, } w_r(x) \text{ is odd}\}$.

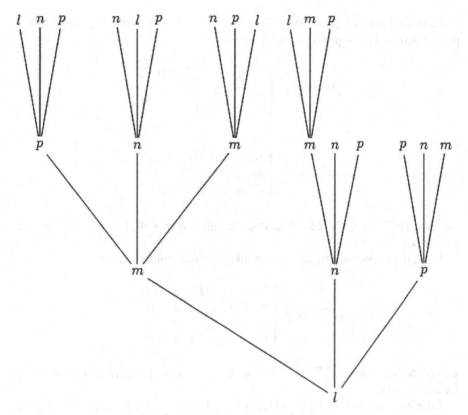

Fig. 1 Representation of the $G_k^{(6)}$ periodic configuration $\varphi_6^{(lmnp)}(x)$ on the Cayley tree of order $k = 3$

For given σ_b, we have

$$
\sigma_b(x) = \begin{cases}
l, & \text{if } x \in H_0^{(6)} \cup H_7^{(6)} \cap b, \\
m, & \text{if } x \in H_1^{(6)} \cup H_6^{(6)} \cap b, \\
n, & \text{if } x \in H_2^{(6)} \cup H_5^{(6)} \cap b, \\
p, & \text{if } x \in H_3^{(6)} \cup H_4^{(6)} \cap b.
\end{cases}
$$

For any $l, m, n, p \in \Phi, l \neq m, l \neq n, l \neq p, m \neq n, m \neq p, n \neq p$, continue the bounded configuration $\sigma_b \in C_6$ to the entire lattice Γ^3 (which is denoted by $\varphi_6^{(lmnp)}$, (see Fig. 1)) as

$$
\varphi_6^{(lmnp)}(x) = \begin{cases}
l, & \text{if } x \in H_0^{(6)} \cup H_7^{(6)}, \\
m, & \text{if } x \in H_1^{(6)} \cup H_6^{(6)}, \\
n, & \text{if } x \in H_2^{(6)} \cup H_5^{(6)}, \\
p, & \text{if } x \in H_3^{(6)} \cup H_4^{(6)}.
\end{cases}
$$

Case C_7. Let $G_k/G_k^{(4)} = \{H_0^{(7)}, H_1^{(7)}, H_2^{(7)}, H_3^{(7)}\}$ be quotient group. In [8] it was proved that for $G_k^{(4)}$ – periodic configurations

$$\varphi_7^{(lmn)}(x) = \begin{cases} l, & \text{if } x \in H_0^{(7)} \cap H_1^{(7)}, \\ m, & \text{if } x \in H_2^{(7)}, \\ n, & \text{if } x \in H_3^{(7)}, \end{cases}$$

and

$$\psi_7^{(lmn)}(x) = \begin{cases} l, & \text{if } x \in H_0^{(7)}, \\ m, & \text{if } x \in H_1^{(7)}, \\ n, & \text{if } x \in H_2^{(7)} \cap H_3^{(7)}, \end{cases}$$

one has: $(\varphi_7^{(lmn)})_{b'} \in C_7$, $(\psi_7^{(lmn)})_{b'} \in C_7$ for all $l, m, n \in \Phi, l \neq m, l \neq n, m \neq n$ and for any $b' \in M$.

In [9] it was proved that $H_{\{1,2,3\}}$ – weakly periodic configurations

$$\xi_7^{(lmn)}(x) = \begin{cases} l, & \text{if } x_{\downarrow} \in H_0, \ x \in H_0 \\ m, & \text{if } x_{\downarrow} \in H_0, \ x \in H_1 \\ n, & \text{if } x_{\downarrow} \in H_1, \ x \in H_0 \\ l, & \text{if } x_{\downarrow} \in H_1, \ x \in H_1, \end{cases}$$

satisfy the following: $(\xi_7^{(lmn)})_{b'} \in C_7$ for all $l, m, n \in \Phi, l \neq m, l \neq n, m \neq n$ and for any $b' \in M$.

Case C_8. Let $G_3^{(8)} = H_{\{i,j\}} \cap H_{\{k\}} \cap H_{\{r\}}$, $i, j, k, r \in \{1, 2, 3, 4\}, i \neq j, i \neq k$, $i \neq r, j \neq k, j \neq r, k \neq r$. We note (see [2]) that $G_3^{(8)}$ is a normal index-eight subgroup in G_3, and $G_3/G_3^{(8)} = \{H_0^{(8)}, H_1^{(8)}, \ldots, H_7^{(8)}\}$ is quotient group, where
$H_0^{(8)} = G_3^{(8)} = \{x \in G_3 : w_i(x) + w_j(x) \text{ is even}, w_k(x) \text{ is even}, w_r(x) \text{ is even}\}$,
$H_1^{(8)} = \{x \in G_3 : w_i(x) + w_j(x) \text{ is even}, w_k(x) \text{ is even}, w_r(x) \text{ is odd}\}$,
$H_2^{(8)} = \{x \in G_3 : w_i(x) + w_j(x) \text{ is even}, w_k(x) \text{ is odd}, w_r(x) \text{ is even}\}$,
$H_3^{(8)} = \{x \in G_3 : w_i(x) + w_j(x) \text{ is even}, w_k(x) \text{ is odd}, w_r(x) \text{ is odd}\}$,
$H_4^{(8)} = \{x \in G_3 : w_i(x) + w_j(x) \text{ is odd}, w_k(x) \text{ is even}, w_r(x) \text{ is even}\}$,
$H_5^{(8)} = \{x \in G_3 : w_i(x) + w_j(x) \text{ is odd}, w_k(x) \text{ is even}, w_r(x) \text{ is odd}\}$,
$H_6^{(8)} = \{x \in G_3 : w_i(x) + w_j(x) \text{ is odd}, w_k(x) \text{ is odd}, w_r(x) \text{ is even}\}$,
$H_7^{(8)} = \{x \in G_3 : w_i(x) + w_j(x) \text{ is odd}, w_k(x) \text{ is odd}, w_r(x) \text{ is odd}\}$.

In this case, for any $l, m, n, p \in \Phi, l \neq m, l \neq n, l \neq p, m \neq n, m \neq p, n \neq p$ we define the configuration $\varphi_8^{(lmnp)}$ as

$$\varphi_8^{(lmnp)}(x) = \begin{cases} l, & \text{if } x \in H_0^{(8)} \cup H_7^{(8)}, \\ m, & \text{if } x \in H_1^{(8)} \cup H_6^{(8)}, \\ n, & \text{if } x \in H_2^{(8)} \cup H_5^{(8)}, \\ p, & \text{if } x \in H_3^{(8)} \cup H_4^{(8)}. \end{cases}$$

We thus obtain a periodic configuration $\varphi_8^{(lmnp)}$ with the period $p = 8$ such that $(\varphi_8^{(lmnp)})_b = \sigma_b$, $(\varphi_8^{(lmnp)})_{b'} \in C_8$ for any $b' \in M$.

Case C_9. We consider a normal subgroup $\mathcal{H}_0 \in G_3$ (see [2]) of infinite index constructed as follows. Let the mapping $\pi_0 : \{a_1, a_2, a_3, a_4\} \to \{e, a_1, a_2\}$ be defined by

$$\pi_0(a_i) = \begin{cases} a_i, & \text{if } i = 1, 2 \\ e, & \text{if } i \neq 1, 2. \end{cases}$$

Consider

$$f_0(x) = f_0(a_{i_1} a_{i_2} \ldots a_{i_m}) = \pi_0(a_{i_1})\pi_0(a_{i_2}) \ldots \pi_0(a_{i_m}).$$

Then it is easy to see that f_0 is a homomorphism and hence $\mathcal{H}_0 = \{x \in G_3 : f_0(x) = e\}$ is a normal subgroup of infinity index.

Now we consider the factor group

$$G_3/\mathcal{H}_0 = \{\mathcal{H}_0, \mathcal{H}_0(a_1), \mathcal{H}_0(a_2), \mathcal{H}_0(a_1 a_2), \mathcal{H}_0(a_2 a_1), \ldots\},$$

where $\mathcal{H}_0(y) = \{x \in G_3 : f_0(x) = y\}$. We introduce the notations

$$\mathcal{H}_n = \mathcal{H}_0(\underbrace{a_1 a_2 \ldots}_{n}), \quad \mathcal{H}_{-n} = \mathcal{H}_0(\underbrace{a_2 a_1 \ldots}_{n}).$$

In this notation, the factor group can be represented as

$$G_3/\mathcal{H}_0 = \{\ldots, \mathcal{H}_{-2}, \mathcal{H}_{-2}, \mathcal{H}_0, \mathcal{H}_1, \mathcal{H}_2, \ldots\}.$$

It is known (see [6]), that for $x \in \mathcal{H}_n$ we have $|S_1(x) \cap \mathcal{H}_{n-1}| = 1$, $|S_1(x) \cap \mathcal{H}_n| = k - 1$, $|S_1(x) \cap \mathcal{H}_{n+1}| = 1$.

Consider the following configuration

$$\varphi_9^{(lm)}(x) = \begin{cases} 2nl, & \text{if } x \in \mathcal{H}_n, n \neq 0, \\ 0, & \text{if } x \in \mathcal{H}_0, \\ (2n - 1)m, & \text{if } x \in \mathcal{H}_{-n}, n \neq 0, \end{cases}$$

where $l, m \in \Phi, l \neq m, n = 1, 2, 3 \ldots$

We thus obtain a periodic configuration $\varphi_9^{(lm)}$ with the infinity period, such that $(\varphi_9^{(lm)})_b = \sigma_b \in C_9$, and $(\varphi_9^{(lm)})_{b'} \in C_9$ for any $b' \in M$.

Case C_{10}. Let S_3 be the group of third-order permutations. We choose $\pi_0, \pi_1, \pi_2 \in S_3$ as

$$\pi_0 = \begin{pmatrix} 1\ 2\ 3 \\ 1\ 2\ 3 \end{pmatrix}, \pi_1 = \begin{pmatrix} 1\ 2\ 3 \\ 1\ 3\ 2 \end{pmatrix}, \pi_2 = \begin{pmatrix} 1\ 2\ 3 \\ 3\ 2\ 1 \end{pmatrix}. \tag{5}$$

It is easily seen that $\pi_0 = \pi_1^2 = \pi_2^2$.

We consider the map $u : \{a_1, a_2, a_3, a_4\} \to \{\pi_1, \pi_2\}$

$$u(a_i) = \begin{cases} \pi_1, i = 1, 2; \\ \pi_2, i = 3, 4 \end{cases} \tag{6}$$

and assume that the function $f : G_3 \to S_3$ is defined as

$$f(x) = f(a_{i_1} a_{i_2} \ldots a_{i_n}) = u(a_{i_1}) \ldots u(a_{i_n}).$$

Let

$$\pi_3 = \begin{pmatrix} 1\,2\,3 \\ 3\,1\,2 \end{pmatrix}, \pi_4 = \begin{pmatrix} 1\,2\,3 \\ 2\,3\,1 \end{pmatrix}, \pi_5 = \begin{pmatrix} 1\,2\,3 \\ 2\,1\,3 \end{pmatrix}.$$

We note (see [14]) that $H_{10} = \{x \in G_3 : f(x) = \pi_0\}$ is a normal index-six subgroup. Let $G_3/H_{10} = \{\aleph_0, \ldots, \aleph_5\}$ be the quotient group, where

$$\aleph_i = \{x \in G_3 : f(x) = \pi_i\}, i = \overline{0, 5}.$$

In this case, we define the configuration

$$\varphi_{10}^{(l,m,n)}(x) = \begin{cases} l, & x \in \aleph_0 \cup \aleph_5, \\ m, & x \in \aleph_1 \cup \aleph_4, \\ n, & x \in \aleph_2 \cup \aleph_3, \end{cases}$$

where $l, m, n \in \Phi, l \neq m, l \neq n, m \neq n$.

We thus obtain a periodic configuration $\varphi_{10}^{(l,m,n)}$ with the period six, such that $(\varphi_{10}^{(l,m,n)})_b = \sigma_b \in C_{10}, (\varphi_{10}^{(l,m,n)})_{b'} \in C_{10}$ for any $b' \in M$.

Case C_{11}. Let S_3 be the group of third-order permutations. It is easily seen that $\pi_0 = \pi_1^2 = \pi_5^2$.

We consider the map $u : \{a_1, a_2, a_3, a_4\} \to \{\pi_1, \pi_5\}$

$$u(a_i) = \begin{cases} \pi_5, i = 1, 2; \\ \pi_1, i = 3, 4, \end{cases} \tag{7}$$

and assume that the function $f : G_3 \to S_3$ is defined as

$$f(x) = f(a_{i_1} a_{i_2} \ldots a_{i_n}) = u(a_{i_1}) \ldots u(a_{i_n}).$$

We note (see [14]) that $H_{11} = \{x \in G_3 : f(x) = \pi_0\}$ is a normal index-six subgroup. Let $G_3/H_{11} = \{\aleph_0, \ldots, \aleph_5\}$ be the quotient group, where

$$\aleph_i = \{x \in G_3 : f(x) = \pi_i\}, i = \overline{0, 5}.$$

In this case, we define the configuration

$$\varphi_{11}^{(l,m,n)}(x) = \begin{cases} l, & x \in \aleph_0 \cup \aleph_2, \\ m, & x \in \aleph_4 \cup \aleph_5, \\ n, & x \in \aleph_1 \cup \aleph_3, \end{cases}$$

where $l, m, n \in \Phi, l \neq m, l \neq n, m \neq n$.

We thus obtain a periodic configuration $\varphi_{11}^{(l,m,n)}$ with the period six, such that $(\varphi_{11}^{(l,m,n)})_b = \sigma_b \in C_{11}, (\varphi_{11}^{(l,m,n)})_{b'} \in C_{11}$ for any $b' \in M$.

Case C_{12}. Let $\mathcal{U} = \{(a_1 a_2)^n \in G_3 : n \in \mathbf{Z}\}$. It is easy to see, that $\mathcal{U}$ is subgroup of the group G_3. Consider the set of right cosets $G_3/\mathcal{U} = \{\mathcal{U}, \mathcal{U} a_1, \ldots, \mathcal{U} a_{k+1}, \mathcal{U} a_1 a_2, \ldots\}$ of $\mathcal{U}$ in G_3. We introduce the notations

$$H_0 = \mathcal{U}, \ H_1 = \mathcal{U} a_1, \ldots, H_{k+1} = \mathcal{U} a_{k+1}, H_{k+2} = \mathcal{U} a_1 a_2, \ldots.$$

In this notation, the set of right coset can be represented as

$$G_3/\mathcal{U} = \{H_0, H_1, \ldots H_{k+1}, H_{k+2}, \ldots\}.$$

Consider the following configuration: $\varphi_{12}^l(x) = l + i$, if $x \in H_i$ for all $i = 0, 1, 2, \ldots$ and for any $l \in \Phi$.

Let $x \in H_n$, then $\varphi_{12}^l(x) = l + n$ and if $H_n = \mathcal{U} a_{j_1} a_{j_2} \ldots a_{j_n}$, then for all $y \in S_1(x)$ we have $y \in \mathcal{U} a_{j_1} a_{j_2} \ldots a_{j_n} a_t$, $t = 1, 2, 3, 4$. By construction of configuration we have $\varphi_{12}^l(y) \neq \varphi_{12}^l(x)$ and $\varphi_{12}^l(y_1) \neq \varphi_{12}^l(y_2)$ for all $y, y_1, y_2 \in S_1(x), y_1 \neq y_2$.

We thus obtain a $\mathcal{U}$−periodic configuration φ_{12}^l with the infinity period, such that $(\varphi_{12}^l)_{b'} \in C_{12}$ for any $b' \in M$.

We set $B = A_1 \cap A_2, B_0 = A_1 \cap A_5, B_1 = A_2 \cap A_9, B_2 = A_9 \cap A_6, B_3 = A_6 \cap A_{12}, \widetilde{A_1} = A_1 \setminus (B \cup B_0), \widetilde{A_2} = A_2 \setminus (B_0 \cup B_1), \widetilde{A_5} = A_5 \setminus (B_0 \cup A_7), \widetilde{A_6} = A_6 \setminus (B_2 \cup B_3), \widetilde{A_9} = A_9 \setminus (B_1 \cup B_2)$ and $\widetilde{A_{12}} = A_{12} \setminus (B_3 \cup A_7)$. Let $GS(H)$ be the set of all ground states, and let $GS_p(H)$ be the set of all periodic ground states.

Remark 1 (1) Note that,

(i) If $q \geq 3$ then the ground states $\sigma(x), \varphi_2^{(lm)}, \varphi_3^{(lm)}, \varphi_4^{(lm)}, \varphi_5^{(lm)}, \varphi_7^{(lmp)}, \psi_7^{(lmn)}, \xi_7^{(lmn)}, \varphi_{10}^{(lmn)}, \varphi_{11}^{(lmn)}$ described in Theorem 1 are ground states for the q state Potts model on the Cayley tree of order three.

(ii) If $q \geq 4$ then the ground states $\varphi_6^{(lmnp)}, \varphi_8^{(lmnp)}$ (described in Theorem 1) are ground states for the q state Potts model on the Cayley tree of order three.

(iii) The ground states $\varphi_9^{(lm)}, \varphi_{12}^l$ are ground states only for the Potts model with countable set of spin values on the Cayley tree of order three.

(2) In this paper we considered the case $k = 3$. If one considers the case $k \geq 4$, the ground states described in Theorem 1 may not be ground state, and this class of ground states may be extended. Besides the set A_i in (4) will be different.

Theorem 2 *A. If $J = (0, 0)$, then $GS(H) = \Omega$.*

B. 1. If $J \in \widetilde{A_1}$, then $GS_p(H) = \{\varphi^{(i)} : i \in \Phi\}$.

2. If $J \in \widetilde{A_2}$, then $GS_p(H) = \{\varphi_2^{(lm)} : l, m \in \Phi, l \neq m\}$.

3. If $J \in \widetilde{A_5}$, then $GS_p(H) = \{\varphi_5^{(lm)} : l, m \in \Phi, l \neq m\}$.

4. If $J \in \widetilde{A_6}$, then $GS_p(H) = \{\varphi_6^{(lmnp)} : l, m, n, p \in \Phi, l \neq m, l \neq n, l \neq p, m \neq n, m \neq p, n \neq p\}$.

5. If $J \in \widetilde{A_9}$, then $GS_p(H) = \{\varphi_9^{(lm)} : l, m \in \Phi, l \neq m\}$.

6. If $J \in \widetilde{A_{12}}$, then $GS_p(H) = \{\varphi_{12}^l : l \in \Phi\}$.

C. 1. If $J \in B \setminus \{(0, 0)\}$, then $GS_p(H) = \{\varphi^{(i)}, \varphi_2^{(lm)} : i, l, m \in \Phi, l \neq m\}$.

2. If $J \in B_0 \setminus \{(0, 0)\}$, then $GS_p(H) = \{\varphi^{(i)}, \varphi_5^{(lm)} : i, l, m \in \Phi, l \neq m\}$.

3. If $J \in B_1 \setminus \{(0, 0)\}$, then $GS_p(H) = \{\varphi_2^{(lm)}, \varphi_9^{(lm)} : i, l, m \in \Phi, l \neq m\}$.

4. If $J \in B_2 \setminus \{(0, 0)\}$ then $GS_p(H) = \{\varphi_6^{(lmnp)}, \varphi_9^{(lm)} : l, m, n, p \in \Phi, l \neq m, l \neq n, l \neq p, m \neq n, m \neq p, n \neq p\}$.

5. If $J \in B_3 \setminus \{(0, 0)\}$ then $GS_p(H) = \{\varphi_6^{(lmnp)}, \varphi_{12}^l : l, m, n, p \in \Phi, l \neq m, l \neq n, l \neq p, m \neq n, m \neq p, n \neq p\}$.

6. If $J \in A_8$, then periodic configuration $\varphi_5^{(lm)}, \xi_7^{(lmn)}(x), \psi_7^{(lmn)}(x), \varphi_8^{(lmnp)}(x), \varphi_{12}^l$ are periodic ground states, and weakly periodic configuration $\xi_7^{(lmn)}(x)$ is weakly periodic ground state, where $l, m, n, p \in \Phi, l \neq m, l \neq n, l \neq p, m \neq n, m \neq p, n \neq p$.

Proof Case A is trivial. In cases B and C, for a given configuration σ_b for which the energy $U(\sigma_b)$ is minimal, we can use Theorem 1 to construct the periodic configurations.

Remark 2 Since the set Φ is countable, it follows that the periodic and weakly periodic ground states described in Theorem 2 are countable.

Acknowledgements The authors thank Professor U. Rozikov for useful discussions. Also the authors thank for anonyme reviewers for useful comments.

References

1. Mukhamedov, F., Rozikov, U., Mendes, F.F.: On contour arguments for the three state Potts model with competing interactions on a semi-infinite Cayley tree. J. Math. Phys. **48**, 013301 (2007). https://doi.org/10.1063/1.2408398
2. Rozikov, U.A.: Gibbs measures on Cayley trees. World Scientific (2013). ISBN-13: 978-9814513371, ISBN-10: 9814513377
3. Rozikov, U.A.: A constructive description of ground states and gibbs measures for Ising model with two-step interactions on Cayley tree. J. Stat. Phys. **122**, 217–235 (2006)
4. Rahmatullaev, M.M.: Desription of weak periodic ground states of Ising model with competing interactions on Cayley tree. Appl. Math. Inf. Sci. **4**(2), 237–241 (2010)
5. Botirov, G.I., Rozikov, U.A.: Potts model with competing interactions on the Cayley tree: the contour method. Theor. Math. Phys. **153**, 1423–1433 (2007)
6. Rozikov, U.A., Ishankulov, F.T.: Description of periodic p-harmonic functions on Cayley tree NoDEA. Nonlinear Differ. Equ. Appl. **17**(2), 153–160 (2010)
7. Rozikov, U.A., Rahmatullaev, M.M.: Weakly periodic ground states and Gibbs measures for the Ising model with competing interactions on the Cayley tree. Theor. Math. Phys. **160**, 1292–1300 (2009)

8. Rahmatullaev, M.M., Rasulova, M.A.: Periodic and weakly periodic ground states for the Potts model with competing interactions on the Cayley tree. Sib. Adv. Math. **26**(3), 215–229 (2016)
9. Rahmatullaev, M.M.: Weakly periodic Gibbs measures and ground states for the Potts model with competing interactions on the Cayley tree. Theor. Math. Phys. **176**, 1236–1251 (2013)
10. Ganikhodjaev, N.N., Mukhamedov, F.M., Mendes, J.F.F.: On the three state Potts model with competing interactions on the Bethe lattice. J. Stat. Mech. P08012, 29p (2006)
11. Ganikhodjaev, N., Mukhamedov, F., Pah, C.H.: Phase diagram of the three states Potts model with next nearest neighbour interactions on the Bethe lattice. Phys. Lett. A **373**, 33–38 (2008)
12. Baxter, R.J.: Exactly Solved Models in Statistical Mechanics. Academic Press, London (1982)
13. Gankhodjaev, N.N.: Group represantation and automorphisms of the Cayley tree. Dokl. Akad. Nauk Resp. Uzbekistan **4**, 3–6 (1994)
14. Ganikhodzhaev, N.N., Rozikov, U.A.: Description of peridoic extreme Gibbs measures of some lattice models on the Cayley tree. Theor. Math. Phys. **111**, 480–486 (1997)

Isomorphic Classification of ∗-Algebras of Log-Integrable Measurable Functions

R. Z. Abdullaev and V. I. Chilin

Abstract Let (Ω, μ) be a σ-finite measure space, and let $L_0(\Omega, \mu)$ be the ∗-algebra of all complex (real) valued measurable functions on (Ω, μ). The ∗-subalgebra $L_{\log}(\Omega, \mu) = \{f \in L_0(\Omega, \mu) : \int_{\Omega} \log(1 + |f|)d\mu < +\infty\}$ of $L_0(\Omega, \mu)$ is called the algebra of log-integrable measurable functions on (Ω, μ). Using the notion of passport of a normed Boolean algebra, we give the necessary and sufficient conditions for a ∗-isomorphism of two algebras of log-integrable measurable functions.

Keywords Passport of Boolean algebra · Isomorphism of Boolean algebras · Equivalent measures · Log-integrable functions

1 Introduction

Let $(\Omega, \mathscr{A}, \mu)$ be a measure space with σ-finite measure μ and let $\mathscr{L}_{\log}(\Omega, \mathscr{A}, \mu)$ be the symmetric function space consisting of complex (real) measurable functions f on $(\Omega, \mathscr{A}, \mu)$ such that $\int \log(1 + |f|)\,d\mu < \infty$. It is known that this symmetric function space is a ∗-subalgebra in an ∗-algebra $\mathscr{L}_0(\Omega, \mathscr{A}, \mu)$ of all complex measurable functions on $(\Omega, \mathscr{A}, \mu)$ (functions equal μ-almost everywhere are identified) [3]. In addition, a function $\|f\|_{\log} = \int_{\Omega} log(1 + |f|)d\mu$ is an F-norm on $\mathscr{L}_{\log}(\Omega, \mathscr{A}, \mu)$, i.e.

$(i).\ \|f\|_{\log} > 0\ (0 \neq f \in \mathscr{L}_{\log}(\Omega, \mathscr{A}, \mu));$
$(ii).\ \|\alpha f\|_{\log} \leq \|f\|_{\log}\ (f \in \mathscr{L}_{\log}(\Omega, \mathscr{A}, \mu),\ |\alpha| \leq 1);$
$(iii).\ \lim_{\alpha \to 0} \|\alpha f\|_{\log} = 0\ (f \in \mathscr{L}_{\log}(\Omega, \mathscr{A}, \mu));$
$(iv).\ \|f + g\|_{\log} \leq \|f\|_{\log} + \|g\|_{\log}\ (f, g \in \mathscr{L}_{\log}(\Omega, \mathscr{A}, \mu)).$

R. Z. Abdullaev
Uzbek State University of World Languages, Tashkent, Uzbekistan
e-mail: arustambay@yandex.com

V. I. Chilin (✉)
National University of Uzbekistan, Tashkent, Uzbekistan
e-mail: vladimirchil@gmail.com

© Springer Nature Switzerland AG 2018
Z. Ibragimov et al. (eds.), *Algebra, Complex Analysis, and Pluripotential Theory*,
Springer Proceedings in Mathematics & Statistics 264,
https://doi.org/10.1007/978-3-030-01144-4_6

With respect to the metric $\rho(f, g) = \|f - g\|_{\log}$ the $*$-algebra $\mathscr{L}_{\log}(\Omega, \mathscr{A}, \mu)$ is a non-locally-convex complete topological $*$-algebra [3, Corrollary 2.7].

In the case when the measure space $(\Omega, \mathscr{A}, m)$ is the unit circle in the complex plane endowed with Lebesgue measure m, boundary values of Nevanlinna functions are in the symmetric function space $\mathscr{L}_{\log}(\Omega, \mathscr{A}, m)$ and the map sending a Nevanlinna function to its boundary values provides an injective and continuous algebraic homomorphism from the Nevanlinna class to $\mathscr{L}_{\log}(\Omega, \mathscr{A}, m)$. Since the Nevanlinna class is not well behaved under the usual metric, it is naturally to study the topological structure on the Nevanlinna using F-norm $\|\cdot\|_{\log}$. Such study was carried out in the paper [3].

Taking into account the importance of $*$-algebras $\mathscr{L}_{\log}(\Omega, \mathscr{A}, \mu)$ in the theory of functions of a complex variable, it is necessary to describe all these $*$-algebras associated with different measures up to their $*$-isomorphisms. This paper is devoted to solving this problem. Using the notion of the passport of a normed Boolean algebra, we give the necessary and sufficient conditions ensuring the existence of a $*$-isomorphism from a $*$-algebra $L_{\log}(\Omega_1, \mu_1)$ onto $*$-algebra $L_{\log}(\Omega_2, \mu_2)$. In the proof of this criterion we use the method of papers [1, 2], which give a description of all $*$-isomorphisms of the Arens algebras of measurable functions.

2 Preliminaries

Let $(\Omega, \mathscr{A}, \mu)$ be a measure space with σ-finite measure μ and let $\nabla = \nabla_\mu$ be the complete Boolean algebra of all equivalence classes $e = [A]$ of sets equal μ-almost everywhere. It is known that $\widehat{\mu}(e) = \mu(A)$ is a strictly positive σ-finite measure on ∇_μ. Further the measure $\widehat{\mu}$ will be denoted by μ. Denote by $\mathscr{L}_0(\nabla) = \mathscr{L}_0(\nabla, \mu)$ the algebra of μ-equivalence classes of μ-(a.e.) finite complex (real)-valued measurable functions on $(\Omega, \mathscr{A}, \mu)$. Let $\mathscr{L}_p(\nabla, \mu) \subset \mathscr{L}_0(\nabla)$, $1 \leq p \leq \infty$, be the L_p-space equipped with the standard norm $\|\cdot\|_p$.

Let $\varphi : \nabla \to \nabla$ be an arbitrary automorphism of the Boolean algebra ∇. It is clear that $\lambda(e) := \mu(\varphi(e))$, $e \in \nabla$ is a strictly positive countably additive σ-finite measure on the Boolean algebra ∇. Denote by Φ the $*$-isomorphism of the $*$-algebra $\mathscr{L}_0(\nabla)$ such that $\varphi(e) = \Phi(e)$ for all $e \in \nabla$. The restriction of Φ on the C^*-algebra $\mathscr{L}_\infty(\nabla)$ is a $*$-isomorphism of $\mathscr{L}_\infty(\nabla)$.

Since

$$\int_\Omega \sum_{i=1}^n |c_i|^p e_i \, d\lambda = \sum_{k=i}^n |c_i|^p \lambda(e_i) = \sum_{k=i}^n |c_i|^p \mu(\varphi(e_i)) = \int_\Omega \Phi\left(\sum_{i=1}^n |c_i|^p e_i\right) d\mu,$$

for all $e_i \in \nabla$, $\mu(e_i) < \infty$, $e_i e_j = 0$, $i \neq j$, $c_i \in \mathbf{C}$, $i, j = 1, \ldots, n$, $n \in \mathbf{N}$, $1 \leq p < \infty$, where $\mathbf{C}$ is the field of complex numbers, $\mathbf{N}$ is the set of all natural numbers, it follows that

$$\int_{\Omega} |f|^P d\lambda = \int_{\Omega} \Phi(|f|^P) d\mu, \quad \int_{\Omega} \Phi^{-1}(|g|^P) d\lambda = \int_{\Omega} |g|^P d\mu$$

for all $f \in \mathscr{L}_p(\nabla, \lambda)$, $g \in \mathscr{L}_p(\nabla, \mu)$. This means that Φ is a surjective isometry from $\mathscr{L}_p(\nabla, \lambda)$ onto $\mathscr{L}_p(\nabla, \mu)$.

Let ∇ be a non-atomic complete Boolean algebra, that is, a Boolean algebra ∇ has not atoms. Let $\nabla_e = \{g \in \nabla : g \leq e\}$, where $0 \neq e \in \nabla$. By $\tau(\nabla_e)$ denote the minimal cardinality of a set that is dense in ∇_e with respect to the order topology ((o)-topology). A non-atomic complete Boolean algebra ∇ is said to be homogeneous if $\tau(\nabla_e) = \tau(\nabla_g)$ for any nonzero $e, g \in \nabla$. The cardinality $\tau(\nabla)$ is called the weight of a homogeneous Boolean algebra ∇ (see, for example [4, Chap. VII]).

Let $\mathbf{1}$ be the unit in a non-atomic Boolean algebra ∇ and $\mu(\mathbf{1}) < \infty$. Then ∇ is a direct product of homogeneous Boolean algebras ∇_{e_n}, $e_n \cdot e_m = 0$, $n \neq m$, $\tau_n = \tau(\nabla_{e_n}) < \tau_{n+1}$, $n, m \in \mathbf{N}$ [4, Chap. VII, Sect. 2, p. 272]).

Set $\mu_n = \mu(e_n)$. The matrix $\begin{pmatrix} \tau_1 & \tau_2 & \cdots \\ \mu_1 & \mu_2 & \cdots \end{pmatrix}$ is called *the passport* of a Boolean algebra (∇, μ).

The following theorem gives a classification of Boolean algebras with finite measure [4, Chap. VII, Sect. 2, Theorems 5 and 6].

Theorem 1 (a). *Let μ and ν be probability measures on the non-atomic complete Boolean algebras ∇_1 and ∇_2 respectively. Let $\begin{pmatrix} \tau_1^{(1)} & \tau_2^{(1)} & \cdots \\ \mu_1 & \mu_2 & \cdots \end{pmatrix}$ be the passport of Boolean algebras (∇_1, μ) and $\begin{pmatrix} \tau_1^{(2)} & \tau_2^{(2)} & \cdots \\ \nu_1 & \nu_2 & \cdots \end{pmatrix}$ be the passport of Boolean algebras (∇_2, ν).*

The following conditions are equivalent:

(i). There exists an isomorphism $\varphi : \nabla_1 \to \nabla_2$ such that $\mu(e) = \nu(\varphi(e))$ for all $e \in \nabla_1$;

(ii). $\tau_n^{(1)} = \tau_n^{(2)}$ and $\mu_n = \nu_n$ for all $n \in \mathbf{N}$.

(b). Let ∇_1 and ∇_2 be non-atomic complete Boolean algebras. Then the Boolean algebras ∇_1 and ∇_2 are isomorphic if and only if the upper rows of their passports are equal.

Corollary 1 ([4, Chap. VII, Sect. 2, Theorem 5 and Corollary]) *Let μ and ν be probability measures on non-atomic homogeneous Boolean algebras ∇_1 and ∇_2 respectively. If $\tau(\nabla_1) = \tau(\nabla_2)$ then there exists an isomorphism φ from ∇_1 onto ∇_2 such that $\nu(\varphi(e)) = \mu(e)$ for all $e \in \nabla_1$.*

Corollary 2 *Let μ and ν be finite measures on the non-atomic homogeneous Boolean algebras ∇_1 and ∇_2 respectively. If $\tau(\nabla_1) = \tau(\nabla_2)$ then there exists an isomorphism φ from ∇_1 onto ∇_2 such that $\nu(\varphi(e)) = \frac{\nu(\mathbf{1}_{\nabla_2})}{\mu(\mathbf{1}_{\nabla_1})} \cdot \mu(e)$, where $\mathbf{1}_{\nabla_i}$ is the unit in the Boolean algebra ∇_i, $i = 1, 2$.*

Proof If $\mu_1(e) = \frac{\mu(e)}{\mu(\mathbf{1}_{\nabla_1})}$, $e \in \nabla_1$ and $\nu_1(q) = \frac{\nu(q)}{\nu(\mathbf{1}_{\nabla_2})}$, $q \in \nabla_2$ then μ_1 and ν_1 are probability measures on the non-atomic homogeneous Boolean algebras ∇_1 and ∇_2

respectively. By Corollary 1 we have that exists an isomorphism φ from ∇_1 onto ∇_2 such that $\nu_1(\varphi(e)) = \mu_1(e)$ for all $e \in \nabla_1$, i.e. $\nu(\varphi(e)) = \frac{\nu(1_{\nabla_2})}{\mu(1_{\nabla_1})} \cdot \mu(e)$, $e \in \nabla_1$. $\square$

Corollary 3 *Let μ and ν be σ-finite measures on the non-atomic homogeneous Boolean algebras ∇_1 and ∇_2 respectively. If $\tau(\nabla_1) = \tau(\nabla_2)$ then there exists an isomorphism φ from ∇_1 onto ∇_2 such that $\nu(\varphi(e)) = \mu(e)$ for all $e \in \nabla_1$.*

Proof Since ∇_1 (respectively, ∇_2) is non-atomic homogeneous Boolean algebra, it follows that there exists a countable partition $\{e_n\}_{n=1}^{\infty}$ (respectively, $\{q_n\}_{n=1}^{\infty}$) of the unit of the Boolean algebra ∇_1 (respectively, ∇_2) such that $\tau(e_n \cdot \nabla_1) = \tau(\nabla_1)$ and $\mu(e_n) = 1$ (respectively, $\tau(q_n \cdot \nabla_2) = \tau(\nabla_2)$ and $\nu(q_n) = 1$ for all $n \in \mathbf{N}$. By Corollary 1 there exists an isomorphism φ_n from $e_n \cdot \nabla_1$ onto $q_n \cdot \nabla_2$ such that $\nu(\varphi_n(e)) = \mu(e)$ for all $e \in e_n \cdot \nabla_1$, $n \in \mathbf{N}$.

Since the Boolean algebra ∇_2 is a complete Boolean algebra it follows that there exists $\varphi(e) := \sup_{n \in \mathbf{N}} \varphi_n(e_n \cdot e) \in \nabla_2$ for each $e \in \nabla_1$. It is clear that a map $\varphi : \nabla_1 \rightarrow \nabla_2$ is a $*$-isomorphism from ∇_1 onto ∇_2 such that $\nu(\varphi(e)) = \mu(e)$ for all $e \in \nabla_1$. $\square$

3 Isomorphic Classification of $*$-Algebras of Log-Integrable Measurable Functions in the Case of Finite Measures

Let μ be a strictly positive σ-finite measure on the Boolean algebra ∇ and let

$$\mathscr{L}_{\log}(\nabla, \mu) = \{f \in \mathscr{L}_0(\nabla, \mu) : log(1 + |f|) \in \mathscr{L}_1(\nabla, \mu)\}$$

be the $*$-algebra of all log-integrable measurable functions (see [3]).

Let μ and ν be a strictly positive σ-finite measures on a Boolean algebra ∇. Denote by $\frac{d\nu}{d\mu}$ the Radon–Nikodym derivative of the measure ν with respect to the measure μ. It is known that $\frac{d\nu}{d\mu} \in \mathscr{L}_0(\nabla)$ and

$$f \in \mathscr{L}_1(\nabla, \nu) \Leftrightarrow f \cdot \frac{d\nu}{d\mu} \in \mathscr{L}_1(\nabla, \mu),$$

in addition,

$$\int_{\Omega} f d\nu = \int_{\Omega} f \cdot \left(\frac{d\nu}{d\mu}\right) d\mu.$$

Since μ and ν are the strictly positive measures it follows that there exists the inverse function $(\frac{d\nu}{d\mu})^{-1} = \frac{d\mu}{d\nu}$.

Proposition 1 *The following conditions are equivalent:*
   (i). $\frac{d\nu}{d\mu} \in \mathscr{L}_\infty(\nabla)$;
   (ii). $\mathscr{L}_{\log}(\nabla, \mu) \subset \mathscr{L}_{\log}(\nabla, \nu)$.

Proof (i)⇒(ii). Let $\frac{d\nu}{d\mu} \in \mathscr{L}_\infty(\nabla)$. If $f \in \mathscr{L}_{\log}(\nabla, \mu)$, then $\int_\Omega \log(1 + |f|)d\mu < \infty$, and we we have that

$$\int_\Omega \log(1 + |f|)d\nu = \int_\Omega \log(1 + |f|)\frac{d\nu}{d\mu}d\mu$$
$$\leq \|\frac{d\nu}{d\mu}\|_\infty \int_\Omega \log(1 + |f|)d\mu < \infty.$$

Consequently, $f \in \mathscr{L}_{\log}(\nabla, \nu)$, and therefore, $\mathscr{L}_{\log}(\nabla, \mu) \subset \mathscr{L}_{\log}(\nabla, \nu)$.

(ii)⇒(i). Let $\mathscr{L}_{\log}(\nabla, \mu) \subset \mathscr{L}_{\log}(\nabla, \nu)$. Suppose that $\frac{d\nu}{d\mu} \notin \mathscr{L}_\infty(\nabla)$. In this case there exists a sequence of natural numbers $n_k \uparrow \infty$, $n_k > k$, such that $\mu(e_{n_k}) > 0$ for all $k \in \mathbf{N}$, where $e_{n_k} = \{n_k \leq \frac{d\nu}{d\mu} < n_k + 1\}$.

Let $g = \sum_{k=1}^\infty \frac{1}{k^2 \cdot \mu(e_{n_k})} \cdot e_{n_k}$ and let $f = e^g - 1$ (we assume that the base of logarithm in the definition of algebra $\mathscr{L}_{\log}(\nabla, \mu)$ is the Euler's Number e). Then

$$\int_\Omega \ln(1 + |f|)d\mu = \sum_{k=1}^\infty \frac{\mu(e_{n_k})}{k^2\mu(e_{n_k})} = \sum_{k=1}^\infty \frac{1}{k^2} < \infty, \tag{1}$$

In the same time,

$$\int_\Omega \ln(1 + |f|)d\nu = \int_\Omega \ln(1 + |f|)\frac{d\nu}{d\mu}d\mu \geq \sum_{k=1}^\infty \frac{n_k \cdot \mu(e_{n_k})}{k^2\mu(e_{n_k})} =$$
$$\sum_{k=1}^\infty \frac{n_k}{k^2} \geq \sum_{k=1}^\infty \frac{1}{k} = \infty. \tag{2}$$

From (1) and (2) it follows that $f \in \mathscr{L}_{\log}(\nabla, \mu)$ and $f \notin \mathscr{L}_{\log}(\nabla, \nu)$, that is $\mathscr{L}_{\log}(\nabla, \mu)$ is not a subset of $\mathscr{L}_{\log}(\nabla, \nu)$. This contradicts the assumption. Consequently, the function $\frac{d\nu}{d\mu}$ belongs to $\mathscr{L}_\infty(\nabla)$. □

The following corollary follows from Proposition 1.

Corollary 4 *The following conditions are equivalent:*
  (i). $\frac{d\nu}{d\mu} \in \mathscr{L}_\infty(\nabla)$ *and* $\frac{d\mu}{d\nu} \in \mathscr{L}_\infty(\nabla)$;
  (ii). $\mathscr{L}_{\log}(\nabla, \mu) = \mathscr{L}_{\log}(\nabla, \nu)$.

Let μ be a strictly positive σ-finite measure on a complete Boolean algebra ∇, and let $0 \neq e \in \nabla$, $e \neq \mathbf{1}$. Consider the Cartesian product

$$\mathscr{L}_{\log}(\nabla_e, \mu) \times \mathscr{L}_{\log}(\nabla_{1-e}, \mu) = \{(f, g) : f \in \mathscr{L}_{\log}(\nabla_e, \mu), \ g \in \mathscr{L}_{\log}(\nabla_{1-e}, \mu)\}$$

of ∗-algebras $\mathscr{L}_{\log}(\nabla_e, \mu)$ and $\mathscr{L}_{\log}(\nabla_{1-e}, \mu)$ with coordinate-wise algebraic operations. It is clear that $\mathscr{L}_{\log}(\nabla_e, \mu) \times \mathscr{L}_{\log}(\nabla_{1-e}, \mu)$ is a ∗-algebra.
  Define a map $\Phi : \mathscr{L}_{\log}(\nabla, \mu) \to \mathscr{L}_{\log}(\nabla_e, \mu) \times \mathscr{L}_{\log}(\nabla_{1-e}, \mu)$ by $\Phi(f) = (e \cdot f, (1 - e) \cdot f)$. Obviously, Φ is a ∗-homomorphism from $\mathscr{L}_{\log}(\nabla, \mu)$ into $\mathscr{L}_{\log}(\nabla_e, \mu) \times \mathscr{L}_{\log}(\nabla_{1-e}, \mu)$. If $\Phi(f) = \Phi(h)$, then $e \cdot f = e \cdot h$ and $(1 - e) \cdot f = (1 - e) \cdot h$, i.e. $f = h$.

If $(f, g) \in \mathscr{L}_{\log}(\nabla_e, \mu) \times \mathscr{L}_{\log}(\nabla_{1-e}, \mu)$, then $h = f + g \in \mathscr{L}_{\log}(\nabla, \mu)$ and $\Phi(h) = (f, g)$, i.e. Φ is a surjection.

Thus, the following proposition holds.

Proposition 2 *If μ is a strictly positive σ-finite measure on complete Boolean algebra ∇, and $0 \neq e \in \nabla$, $e \neq \mathbf{1}$, then $\Phi(f) = (e \cdot f, (\mathbf{1} - e) \cdot f)$ is a $*$-isomorphism from $\mathscr{L}_{\log}(\nabla, \mu)$ onto $\mathscr{L}_{\log}(\nabla_e, \mu) \times \mathscr{L}_{\log}(\nabla_{1-e}, \mu)$.*

Let μ (respectively, ν) be a strictly positive σ-finite measure on a complete Boolean algebra ∇_1 (respectively, ∇_2), let $\varphi : \nabla_1 \rightarrow \nabla_2$ be an isomorphism from ∇_1 onto ∇_2, and let $\Phi : \mathscr{L}_0(\nabla_1) \rightarrow \mathscr{L}_0(\nabla_2)$ be an $*$-isomorphism such that $\varphi(e) = \Phi(e)$ for all $e \in \nabla_1$. It is clear that $\lambda(\varphi(e))$, $e \in \nabla_1$, is a strictly positive σ-finite measure on the Boolean algebra ∇_2.

Proposition 3 $\Phi(\log(1 + |f|)) = \log(1 + \Phi(|f|))$ *for all $f \in \mathscr{L}_0(\nabla_1)$ and* $\Phi((\mathscr{L}_{\log}(\nabla_1, \mu)) = \mathscr{L}_{\log}(\nabla_2, \lambda)$.

Proof The restriction Ψ of $*$-isomorphism Φ on the C^*-algebra $\mathscr{L}_\infty(\nabla_1)$ is a $*$-isomorphism from C^*-algebra $\mathscr{L}_\infty(\nabla_1)$ onto C^*-algebra $\mathscr{L}_\infty(\nabla_2)$. Then for any $g \in \mathscr{L}_\infty(\nabla_1)$ and every continuous function $u : [0, +\infty) \rightarrow \mathbf{R}$ the equality

$$\Psi(u \circ |g|) = u \circ \Psi(|g|)$$

holds. The function $u(t) = \log(1 + t)$ is continuous on the interval $[0, +\infty)$. Therefore,

$$\Phi(\log(1 + |g|)) = \Psi(\log(1 + |g|)) = \log(1 + \Psi(|g|) = \log(1 + \Phi(|g|))$$

for all $g \in \mathscr{L}_\infty(\nabla_1)$. If $f \in \mathscr{L}_0(\nabla_1)$, then setting $g_n = |f| \cdot \{|f| \leq n\}$, we obtain that $g_n \in \mathscr{L}_\infty(\nabla_1)$, $n \in \mathbf{N}$, $0 \leq g_n \uparrow |f|$ and $\log(1 + g_n) \uparrow \log(1 + |f|)$.

Since $\Phi : \mathscr{L}_0(\nabla_1) \rightarrow \mathscr{L}_0(\nabla_2)$ is an $*$-isomorphism it follows that $\Phi : \mathscr{L}_0(\nabla_1)^h \rightarrow \mathscr{L}_0(\nabla_2)^h$ is a bijective positive linear map, where $\mathscr{L}_0(\nabla)^h := \{f \in \mathscr{L}_0(\nabla) : f = \bar{f}\}$. Consequently, $\Phi : \mathscr{L}_0(\nabla_1)^h \rightarrow \mathscr{L}_0(\nabla_2)^h$ is an order isomorphism, in particular, $\Phi(h_n) \uparrow \Phi(h)$ if h_n, $h \in \mathscr{L}_0(\nabla_1)^h$, $h_n \uparrow h$. Thus $\Phi(g_n) \uparrow \Phi(|f|)$ and $\log(1 + \Phi(g_n)) \uparrow \log(1 + \Phi(|f|))$. Therefore,

$$\log(1 + \Phi(g_n)) = \Phi(\log(1 + g_n)) \uparrow \Phi(\log(1 + |f|)).$$

Hence

$$\Phi(\log(1 + |f|)) = \log(1 + \Phi(|f|))$$

for all $f \in \mathscr{L}_0(\nabla_1)$.

By the definition of the $*$-algebra $\mathscr{L}_{\log}(\nabla_1, \mu)$ we have that $f \in \mathscr{L}_{\log}(\nabla_1, \mu)$ if and only if $f \in \mathscr{L}_0(\nabla_1)$ and $\log(1 + |f|) \in \mathscr{L}_1(\nabla_1, \mu)$. Consequently, using the equalities

$$\Phi(\mathscr{L}_1(\nabla_1, \mu)) = \mathscr{L}_1(\nabla_2, \lambda)$$

and

$$\Phi(log(1 + |f|)) = log(1 + \Phi(|f|)), \ f \in \mathscr{L}_0(\nabla_1),$$

we obtain that

$$log(1 + |\Phi(f)|) = log(1 + \Phi(|f|)) \in \mathscr{L}_1(\nabla_2, \lambda),$$

that is $\Phi(f) \in \mathscr{L}_{log}(\nabla_2, \lambda)$ for all $f \in \mathscr{L}_{log}(\nabla_1, \mu)$. Similarly, using the inverse ∗-isomorphism Φ^{-1}, we obtain that $\Phi^{-1}(h) \in \mathscr{L}_{log}(\nabla_1, \mu)$ for all $h \in \mathscr{L}_{log}(\nabla_2, \lambda)$. Therefore $\Phi(\mathscr{L}_{log}(\nabla_1, \mu)) = \mathscr{L}_{log}(\nabla_2, \lambda)$. □

Let μ (respectively, ν) be a strictly positive σ-finite measure on a complete Boolean algebra ∇_1 (respectively, ∇_2). The measures μ and ν are called log-*equivalent* if there exists an isomorphism $\varphi : \nabla_1 \to \nabla_2$ such that $\mathscr{L}_{log}(\nabla_2, \nu) = \mathscr{L}_{log}(\nabla_2, \mu \circ \varphi^{-1})$, where $(\mu \circ \varphi^{-1})(e) = \mu(\varphi^{-1}(e))$, $e \in \nabla_2$.

As noted above, $\lambda = \mu \circ \varphi^{-1}$ is strictly positive σ-finite measure on the Boolean algebra ∇_2. Therefore, by Corollary 4, equality $\mathscr{L}_{log}(\nabla_2, \nu) = \mathscr{L}_{log}(\nabla_2, \mu \circ \varphi^{-1})$ is equivalent to $\frac{d\nu}{d\lambda} \in \mathscr{L}_\infty(\nabla_2)$ and $\frac{d\lambda}{d\nu} \in \mathscr{L}_\infty(\nabla_2)$.

Theorem 2 *Let μ (respectively, ν) be a strictly positive σ-finite measure on a complete Boolean algebra ∇_1 (respectively, ∇_2). Then the ∗-algebras $\mathscr{L}_{log}(\nabla_1, \mu)$ and $\mathscr{L}_{log}(\nabla_2, \nu)$ are ∗-isomorphic if and only if the measures μ and ν are log-equivalent.*

Proof Let measures μ and ν be log-equivalent, i.e. there exists an isomorphism $\varphi : \nabla_1 \to \nabla_2$ such that $\mathscr{L}_{log}(\nabla_2, \nu) = \mathscr{L}_{log}(\nabla_2, \mu \circ \varphi^{-1})$. Let $\Phi : \mathscr{L}_0(\nabla_1) \to \mathscr{L}_0(\nabla_2)$ be a ∗-isomorphism, for which $\varphi(e) = \Phi(e)$ for all $e \in \nabla$. By Proposition 3 we have that

$$\Phi(\mathscr{L}_{log}(\nabla_1, \mu)) = (\mathscr{L}_{log}(\nabla_2, \mu \circ \varphi^{-1})) = \mathscr{L}_{log}(\nabla_2, \nu).$$

This means that the ∗-algebras $\mathscr{L}_{log}(\nabla_1, \mu)$ and $\mathscr{L}_{log}(\nabla_2, \nu)$ are ∗-isomorphic.

Conversely, suppose that the ∗-algebras $\mathscr{L}_{log}(\nabla_1, \mu)$ and $\mathscr{L}_{log}(\nabla_2, \nu)$ are ∗-isomorphic, that is there exists a ∗-isomorphism

$$\Psi : \mathscr{L}_{log}(\nabla_1, \mu) \to \mathscr{L}_{log}(\nabla_2, \nu).$$

Let $\{e_n\}_{n=1}^\infty$ be a partition of a unit of the Boolean algebra ∇_1 such that $\mu(e_n) < \infty$ and $\nu(\Psi(e_n)) < \infty$ for all $n \in \mathbf{N}$. It is clear that the map $\Psi : \mathscr{L}_{log}(e_n \nabla_1, \mu) \to \mathscr{L}_{log}(\Psi(e_n)\nabla_2, \nu)$ is a ∗-isomorphism. Since $e_n \nabla_1 \subset \mathscr{L}_{log}(e_n \nabla_1, \mu)$ and $\Psi(e_n)\nabla_2 \subset \mathscr{L}_{log}(\Psi(e_n)\nabla_2, \nu)$ it follows that the restriction φ_n of the ∗-isomorphism $\Psi : \mathscr{L}_{log}(e_n \nabla_1, \mu) \to \mathscr{L}_{log}(\Psi(e_n)\nabla_2, \nu)$ on the Boolean algebra $e_n \nabla_1$ is an isomorphism from $e_n \nabla_1$ onto $\Psi(e_n)\nabla_2$. Define the map $\varphi : \nabla_1 \to \nabla_2$ by the formula

$$\varphi(e) = \sup_{n \geq 1} \varphi_n(e \cdot e_n), \ e \in \nabla_1.$$

It is clear that φ is an isomorphism from ∇_1 onto ∇_2 and $\Psi(e) = \varphi(e)$ for all $e \in \nabla_1$. It means that (see Proposition 3)

$$\mathscr{L}_{\log}(\nabla_2, \mu \circ \varphi^{-1}) = \Psi(\mathscr{L}_{\log}(\nabla_1, \mu)) = \mathscr{L}_{\log}(\nabla_2, \nu),$$

that is the measures μ and ν are log-equivalent. □

Now we give a criterion for $*$-isomorphism of $*$-algebras log-integrable measurable functions, associated with the non-atomic finite measures.

Let μ and ν be strictly positive finite measures on non-atomic complete Boolean algebras ∇_1 and ∇_2. Since $\mathscr{L}_{\log}(\nabla, \mu) = \mathscr{L}_{\log}(\nabla, \frac{\mu}{\mu(1)})$, we can assume that both measures are probability measures, i.e. $\mu(1_{\nabla_1}) = 1 = \nu(1_{\nabla_2})$.

Theorem 3 *Let μ and ν be probability measures on non-atomic complete Boolean algebras ∇_1 and ∇_2 respectively. Let $\begin{pmatrix} \tau_1^{(1)} & \tau_2^{(1)} & \cdots \\ \mu_1 & \mu_2 & \cdots \end{pmatrix}, \begin{pmatrix} \tau_1^{(2)} & \tau_2^{(2)} & \cdots \\ \nu_1 & \nu_2 & \cdots \end{pmatrix}$ be the passports of (∇_1, μ) and (∇_2, ν). The following conditions are equivalent:*

(i). The $*$-algebras $\mathscr{L}_{\log}(\nabla_1, \mu)$ and $\mathscr{L}_{\log}(\nabla_2, \nu)$ are $*$-isomorphic;

(ii). The upper rows of passports of (∇_1, μ) and (∇_2, ν) are equal and the sequences $\{\frac{\mu_n}{\nu_n}\}$ and $\{\frac{\nu_n}{\mu_n}\}$ are bounded.

Proof $(i) \Rightarrow (ii)$. Let $\Psi : \mathscr{L}_{\log}(\nabla_1, \mu) \to \mathscr{L}_{\log}(\nabla_2, \nu)$ be a $*$-isomorphism. Then the restriction φ of the $*$-isomorphism Ψ on the Boolean algebra ∇_1 is an isomorphism from ∇_1 onto ∇_2. By Theorem 1 we have that the upper rows of passports of (∇_1, μ) and (∇_2, ν) are equal.

Let $\{e_n\}_{n=1}^{\infty}$ be a partition of the unit 1_{∇_1} of the Boolean algebra ∇_1 such that $e_n \nabla_1$ is a homogeneous Boolean algebra, $\tau_n^{(1)} = \tau(e_n \nabla_1) < \tau_{n+1}^{(1)}$, and $\mu_n = \mu(e_n)$, $n \in \mathbf{N}$. Set $q_n = \varphi(e_n)$. It is clear that $\{q_n\}_{n=1}^{\infty}$ is a partition of the unit 1_{∇_2} of the Boolean algebra ∇_2 such that $q_n \nabla_2$ is a homogeneous Boolean algebra, $\tau_n^{(2)} = \tau(q_n \nabla_2) < \tau_{n+1}^{(2)}$, and $\nu_n = \nu(q_n)$, $n \in \mathbf{N}$.

By Proposition 3 for a probability measure $\lambda(\varphi(e)) = \mu(e)$, $e \in \nabla_1$, on ∇_2 we have that

$$\mathscr{L}_{\log}(\nabla_2, \nu) = \Psi(\mathscr{L}_{\log}(\nabla_1, \mu)) = \mathscr{L}_{\log}(\nabla_2, \lambda).$$

Using now Corollary 4 we get that $\frac{d\nu}{d\lambda} \in \mathscr{L}_{\infty}(\nabla_2)$ and $\frac{d\lambda}{d\nu} \in \mathscr{L}_{\infty}(\nabla_2)$.

Consequently,

$$\nu_n = \nu(q_n) = \int\limits_{q_n} \frac{d\nu}{d\lambda} d\lambda \leq \|\frac{d\nu}{d\lambda}\|_{\infty} \cdot \lambda(q_n) = \|\frac{d\nu}{d\lambda}\|_{\infty} \cdot \mu(e_n)$$

for all $n \in \mathbf{N}$. Therefore the sequence $\{\frac{\nu_n}{\mu_n}\}$ is bounded. Similarly, it is proved that the sequence $\{\frac{\mu_n}{\nu_n}\}$ is also bounded.

$(ii) \Rightarrow (i)$. Let the upper rows of passports of (∇_1, μ) and (∇_2, ν) be equal and the sequences $\{\frac{\mu_n}{\nu_n}\}$ and $\{\frac{\nu_n}{\mu_n}\}$ be bounded. By Theorem 1 we have that there exists an isomorphism $\varphi : \nabla_1 \to \nabla_2$. Let $\{e_n\}_{n=1}^{\infty}$ and $q_n = \varphi(e_n)$, $n \in \mathbf{N}$ be the same as in the proof of implication $(i) \Rightarrow (ii)$. Consider on ∇_2 the probability measure $\gamma(q) = \sum_{n=1}^{\infty} \mu_n \nu_n^{-1} \nu(q_n q)$, $q \in \nabla_2$. Since the passports of Boolean algebras (∇_1, μ)

and (∇_2, γ) are equal, it follows by Theorem 1 that there exists an isomorphism $\psi : \nabla_1 \to \nabla_2$ such that $\mu(e) = \gamma(\psi(e))$ for all $e \in \nabla_1$.

Let $\Psi : \mathscr{L}_0(\nabla_1, \mu) \to \mathscr{L}_0(\nabla_2, \gamma)$ be an *-isomorphism such that $\psi(e) = \Psi(e)$ for all $e \in \nabla_1$. By Proposition 3 we have that $\Psi(\mathscr{L}_{\log}(\nabla_1, \mu)) = \mathscr{L}_{\log}(\nabla_2, \gamma)$. Since $\gamma(q) = \sum_{n=1}^{\infty} \mu_n \nu_n^{-1} \nu(q_n q)$, $q \in \nabla_2$ it follows that

$$\frac{d\gamma}{d\nu} = \sum_{n=1}^{\infty} \mu_n \nu_n^{-1} q_n \in \mathscr{L}_\infty(\nabla_2).$$

On the other hand, if $q \in q_n \nabla_2$ then $\nu(q) = \nu(q_n q) = \nu_n \mu_n^{-1} \gamma(q)$, that is

$$\nu(q) = \sum_{n=1}^{\infty} \nu_n \mu_n^{-1} \gamma(q_n q), \quad q \in \nabla_2.$$

Consequently,

$$\frac{d\nu}{d\gamma} = \sum_{n=1}^{\infty} \nu_n \mu_n^{-1} q_n \in \mathscr{L}_\infty(\nabla_2).$$

By Corollary 4 we have that $\mathscr{L}_{\log}(\nabla_2, \nu) = \mathscr{L}_{\log}(\nabla_2, \gamma) = \Psi(\mathscr{L}_{\log}(\nabla_1, \mu))$. $\quad\square$

4 Isomorphic Classification of *-Algebras of Log-Integrable Measurable Functions in the Case of the σ-Finite Measures

Let ∇ be a non-atomic complete Boolean algebra and let μ be a strictly positive σ-finite measure on ∇. Then ∇ has a countable type, i.e. every partition of the unit $\mathbf{1}$ of the Boolean algebra ∇ is no more countably many [4, Chap. I, Sect. 6, Sect. 3]. Consequently, a Boolean algebra ∇ has no more than countably homogeneous components.

Let ∇ be a non-atomic complete Boolean algebra. Then ∇ is a direct product of homogeneous Boolean algebras ∇_{e_n}, $e_n \cdot e_m = 0$, $n \neq m$, $\tau_n = \tau(\nabla_{e_n}) < \tau_{n+1}$, $n, m \in \mathbf{N}$ [4, chap. VII, Sect. 2, p. 272]. Suppose that $\mu_n = \mu(e_n) < \infty$ for all $n \in \mathbf{N}$. Since $\mu(\mathbf{1}) = \infty$ it follows that $\sup_{n \in \mathbf{N}} \mu_n = \infty$.

Theorem 4 *Let μ and ν be strictly positive σ-finite (non finite) measures on non-atomic complete Boolean algebras ∇_1 and ∇_2 respectively. Let $\begin{pmatrix} \tau_1^{(1)} & \tau_2^{(1)} & \cdots \\ \mu_1 & \mu_2 & \cdots \end{pmatrix}$ and $\begin{pmatrix} \tau_1^{(2)} & \tau_2^{(2)} & \cdots \\ \nu_1 & \nu_2 & \cdots \end{pmatrix}$ be the passports of (∇_1, μ) and (∇_2, ν). Suppose that $\mu_n < \infty$ and $\nu_n < \infty$ for all $n \in \mathbf{N}$. Then the following conditions are equivalent:*

(*i*). *The $*$-algebras $\mathscr{L}_{\log}(\nabla_1, \mu)$ and $\mathscr{L}_{\log}(\nabla_2, \nu)$ are $*$-isomorphic;*

(*ii*). *The upper rows of passports of (∇_1, μ) and (∇_2, ν) are equal and the sequences $\{\frac{\mu_n}{\nu_n}\}$ and $\{\frac{\nu_n}{\mu_n}\}$ are bounded.*

Proof The proof of the implication $(i) \Rightarrow (ii)$ is the same as in Theorem 3.

$(ii) \Rightarrow (i)$. Let the upper rows of passports of (∇_1, μ) and (∇_2, ν) be equal and the sequences $\{\frac{\mu_n}{\nu_n}\}$ and $\{\frac{\nu_n}{\mu_n}\}$ are bounded. Using Corollaries 1 and 2, we have that there exists an isomorphism $\varphi : \nabla_1 \to \nabla_2$ such that $\frac{\nu(q_n \cdot \varphi(e))}{\nu_n} = \frac{\mu(e \cdot e_n)}{\mu_n}$ for all $n \in \mathbf{N}$, where $\{e_n\}_{n=1}^{\infty}$ and $q_n = \varphi(e_n)$, $n \in \mathbf{N}$, are the same as in the proof of implication $(i) \Rightarrow (ii)$ in Theorem 3.

Consider on ∇_2 the strictly positive σ-finite measure $\lambda(\varphi(e)) = \mu(e)$, $e \in \nabla_1$. Let $\Phi : \mathscr{L}_0(\nabla_1, \mu) \to \mathscr{L}_0(\nabla_2, \lambda)$ be a $*$-isomorphism such that $\varphi(e) = \Phi(e)$ for all $e \in \nabla_1$. By Proposition 3 we have that $\Phi((\mathscr{L}_{\log}(\nabla_1, \mu)) = \mathscr{L}_{\log}(\nabla_2, \lambda)$, in addition, $\lambda(q_n) = \mu_n$ for all $n \in \mathbf{N}$. Consequently,

$$\frac{d\lambda}{d\nu} = \sum_{n=1}^{\infty} \frac{\mu_n}{\nu_n} q_n \in \mathscr{L}_{\infty}(\nabla_2),$$

and

$$\frac{d\nu}{d\lambda} = \sum_{n=1}^{\infty} \frac{\nu_n}{\mu_n} q_n \in \mathscr{L}_{\infty}(\nabla_2).$$

By Corollary 4 we have that $\mathscr{L}_{\log}(\nabla_2, \nu) = \mathscr{L}_{\log}(\nabla_2, \lambda) = \Phi(\mathscr{L}_{\log}(\nabla_1, \mu))$. $\square$

Theorem 5 *Let μ and ν be strictly positive σ-finite (non finite) measures on non-atomic complete Boolean algebras ∇_1 and ∇_2 respectively. Let $\begin{pmatrix} \tau_1^{(1)} & \tau_2^{(1)} & \cdots \\ \mu_1 & \mu_2 & \cdots \end{pmatrix}$ and $\begin{pmatrix} \tau_1^{(2)} & \tau_2^{(2)} & \cdots \\ \nu_1 & \nu_2 & \cdots \end{pmatrix}$ be the passports of (∇_1, μ) and (∇_2, ν). Suppose that $\mu_n = \nu_n = \infty$ for all $n \in \mathbf{N}$. Then the following conditions are equivalent:*

(*i*). *The $*$-algebras $\mathscr{L}_{\log}(\nabla_1, \mu)$ and $\mathscr{L}_{\log}(\nabla_2, \nu)$ are $*$-isomorphic;*

(*ii*). *The upper rows of passports of (∇_1, μ) and (∇_2, ν) are equal.*

The proof of Theorem 5 uses Corollary 3 and repeats the proof of Theorems 3 and 4.

Let now ∇ be an arbitrary non-atomic Boolean algebra ∇, and let μ be a σ-finite measure on the Boolean algebra ∇. It is known that ∇ is a direct product of homogeneous Boolean algebras ∇_{e_n}, where $e_n \cdot e_m = 0$, $n \neq m$, $\tau_n = \tau(\nabla_{e_n}) < \tau_{n+1}$, $n, m \in \mathbf{N}$, and $\sup_{n \geq 1} = \mathbf{1}$ [4, Chap. VII, Sect. 2, Theorem 5]). We denote by $\nabla_{e_{s_i}}$ those homogeneous components of the Boolean algebra ∇, for which

$$\tau_{s_i} < \tau_{s_{i+1}}, \quad \mu_{s_i} = \mu(e_{s_i}) < \infty, \quad i \in \mathbf{N},$$

and denote by $\nabla_{e_{m_i}}$ those homogeneous components of the Boolean algebra ∇, for which

$$\tau_{m_i} < \tau_{m_{i+1}}, \quad \mu_{m_i} = \mu(e_{m_i}) = \infty, \quad i \in \mathbf{N}.$$

The matrix $\begin{pmatrix} \tau_{s_1} & \tau_{s_2}, & \dots & \tau_{m_1} & \tau_{m_2}, \dots \\ \mu_{s_1} & \mu_{s_2}, & \dots \infty, & \infty, \dots \end{pmatrix}$ is called *the passport* of the Boolean algebra (∇, μ).

Using now Theorems 1, 3, 4, 5, Proposition 3 and the equalities $e = \sup\limits_{i \in \mathbf{N}} e_{s_i}$ and $1 - e = \sup\limits_{i \in \mathbf{N}} e_{m_i}$, we get the following criterion for the existence of a ∗-isomorphism from an ∗-algebra $L_{\log}(\nabla_1, \mu)$ onto ∗-algebra $L_{\log}(\nabla_2, \nu)$.

Theorem 6 *Let μ and ν be strictly positive σ-finite measures on non-atomic complete Boolean algebras ∇_1 and ∇_2 respectively. Let*
$$\begin{pmatrix} \tau_{s_1}^{(1)} & \tau_{s_2}^{(1)}, & \dots & \tau_{m_1}^{(1)} & \tau_{m_2}^{(1)}, \dots \\ \mu_{s_1} & \mu_{s_2}, & \dots \infty, & \infty, \dots \end{pmatrix} \text{ and } \begin{pmatrix} \tau_{s_1}^{(2)} & \tau_{s_2}^{(2)}, & \dots & \tau_{m_1}^{(2)} & \tau_{m_2}^{(2)}, \dots \\ \nu_{s_1} & \nu_{s_2}, & \dots \infty, & \infty, \dots \end{pmatrix} \text{ be the passports}$$
of (∇_1, μ) and (∇_2, ν). Then the following conditions are equivalent:

(i). The ∗-algebras $\mathscr{L}_{\log}(\nabla_1, \mu)$ and $\mathscr{L}_{\log}(\nabla_2, \nu)$ are ∗-isomorphic;

(ii). The upper rows of passports of (∇_1, μ) and (∇_2, ν) are equal and the sequences $\{\frac{\mu_{s_i}}{\nu_{s_i}}\}$ and $\{\frac{\nu_{s_i}}{\mu_{s_i}}\}$ are bounded.

References

1. Abdullayev, R.Z.: Isomorphisms of Arens algebras. Sib. J. Ind. Math. **1**(2), 3–15 (1998). (Russian)
2. Abdullaev, R.Z., Chilin, V.I.: Arens algebras, associated with commutative von Neumann algebras. Ann. Math. Blaise Pascal **5**(1), 1–12 (1998)
3. Dykema, K., Sukochev, F., Zanin, D.: Algebras of log-integrable functions and operators. Complex Anal. Oper. Theory **10**(8), 1775–1787 (2016)
4. Vladimirov, D.A.: Boolean Algebras. Nauka Publications, Moscow (1969). [in Russian]

Matrix Differential Equations
for Pseudo-orthogonal Groups

V. I. Chilin and K. K. Muminov

Abstract We consider a system of matrix differential equations whose nondegenerate solutions are $O(n, p, R)$-equivalent, where $O(n, p, R)$ is the pseudo-orthogonal group of invertible linear transformations of R^n. We show that the class of first columns of the set of matrices that are nondegenerate solutions of this system coincides with the class of $O(n, p, R)$-equivalent paths in R^n.

Keywords Pseudo-orthogonal group · Regular path · Equivalence of paths
Matrix differential equation

1 Introduction and Preliminaries

Let R^n be the n-dimensional real linear space and let G be a subgroup in the group $GL(n, R)$ of invertible linear transformations of R^n. One of the problems of differential geometry is finding necessary and sufficient conditions for $\overrightarrow{y}(t) = g \overrightarrow{x}(t)$ for some $g \in G$ and for all t (G-equivalence of paths $\overrightarrow{x}(t)$, $\overrightarrow{y}(t)$), where paths $\overrightarrow{x}(t), \overrightarrow{y}(t) \subset R^n$. Finding explicit forms of rational bases of differential fields of G-invariant differential rational functions makes possible to establish effective criteria of G-equivalence of paths. This approach was used for G a linear, special linear, orthogonal, pseudo-orthogonal, or a symplectic subgroup in $GL(n, R)$ (for a survey of these results see, for example, [4, 7]).

Use of the theory of differential fields of G-invariant differential rational functions to obtain criteria of G-equivalence of paths is connected with presence of special matrix equations. These matrix equations define systems of matrix differential equations whose solutions are matrices associated with classes of G-equivalent paths (see, for example, [7]). There arises the problem of description of the systems of

V. I. Chilin (✉) · K. K. Muminov
National University of Uzbekistan, Tashkent, Uzbekistan
e-mail: vladimirchil@gmail.com

K. K. Muminov
e-mail: m.muminov@rambler.ru

© Springer Nature Switzerland AG 2018   85
Z. Ibragimov et al. (eds.), *Algebra, Complex Analysis, and Pluripotential Theory*,
Springer Proceedings in Mathematics & Statistics 264,
https://doi.org/10.1007/978-3-030-01144-4_7

differential equations whose solutions recover classes of G-equivalent paths. In the case of orthogonal groups, this problem was solved by R. Aripov and J. Khadzhiev [1]. In the case of symplectic groups, the solution of this problem was obtained by K. Muminov [5, 6].

In the present paper we consider the action of the pseudo-orthogonal $O(n, p, R)$ in the n-dimensional real pseudo-Euclidean space. Using special matrix equations, we construct systems of matrix differential equations whose solutions are matrices associated with the classes of $O(n, p, R)$-equivalent paths.

We assume that an element $\overrightarrow{x} \in R^n$ (respectively, a linear transformation $g \in GL(n, R)$) is represented as an n-dimensional vector-column $\overrightarrow{x} = \{x_j\}_{j=1}^n$ (respectively, as an invertible $n \times n$-matrix $(g_{ij})_{i,j=1}^n$), where x_j, $g_{ij} \in R$, $i, j = 1, \ldots, n$. The action of the group $GL(n, R)$ in the space R^n is defined by multiplication of a matrix g on a vector-column $\overrightarrow{x}$.

Fix a natural number $p \in \{1, \ldots, n - 1\}$ and consider the following bilinear form in R^n.

$$[\overrightarrow{x}, \overrightarrow{y}]_p = x_1 y_1 + \cdots + x_p y_p - x_{p+1} y_{p+1} - \cdots - x_n y_n, \quad \overrightarrow{x} = \{x_j\}_{j=1}^n, \overrightarrow{y} = \{y_j\}_{j=1}^n \in R^n.$$

Denote by J_p the matrix $(J_{ij}^{(p)})_{i,j=1}^n$ such that $J_{jj}^{(p)} = 1$, if $j = 1, \ldots, p$, $J_{jj}^{(p)} = -1$, if $j = p + 1, \ldots, n$, and $J_{ij}^{(p)} = 0$, if $i \neq j$, $i, j = 1, \ldots, n$. It is clear that $[\overrightarrow{x}, \overrightarrow{y}]_p = \overrightarrow{x}^T J_p \overrightarrow{y}$, where $\overrightarrow{x}^T$ is the row-vector transposed to a column-vector $\overrightarrow{x}$.

Let $O(n, p, R)$ be the pseudo-orthogonal subgroup in $GL(n, R)$, that is,

$$O(n, p, R) = \{g \in GL(n, R) : g^T J_p g = J_p\} =$$

$$= \{g \in GL(n, R) : [g\overrightarrow{x}, g\overrightarrow{y}]_p = [\overrightarrow{x}, \overrightarrow{y}]_p, \forall \overrightarrow{x}, \overrightarrow{y} \in R^n\},$$

where g^T is the transposed matrix to a matrix g.

Let us denote by I the interval $(0, 1) \subset R$. A vector function

$$\overrightarrow{x}(t) = \{x_j(t)\}_{j=1}^n : I \to R^n$$

is called a *path* in R^n if each coordinate function $x_j(t) : I \to R$ is infinitely differentiable.

The rth derivation of a path $\overrightarrow{x}(t) = \{x_j(t)\}_{j=1}^n$ is the vector function $\overrightarrow{x}^{(r)}(t) = \{x_j^{(r)}(t)\}_{j=1}^n$, where $x_j^{(r)}(t)$ is the rth derivation of the coordinate function $x_j(t)$, $j = 1, \ldots, n$, $r = 1, 2, \ldots$ It is clear that $\overrightarrow{x}^{(r)}(t)$ is a path for all r.

Let $M(\overrightarrow{x})(t)$ (respectively, $M^{(1)}(\overrightarrow{x})(t)$) be the matrix $(x_i^{(j-1)}(t))_{i,j=1}^n$ (respectively, the matrix $(x_i^{(j)}(t))_{i,j=1}^n$), where $x_i^{(0)}(t) = x_i(t)$, $i = 1, \ldots, n$. A path $\overrightarrow{x}(t)$ is called *strongly regular* if the determinate $\det M(\overrightarrow{x})(t)$ is not equal to zero for all $t \in I$.

Paths $\overrightarrow{x}(t)$ and $\overrightarrow{y}(t)$ are called $O(n, p, R)$-*equivalent* if there exists an element $g \in O(n, p, R)$ such that $\overrightarrow{y}(t) = g\overrightarrow{x}(t)$ for all $t \in I$. In this case $\overrightarrow{y}^{(j)}(t) =$

$g \overrightarrow{x}^{(j)}(t)$, $j = 1, 2, \ldots$, hence $M(\overrightarrow{y})(t) = g M(\overrightarrow{x})(t)$. It is clear that $M(\overrightarrow{y})(t) = g M(\overrightarrow{x})(t)$, $g \in O(n, p, R)$, implies that the paths $\overrightarrow{x}(t)$ and $\overrightarrow{y}(t)$ are $O(n, p, R)$-equivalent.

We use the following criterion for $O(n, p, R)$-equivalence of strongly regular paths $\overrightarrow{x}(t)$ and $\overrightarrow{y}(t)$.

Theorem 1 *Strongly regular paths $\overrightarrow{x}(t)$ and $\overrightarrow{y}(t)$ are $O(n, p, R)$-equivalent if and only if*

$$\left(M(\overrightarrow{x}) \right)^{-1}(t) M^{(1)}(\overrightarrow{x})(t) = \left(M(\overrightarrow{y}) \right)^{-1} M^{(1)}(\overrightarrow{y})(t)$$

and

$$M^T(\overrightarrow{x})(t) J_p M(\overrightarrow{x})(t) = M^T(\overrightarrow{y})(t) J_p M(\overrightarrow{y})(t)$$

for all $t \in I$.

For a proof of Theorem 1, see [7, Ch. 3, § 3.2]; see also [2].

Using Theorem 1, we obtain necessary and sufficient conditions for $O(n, p, R)$-equivalence of strongly regular paths $\overrightarrow{x}(t)$ and $\overrightarrow{y}(t)$ in terms of equations for the corresponding differential rational functions of the variables $x_i(t)$, $x_i^{(j)}(t)$, $i, j = 1, \ldots, n$.

2 Matrix Differential Equations Generated by the Action of a Pseudo-orthogonal Group

Let $\overrightarrow{x}(t)$ be a strongly regular path, and let

$$A(t) = (M(\overrightarrow{x}(t)))^{-1} M^{(1)}(\overrightarrow{x}(t)) = \left(a_{ij}(t) \right)_{i,j=1}^n.$$

A direct calculation of the elements $a_{ij}(t)$ of the matrix $A(t)$ shows that

$$\{a_{1j}(t)\}_{j=1}^n = \{0, 0, 0, \ldots, 0, a_{1n}(t)\}, \{a_{2j}(t)\}_{j=1}^n = \{1, 0, 0, \ldots, 0, a_{2n}(t)\},$$

$$\{a_{3j}(t)\}_{j=1}^n = \{0, 1, 0, \ldots, 0, a_{3n}(t)\}, \{a_{4j}(t)\}_{j=1}^n = \{0, 0, 1, 0 \ldots, 0, a_{4n}(t)\}, \ldots,$$

$$\{a_{nj}(t)\}_{j=1}^n = \{0, 0, 0, \ldots, 1, a_{nn}(t)\}, \tag{1A}$$

where $a_{jn}(t)$ are infinitely differentiable functions, $t \in I$, $j = 1, \ldots, n$. In addition, $B(t) = M^T(\overrightarrow{x}(t)) J_p M(\overrightarrow{x}(t))$, $1 \le p \le n - 1$, is an invertible symmetric $n \times n$-matrix for which we have

$$B^{(1)}(t) = (M^T(\overrightarrow{x}(t)))^{(1)} J_p M(\overrightarrow{x}(t)) + M^T(\overrightarrow{x}(t))(J_p M(\overrightarrow{x}(t)))^{(1)} =$$

$$= (M^T(\overrightarrow{x}(t)))^{(1)} (M^T(\overrightarrow{x}(t)))^{-1} M^T(\overrightarrow{x}(t)) J_p M(\overrightarrow{x}(t)) +$$

$$+ M^T(\vec{x}(t)) J_p M(\vec{x}(t))(M(\vec{x}(t)))^{-1}(M(\vec{x}(t)))^{(1)} =$$

$$= (M(\vec{x}(t)))^{-1} M^{(1)}(\vec{x}(t)))^T M^T(\vec{x}(t)) J_p M(\vec{x}(t)) +$$

$$+ M^T(\vec{x}(t)) J_p M(\vec{x}(t))(M(\vec{x}(t)))^{-1} M^{(1)}(\vec{x}(t)) = A^T(t) B(t) + B(t) A(t),$$

that is,

$$B^{(1)}(t) = A^T(t) B(t) + B(t) A(t) \tag{1}$$

for all $t \in I$.

Let $\vec{x}(t)$ and $\vec{y}(t)$ be $O(n, p, R)$-equivalent strongly regular paths. By Theorem 1, we have

$$\begin{cases} (M(\vec{x}(t)))^{-1} M^{(1)}(\vec{x}(t)) = (M(\vec{y}(t)))^{-1} M^{(1)}(\vec{y}(t)); \\ M^T(\vec{x}(t)) J_p M(\vec{x}(t)) = M^T(\vec{y}(t)) J_p M(\vec{y}(t)), \end{cases} \tag{2}$$

for all $t \in I$. Consequently, the matrices $M(\vec{x}(t))$ and $M(\vec{y}(t))$ are solutions of the system of matrix differential equations

$$\begin{cases} X^{(1)}(t) = X(t) A(t) \\ X^T(t) J_p X(t) = B(t), \end{cases} \tag{3}$$

where the matrix $X(t) = (x_{jk}(t))_{j,k=1}^n$ is unknown, $X^{(1)}(t) = (x_{jk}^{(1)}(t)))_{j,k=1}^n$, $x_{jk} :$ $I \to R$, $j, k = 1, \ldots, n$; in addition, the matrix $A(t)$ has the form (1A), and the matrix $B(t)$ is an invertible symmetric matrix satisfying Eq. (1).

A solution $X(t)$ of the system (3) will be called *nondegenerate* if the determinant of the matrix $X(t)$ does not vanish for all $t \in I$. We say that two solutions $X(t)$ and $Y(t)$ of the system (3) are $O(n, p, R)$-equivalent if there exists an element $g \in O(n, p, R)$ such that $Y(t) = g X(t)$ for all $t \in I$.

Theorem 2 *Let a matrix $A(t)$ have the form (1A), and let a matrix $B(t)$ be an invertible symmetric $n \times n$-matrix satisfying Eq. (1). Then*

(i). The system (3) has a nondegenerate solution $X(t)$. In addition, if $Y(t)$ is another solution of the system (3), then $X(t)$ and $Y(t)$ are $O(n, p, R)$-equivalent;

(ii). If $\widetilde{X}(t)$ is a nondegenerate solution of the system (3), then $\widetilde{Y}(t) = g \widetilde{X}(t)$ is a solution of the system (3) for any $g \in O(n, p, R)$.

Proof (i). If $X(t) = (x_{jk}(t))_{j,k=1}^n$ is a nondegenerate solution of the system (3), then, by (1A), the first equation of the system (3) can be rewritten as follows:

$$\begin{cases} x_{j1}'(t) = x_{j2}(t), \; j = 1, \ldots, n \\ \cdots\cdots\cdots\cdots\cdots\cdots\cdots\cdots\cdots\cdots \\ x_{jn-1}'(t) = x_{jn}(t), \; j = 1, \ldots, n \\ x_{jn}'(t) = \sum_{k=1}^n a_{kn} x_{jk}(t), \; j = 1, \ldots, n. \end{cases} \tag{4}$$

Transforming the last line of the system (4) using the previous lines, we obtain the following system differential equations.

$$\begin{cases} x'_{1n}(t) - a_{1n}(t)x_{11}(t) - a_{2n}(t)x'_{11}(t) - \ldots - a_{nn}(t)x'_{1n-1}(t) = 0 \\ \cdots\cdots\cdots\cdots\cdots\cdots\cdots\cdots\cdots\cdots\cdots\cdots\cdots\cdots\cdots\cdots\cdots\cdots\cdots\cdots\cdots \\ x'_{nn}(t) - a_{1n}(t)x_{n1}(t) - a_{2n}(t)x'_{n1}(t) - \ldots - a_{nn}(t)x'_{nn-1}(t) = 0. \end{cases} \quad (5)$$

Next, it follows from (4) that

$$\begin{cases} x_{12}(t) = x'_{11}(t), \ x_{13}(t) = x'_{12}(t) = x''_{11}(t), \ldots, \\ \quad x_{1n}(t) = x'_{1n-1}(t) = \ldots = x_{11}^{(n-1)}(t), \\ x_{22}(t) = x'_{21}(t), \ x_{23}(t) = x'_{22}(t) = x''_{21}(t), \ldots, \\ \quad x_{2n}(t) = x'_{2n-1}(t) = \ldots = x_{21}^{(n-1)}(t), \\ \cdots\cdots\cdots\cdots\cdots\cdots\cdots\cdots\cdots\cdots\cdots \\ x_{n2}(t) = x'_{n1}(t), \ x_{n3}(t) = x'_{n2}(t) = x''_{n1}(t), \ldots, \\ \quad x_{nn}(t) = x'_{nn-1}(t) = \ldots = x_{n1}^{(n-1)}(t). \end{cases} \quad (6)$$

Using Eq. (6), we rewrite the system (5) in the form

$$\begin{cases} x_{11}^{(n)}(t) - a_{nn}(t)x_{11}^{(n-1)}(t) - \ldots - a_{2n}x'_{11}(t) - a_{1n}(t)x_{11}(t) = 0 \\ \cdots\cdots\cdots\cdots\cdots\cdots\cdots\cdots\cdots\cdots\cdots\cdots\cdots\cdots\cdots\cdots\cdots\cdots\cdots \\ x_{n1}^{(n)}(t) - a_{nn}(t)x_{n1}^{(n-1)}(t) - \ldots - a_{2n}x'_{n1}(t) - a_{1n}(t)x_{n1}(t) = 0. \end{cases}$$

Therefore, if $X(t) = (x_{jk}(t))_{j,k=1}^n$ is a solution of the first equation of the system (3), then $x_{jk}(t) = x_{j1}^{(k-1)}(t)$, $j, k = 1, \ldots, n$; in addition, each coordinate of the vector $\overrightarrow{x}(t) = \{x_{j1}(t)\}_{j=1}^n$ is a solution of the differential equation

$$y^n(t) - a_{nn}(t)y^{(n-1)}(t) - \cdots - a_{2n}(t)y'(t) - a_{1n}(t)y(t) = 0. \quad (7)$$

By [8, §§ 5,18], the Eq. (7) has solutions $\varphi_1(t), \ldots, \varphi_n(t)$, $t \in I$, such that the functions $\varphi_i(t)$ are linearly independent and the determinant of the matrix

$$\begin{pmatrix} \varphi_1(t) & \varphi_1^{(1)}(t) & \ldots & \varphi_1^{(n-1)}(t) \\ \varphi_2(t) & \varphi_2^{(1)}(t) & \ldots & \varphi_2^{(n-1)}(t) \\ \ldots & \ldots & \ldots & \ldots \\ \varphi_n(t) & \varphi_n^{(1)}(t) & \ldots & \varphi_n^{(n-1)}(t) \end{pmatrix}$$

is not equal to zero for all $t \in I$.

Set $\widetilde{x}_{j1}(t) = \varphi_j(t)$, $\widetilde{x}_{jk}(t) = \varphi_j^{(k-1)}(t)$, $j = 1, \ldots, n$, $k = 2, \ldots n$, $t \in I$. It is clear that the matrix $\widetilde{X}(t) = (\widetilde{x}_{jk}(t))_{j,k=1}^n$ is a nondegenerate solution of the first equation of the system (3).

Let $\widetilde{Y}(t) = (y_{jk}(t))_{j,k=1}^n$ be another nondegenerate solution of the first equation of the system (3). Since $\widetilde{Y}^{(1)}(t) = \widetilde{Y}(t)A(t)$ and $\widetilde{X}^{(1)}(t) = \widetilde{X}(t)A(t)$, it follows that

$$(\widetilde{Y}(t)(\widetilde{X}(t))^{-1})^{(1)} = \widetilde{Y}^{(1)}(t)(\widetilde{X}(t))^{-1} + \widetilde{Y}(t)((\widetilde{X}(t))^{-1})^{(1)} = \widetilde{Y}^{(1)}(t)(\widetilde{X}(t))^{-1} -$$

$$-\widetilde{Y}(t)(\widetilde{X}(t))^{-1}\widetilde{X}^{(1)}(t)(\widetilde{X}(t))^{-1} = (\widetilde{Y}(t)A(t) - \widetilde{Y}(t)A(t))(\widetilde{X}(t))^{-1} = 0.$$

This means that $\widetilde{Y}(t)(\widetilde{X}(t))^{-1} = g \in GL(n, R)$. Consequently, the first equation of the system (3) has only one, up to $GL(n, R)$-equivalence, nondegenerate solution $\widetilde{X}(t) = (\widetilde{x}_{jk}(t))_{j,k=1}^{n}$.

Further, if $\widetilde{X}(t)$ is a non-degenerate solution of the first equation of the system (3), $g \in GL(n, R)$, $Z(t) = g\widetilde{X}(t)$, then $Z^{(1)}(t) = g\widetilde{X}^{(1)}(t) = g\widetilde{X}(t)A(t) = Z(t)A(t)$. Consequently,

$$\{g\widetilde{X}(t) : g \in GL(n, R)\} \tag{8}$$

is the set of all solutions of the first equation of the system (3).

It is clear that $W(t) = (\widetilde{X}^T(t))^{-1}B(t)\widetilde{X}^{-1}(t)$ is an invertible nondegenerate matrix. Since $\widetilde{X}(t)$ is a solution of the first equation of the system (3), using (1), we get

$$W^{(1)}(t) = ((\widetilde{X}^T(t))^{-1}B(t)\widetilde{X}^{-1}(t))^{(1)} = ((\widetilde{X}^T(t))^{-1})^{(1)}B(t)\widetilde{X}^{-1}(t) +$$

$$+ (\widetilde{X}^T(t))^{-1}B^{(1)}(t)\widetilde{X}^{-1}(t) + (\widetilde{X}^T(t))^{-1}B(t)(\widetilde{X}^{-1}(t))^{(1)} =$$

$$= -(\widetilde{X}^{-1}(t)\widetilde{X}^{(1)}(t)\widetilde{X}^{-1}(t))^T B(t)\widetilde{X}^{-1}(t) + (\widetilde{X}^T(t))^{-1}B^{(1)}(t)\widetilde{X}^{-1}(t) -$$

$$- (\widetilde{X}^T(t))^{-1}B(t)\widetilde{X}^{-1}(t)\widetilde{X}^{(1)}(t)\widetilde{X}^{-1}(t) =$$

$$= (\widetilde{X}^{-1}(t))^T[-(\widetilde{X}^{(1)}(t))^T(\widetilde{X}^{-1}(t))^T B(t) + B^{(1)}(t) - B(t)\widetilde{X}^{-1}(t)\widetilde{X}^{(1)}(t)]\widetilde{X}^{-1}(t) =$$

$$= (\widetilde{X}^{-1}(t))^T[-A^T(t)B(t) + A^T(t)B(t) + B(t)A(t) - B(t)A(t)]\widetilde{X}^{-1}(t) = 0.$$

Therefore, the matrix $W(t)$ is independent of the parameter $t \in I$, hence $W(t) = h \in GL(n, R)$; in addition,

$$\widetilde{X}^T(t)h\widetilde{X}(t) = B(t), \ t \in I. \tag{9}$$

By Takagi representation for a nondegenerate symmetric matrix h (see, for example, [3, Ch.4,§ 4.4]), we have $h = U^T DU$, where U is a unitary matrix and $D = diag(\lambda_1, \ldots, \lambda_n)$ is the diagonal matrix with $\lambda_j > 0$ on the main diagonal, $j = 1, \ldots, n$. If $D_p = diag(\sqrt{\lambda_1}, \ldots, \sqrt{\lambda_p}, i\sqrt{\lambda_{p+1}}, \ldots, i\sqrt{\lambda_n})$, where $i^2 = -1$, and $g = D_p U$, then $D = D_p^T J_p D_p$ and $h = U^T D_p^T J_p D_p U = g^T J_p g$. Then, using (8) and $\widetilde{Y}(t) = g\widetilde{X}(t)$, we obtain

$$\widetilde{Y}^T(t)J_p\widetilde{Y}(t) = \widetilde{X}^T(t)g^T J_p g\widetilde{X}(t) = \widetilde{X}^T(t)h\widetilde{X}(t) = B(t),$$

hence $\widetilde{Y}(t)$ is a nondegenerate solution of the system (3) (see (8)).

Let $V(t) = (v_{ij}(t))^n_{i,j=1}$ be another nondegenerate solution of the system (3). By (8), there is an element $q \in GL(n, R)$ such that $V(t) = q\widetilde{Y}(t)$. Using the second equation of the system (3), we get

$$\widetilde{Y}^T(t)q^T J_p q \widetilde{Y}(t) = (q\widetilde{Y}(t))^T J_p(q\widetilde{Y}(t)) =$$

$$= (V(t))^T J_p V(t) = B(t) = \widetilde{Y}^T(t)J_p\widetilde{Y}(t).$$

Consequently, $q^T J_p q = J_p$, that is, $q \in O(n, p, R)$.

(ii). If $\widetilde{X}(t)$ is a nondegenerate solution of the system (3), $g \in O(n, p, R)$, $\widetilde{Y}(t) = g\widetilde{X}(t)$, then

$$\widetilde{Y}^{(1)}(t) = g\widetilde{X}^{(1)}(t) = g\widetilde{X}(t)A(t) = \widetilde{Y}(t)A(t),$$

hence $\widetilde{Y}^{(1)}(t) = \widetilde{Y}(t)A(t)$. In addition,

$$\widetilde{Y}^T(t)J_p\widetilde{Y}^T(t) = \widetilde{X}^T(t)g^T J_p g\widetilde{X}(t) = \widetilde{X}^T J_p\widetilde{X}^T(t) = B(t).$$

Therefore $\widetilde{Y}(t)$ is a nondegenerate solution of the system (3).   □

Theorem 2 implies the following.

Corollary 1 *If $X(t)$ is a nondegenerate solution of the system (3) then $\{gX(t) : g \in O(n, p, R)\}$ is the set of all nondegenerate solutions of the system (3).*

Theorem 3 *Let matrices $A(t)$ and $B(t)$ be as in Theorem 2, and let $X(t) = (x_{jk}(t))^n_{j,k=1}$ be a nondegenerate solution of the system (3). Then*

(i). The vector function $\overrightarrow{x}(t) = \{x_{j1}(t)\}^n_{j=1}$, $t \in I$, is a strongly regular path such that $M(\overrightarrow{x}(t)) = X(t)$ for all $t \in I$;

(ii). There is only one strongly regular path $\overrightarrow{x}(t)$, up to $O(n, p, R)$-equivalence, such that the matrix $M(\overrightarrow{x}(t))$ is a solution of the system (3).

Proof Statement (i) follows from the Eq. (6).

(ii). By Theorem 2 and (i), the system (3) has a nondegenerate solution $M(\overrightarrow{x}(t))$ for some strongly regular path $\overrightarrow{x}(t)$. Let $\overrightarrow{y}(t) = \{y_j(t)\}^n_{j=1}$ be another strongly regular path for which the matrix $M(\overrightarrow{y}(t))$ is a solution of the system (3). By Theorem 2, there exists $g \in O(n, p, R)$ such that $M(\overrightarrow{y}(t)) = gM(\overrightarrow{x}(t))$; in particular, $\overrightarrow{y}(t) = g\overrightarrow{x}(t)$ for all $t \in I$. Therefore, the paths $\overrightarrow{x}(t)$ and $\overrightarrow{y}(t)$ are $O(n, p, R)$-equivalent.   □

Theorems 2 and 3 imply the following description of the classes of $O(n, p, R)$-equivalent strongly regular paths in terms of solutions of the system of differential matrix equation (3).

Theorem 4 *(i). Let $\overrightarrow{x}(t)$ be a strongly regular path, $A(t) = (M(\overrightarrow{x}(t)))^{-1}M^{(1)}(\overrightarrow{x}(t))$, $B(t) = M^T(\overrightarrow{x}(t))J_p M(\overrightarrow{x}(t))$. Then*

$$\{Y_g(t) = (y_{jk}^{(g)}(t))_{j,k=1}^n = gM(\overrightarrow{x}(t)) : g \in O(n, p, R)\}$$

is the set of all nondegenerate solutions of the system (3). In addition,

$$\{\overrightarrow{y_g}(t) = \{y_{j1}^{(g)}(t)\}_{j=1}^n : g \in O(n, p, R), \ t \in I\}$$

is the set of all strongly regular paths that are $O(n, p, R)$-equivalent to the path $\overrightarrow{x}(t)$.

(ii). Let matrices $A(t)$ and $B(t)$ be as in Theorem 2. Then the first columns $\overrightarrow{x}(t) = \{x_{j1}(t)\}_{j=1}^n$, $t \in I$, of the set of nondegenerate solutions $X(t) = (x_{jk}(t))_{j,k=1}^n$ of the system (3) is a class of $O(n, p, R)$-equivalent paths.

References

1. Aripov, R.G., Khadzhiev, J.: A complete system of global differential and integral invariants of a curve in Euclidean geometry, Izvestiya Vuzov. Russian Math. **7**(542), 3–16 (2007)
2. Chilin, V.I., Muminov, K.K.: The complete system of differential invariants of a curve in pseudo-Euclidean space. Dyn. syst. **3**(31), 1–2, 135–149 (2013)
3. Horn, R., Johnson, C.: Matrix Analysis. Cambridge University Press, Cambridge (1990)
4. Khadzhiev, J.: Application of the Theory of Invariants to the Differential Geometry of Curves. FAN Publications, Tashkent (1988). (Russian)
5. Muminov, K.K.: Equivalence of paths with respect to the action of the symplectic group, Izvestiya Vuzov. Russian Math. **7**(537), 27–38 (2002)
6. Muminov, K.K.: Equivalence of curves with respect to an action of the symplectic group, Izvestiya Vuzov. Russian Math. **6**, 31–36 (2009)
7. Muminov, K.K., Chilin, V.I.: Equivalence of Curves in Finite-Dimensional Vector Spaces. LAP LAMBEPT Academic Publishing, Germany (2015)
8. Pontryagin, L.S.: Ordinary Differential Equations. London Addison-Wesley, Reading (1962)

Polynomial Estimates over Exponential Curves in $\mathbb{C}^2$

Shirali Kadyrov and Yershat Sapazhanov

Abstract For any complex α with non-zero imaginary part we show that Bernstein-Walsh type inequality holds on the piece of the curve $\{(e^z, e^{\alpha z}) : z \in \mathbb{C}\}$. Our result extends a theorem of Coman–Poletsky [6] where they considered real-valued α.

Keywords Bernstein-Walsh inequality · Several complex variables · Exponential curves

1 Introduction

In pluripotential theory, one is often interested in growth of polynomials of several variables. A classical Bernstein–Walsh inequality [9] gives important implications in this direction. Recently, there has been significant research work carried in obtaining Bernstein-Walsh type inequalities, see e.g [1–8]. We now recall the result of Coman–Poletsky in [6].

Let $\alpha \in (0, 1)\backslash\mathbb{Q}$ and $K \subset \mathbb{C}^2$ be a compact set given by $K = \{(e^z, e^{\alpha z}) : |z| \leq 1\}$. Define

$$E_n(\alpha) = \sup\{\|P\|_{\Delta^2} : P \in \mathscr{P}_n, \|P\|_K \leq 1\}, \tag{1}$$

where $\mathscr{P}_n$ is the space of polynomials in $\mathbb{C}[z, w]$ of degree at most n, Δ^2 is the closed bidisk $\{(z, w) \in \mathbb{C}^2 : |z|, |w| \leq 1\}$, and $\|\cdot\|_{\Delta^2}, \|\cdot\|_K$ are the uniform norms defined on compact sets Δ^2 and K, respectively. Let $e_n(\alpha) := \log E_n(\alpha)$. Then, Coman–Poletsky prove

Theorem 1 *For any Diophantine $\alpha \in (0, 1)$ one has*

S. Kadyrov (✉) · Y. Sapazhanov
Suleyman Demirel University, 040900 Kaskelen, Kazakhstan
e-mail: shirali.kadyrov@sdu.edu.kz

Y. Sapazhanov
e-mail: yershat.sapazhanov@sdu.edu.kz

© Springer Nature Switzerland AG 2018
Z. Ibragimov et al. (eds.), *Algebra, Complex Analysis, and Pluripotential Theory*,
Springer Proceedings in Mathematics & Statistics 264,
https://doi.org/10.1007/978-3-030-01144-4_8

$$\frac{n^2 \log n}{2} - n^2 \le e_n(\alpha) \le \frac{n^2 \log n}{2} + 9n^2 + Cn,$$

for any $n \ge 1$, where constant $C > 0$ depends on α.

Here, term 'Diophantine' comes from the Diophantine approximation theory and is an exponent that tells how well a real number can be approximated by rationals. For a proper definition see [6]. As a consequence of Theorem 1 one gets the Bernstein–Walsh type inequality

$$|P(z, w)| \le \|P\|_K E_n(\alpha) e^{n \log^+ \max\{|z|, |w|\}}, \tag{2}$$

for any $(z, w) \in \mathbb{C}^2$, $P \in \mathscr{P}_n$ and $E_n(\alpha) = e^{e_n(\alpha)}$ is determined by the theorem above.

We note that the inequality (2) holds for any $\alpha \in \mathbb{C}$ and finding the optimal bounds for $e_n(\alpha)$ is what makes it challenging in general. The proof of Theorem 1 makes use of the well-developed continued fraction expansion theory. We note that the theorem considers real-valued α's only. In this note, we aim to extend Theorem 1 to complex α's.

We now state our main result.

Theorem 2 *Let $\alpha = \alpha_1 + i\alpha_2$, $\alpha_1, \alpha_2 \in \mathbb{R}$ be given such that $|\alpha| < 1$ and $\alpha_2 \ne 0$. Then,*
$$\frac{n^2 \log n}{2} - n^2 \le e_n(\alpha) \le \frac{n^2 \log n}{2} + 8n^2 - n \log |\alpha_2|.$$

We remark here that our proof of Theorem 2 closely follows that of [6]. However, in our case we do not need to appeal to continued fraction theory and as a result our proof requires less effort. Nonetheless, it holds true for *all* non-real complex numbers α.

2 Proof of Main Result

In this section we prove our main result Theorem 2. For any real x let $\langle x \rangle$ denote the closest integer to x. We need the following lemma.

Lemma 1 (cf. Lemma 2.4 in [6]) *Let $k, x, y \in \mathbb{Z}$, $x \le y$, $k \ge 1$. Then (with $0^0 := 1$)*

$$\prod_{j=x}^{y} |j - k\alpha| \ge \begin{cases} (\frac{y-x}{2e})^{y-x} & , \text{if } \langle k\alpha_1 \rangle \notin [x, y], \\ (\frac{y-x}{2e})^{y-x} \cdot |k\alpha_2| & , \text{if } x \le \langle k\alpha_1 \rangle \le y \end{cases}$$

Proof We argue as in Lemma 2.4 of [6]. Using Stirling formula

$$e^{7/8} \le \frac{m!}{(\frac{m}{e})^m \sqrt{m}} \le e, \forall m \in \mathbb{N},$$

one gets

$$\prod_{j=1}^{m}\left(j-\frac{1}{2}\right) = \frac{(2m)!}{2^{2m}\cdot m!} \geq \left(\frac{m}{e}\right)^{m}$$

Let $j_0 = \langle k\alpha_1 \rangle$, if $j \neq j_0$ then,

$$|j - k\alpha| \geq |j - k\alpha_1| = |j - j_0| - \frac{1}{2}$$

Hence, when $j_0 < x$, we get

$$\prod_{j=x}^{y} |j - k\alpha| \geq \frac{1}{2}(y-x)!$$

Similarly, if $y < j_0$,

$$\prod_{j=x}^{y} |j - k\alpha| \geq \frac{1}{2}(y-x)!$$

On the other hand, for $x \leq j_0 \leq y$, we obtain

$$\prod_{j=x}^{y} |j - k\alpha| \geq \left(\frac{j_0-x}{e}\right)^{j_0-x} \cdot \left(\frac{y-j_0}{e}\right)^{y-j_0} \cdot |j_0 - k\alpha| \geq \left(\frac{y-x}{2e}\right)^{y-x} \cdot |j_0 - k\alpha|.$$

However, $|j_0 - k\alpha| \geq |k\alpha_2|$, which finishes the proof. $\qquad\square$

It is easy to see that the space $\mathscr{P}_n$ of polynomials in $\mathbb{C}[z, w]$ of degree at most n has dimension equal to $N + 1$, where $N := (n^2 + 3n)/2$. We are now ready to proceed with the proof of Theorem 2.

Proof Our argument closely follows that of [6]. We first obtain the upper estimate for the exponent $e_n(\alpha)$. For a given polynomial $R(\lambda) = \sum_{j=0}^{m} c_j \lambda^j$ of a single variable we let D_R denote the following differential operator

$$D_R = R\left(\frac{d}{dz}\right) = \sum_{j=0}^{m} c_j \frac{d^j}{dz^j}.$$

Then, $\forall \alpha \in \mathbb{C}$, we have

$$D_R(e^{\alpha z})|_{z=0} = \sum_{j=0}^{m} c_j \alpha^j e^{\alpha \cdot 0} = R(\alpha). \tag{3}$$

Let $P(z, w) = \sum_{j+k \leq n} c_{jk} z^j w^k \in \mathscr{P}_n$, $n \geq 1$, be given with $\|P\|_K \leq 1$, where as before $K = \{(e^z, e^{\alpha z}) : |z| \leq 1\}$. We set

$$f(z) := P(e^z, e^{\alpha z}) = \sum_{j+k \leq n} c_{jk} e^{(j+k\alpha)z}.$$

To obtain the upper bound for $e_n(\alpha)$ it suffices to estimate the coefficients c_{lm} from above. To this end, we define the following polynomials R_{lm} of degree N

$$R_{lm}(\lambda) = \prod_{j+k \leq n, (j,k) \neq (l,m)} (\lambda - j - k\alpha) = \sum_{t=0}^{N} a_t \lambda^t.$$

Using (3) we have

$$\begin{aligned} D_{R_{lm}} f(z)|_{z=0} &= \sum_{j+k \leq n} c_{jk} R_{lm}(j + k\alpha) \\ &= c_{lm} R_{lm}(l + m\alpha) \\ &= c_{lm} \beta_{lm} \\ \beta_{lm} &:= \prod_{j+k \leq n, (j,k) \neq (l,m)} (l - j + (m - k)\alpha) \end{aligned}$$

Using Cauchy's estimates $|f^{(t)}(0)| \leq t! \leq N^t$ for $t \leq N$, we arrive at

$$\left| D_{R_{lm}} f(z)|_{z=0} \right| = \left| \sum_{t=0}^{N} a_t f^{(t)}(0) \right| \leq \sum_{t=0}^{N} |a_t| N^t \leq (N + n)^N,$$

where the last inequality follows from Vieta's formulas and the fact that $|j + k\alpha| \leq n$. Therefore

$$\log(|c_{lm} \beta_{lm}|) \leq N \log(N + n) \leq n^2 \log n + 3.7n^2. \tag{4}$$

We now study the lower estimates on $|\beta_{lm}|$ which will lead to upper estimate for c_{lm}. Clearly,

$$|\beta_{lm}| \geq \prod_{k=0, k \neq m}^{n} \prod_{j=0}^{n-k} |l - j + (m - k)\alpha| = A_1 A_2,$$

Where

$$A_1 = \prod_{k=0}^{m-1} \prod_{j=0}^{n-k} |j - l - (m - k)\alpha| = \prod_{k=1}^{m} \prod_{j=-l}^{n-l-m+k} |j - k\alpha|,$$

$$A_2 = \prod_{k=m+1}^{n} \prod_{j=0}^{n-k} |l - j - (k - m)\alpha| = \prod_{k=1}^{n-m} \prod_{j=l+m+k-n}^{l} |j - k\alpha|,$$

From Lemma 1 we see that

$$A_1 \geq \prod_{k=1}^{m} \left(\frac{n-m+k}{2e} \right)^{n-m+k} \cdot |k\alpha_2|,$$

$$A_2 \geq \prod_{k=1}^{n-m} \left(\frac{n-m+k}{2e} \right)^{n-m+k} \cdot |k\alpha_2|.$$

Thus,

$$|\beta_{lm}| \geq A_1 A_2 \geq n! \cdot |\alpha_2|^n \cdot \prod_{k=1}^{n} \left(\frac{k}{2e} \right)^k$$

$$\geq |\alpha_2|^n \cdot e^{-2n^2} \cdot \prod_{k=1}^{n} k^k.$$

Using $\prod_{k=1}^{n} k^k \geq \frac{n^2 \log n}{2} - \frac{n^2}{4}$ (cf. Lemma 2.1 in [5]) we get

$$\log |\beta_{lm}| \geq \frac{n^2 \log n}{2} - \frac{n^2}{4} + n \log |\alpha_2| - 2n^2$$

$$\geq \frac{n^2 \log n}{2} - \frac{9n^2}{4} + n \log |\alpha_2|$$

Thus, using (4) we arrive at

$$\log |c_{lm}| \leq n^2 \log n + 3.7n^2 - \log |\beta_{lm}|$$

$$\leq \frac{n^2}{2} \log n + 5.95n^2 - n \log |\alpha_2|.$$

Clearly, $\|P\|_{\Delta^2} \leq \sum_{j+k \leq n} |c_{jk}| \leq (N+1) \max_{j+k \leq n} |c_{jk}|$. Recalling $N = (n^2 + 3n)/2$ and using (1) we obtain that

$$e_n(\alpha) = \log E_n(\alpha) \leq \log(N+1) + \frac{n^2}{2} \log n + 5.95n^2 - n \log |\alpha_2|,$$

Since $\log(N+1) \leq N = (n^2 + 3n)/2 \leq 2n^2$ for all $n \geq 1$, we get that

$$e_n(\alpha) \leq \frac{n^2}{2} \log n + 8n^2 - n \log |\alpha_2|,$$

which gives the upper estimate.

We now turn to obtaining lower estimate for $e_n(\alpha)$. To this end, we would like to construct a polynomial whose Δ^2-norm is large compared to its K-norm. We want to show that we can pick coefficients of $P(z, w) = \sum_{k+j \leq n} c_{jk} z^k w^j \in \mathscr{P}_n$ such that the Maclaurin series expansion of $f(t) := P(e^t, e^{\alpha t}) = \sum_{j+k \leq n} c_{jk} e^{(k+\alpha j)t}$ is

$f(t) = \sum_{k=N}^{\infty} a_k t^k$, that is, $f(t)$ has zero of order N at 0, where as before dim $\mathscr{P}_n = N + 1$. In other words, we want $f^{(k)}(0) = 0$ for all $0 \leq k \leq N - 1$. Thus, we get a system of N linear equations

$$\sum_{j+k \leq n} c_{jk}(k + \alpha j)^m = 0, \quad m = 0, 1, \ldots, N - 1.$$

List $\{(k + \alpha j) : k + j \leq n\}$ by $\{a_1, a_2, \ldots, a_N\}$, then the system has a solution provided that the following Vandermonde matrix

$$\begin{pmatrix} 1 & 1 & 1 & \ldots & 1 \\ a_1 & a_2 & a_3 & \ldots & a_N \\ a_1^2 & a_2^2 & a_3^2 & \ldots & a_N^2 \\ \vdots & \vdots & \vdots & \ddots & \vdots \\ a_1^{N-1} & a_2^{N-1} & a_3^{N-1} & \ldots & a_N^{N-1} \end{pmatrix}$$

is invertible. Hence, it suffices to show that $a_j \neq a_k$ unless $j = k$. Indeed, $k_1 + \alpha j_1 = k_2 + \alpha j_2$ implies $\alpha_2 j_1 = \alpha_2 j_2$, but $\alpha_2 \neq 0$ so that we must have $j_1 = j_2$. This in turn gives $k_1 = k_2$. So, the system has a solution and we can make sure that $f(t) = P(e^t, e^{\alpha t}) = t^N g(t)$ for some entire holomorphic function $g(t)$. We set $h(t) := f(t)/\|P\|_K$. Then, $\|h\|_\Delta = \|f\|_\Delta/\|P\|_K = 1$ as $\|P\|_K = \sup_{|z| \leq 1} |P(e^z, e^{\alpha z})| = \sup_{|z| \leq 1} |f(t)|$. Fix $r \geq 1$ (to be determined later) and consider $|t| = r$, then, Maximum Modulus Principle for holomorphic functions yields

$$\sup_{|t|=r} |h(t)| = \frac{\sup_{|t|=r} |f(t)|}{\|P\|_K} \geq \frac{r^N \sup_{|t|=r} |g(t)|}{\|P\|_K} = \frac{r^N \sup_{|t|=r} |g(t)|}{\|P\|_K} \geq r^N.$$

Equation (2) gives

$$|f(t)| = |P(e^t, e^{\alpha t})| \leq \|P\|_K E_n(\alpha) e^{n \log^+ \max\{|e^t|, |e^{\alpha t}|\}},$$

Hence,

$$r^N \leq \sup_{|t|=r} |h(t)| \leq E_n(\alpha) e^{nr} \text{ for any } r \geq 1.$$

Now taking $r = N/n$ we get

$$e_n(\alpha) \geq N \log(N/n) - N \geq \frac{n^2}{2} \log n - n^2.$$

This finishes the proof.   □

Acknowledgements The first author would like to thank Dan Coman for a useful discussion.

References

1. Abdullayev, F.G., Ozkartepe, N.P.: An analogue of the Bernstein-Walsh lemma in Jordan regions of the complex plane. J. Inequal. Appl. **1**, 570 (2013)
2. Brudnyi, A.: Bernstein type inequalities for restrictions of polynomials to complex submanifolds of $\mathbb{C}^N$. J. Approx. Theory **225**, 106–147 (2018)
3. Bos, L.P., Brudnyi, A., Levenberg, N.: On Polynomial Inequalities on Exponential Curves in $\mathbb{C}^n$. Constr. Approx. **31**(1), 139–147 (2010)
4. Coman, D., Poletsky, E.A.: Measures of trancendency for entire functions. Mich. Math. J. **51**(3), 575–591 (2003)
5. Coman, D., Poletsky, E.: Bernstein-Walsh inequalities and the exponential curve in $\mathbb{C}^2$. Proc. Am. Math. Soc. **131**(3), 879–887 (2003)
6. Coman, D., Poletsky, E.A.: Polynomial estimates, exponential curves and Diophantine approximation. Math. Res. Lett. **17**(6), 1125–1136 (2010)
7. Kadyrov, S., Lawrence, M.: Bernstein-Walsh inequalities in higher dimensions over exponential curves. Constr. Approx. **44**(3), 327–338 (2016)
8. Neelon, T.: A Bernstein-Walsh type inequality and applications. Can. Math. Bull. **49**(2), 256–264 (2006)
9. Ransford, T.: Potential Theory in the Complex Plane. Cambridge University Press, Cambridge (1995)

Carleman Formula for Matrix Ball of the Third Type

G. Khudayberganov and U. S. Rakhmonov

Abstract In this paper we consider the problem of restoring the values of a holomorphic function in a matrix ball of the third type with respect to values on a part of its boundary.

Keywords Carlemen formula · Matrix ball of the third type · Poisson kernel
Cauchy formula · Lebesgue measure · Cauchy-Szego kernel

1 Introduction and Preliminaries

Let $Z = (Z_1, \ldots, Z_n)$ be vector where Z_j square matrices of order m, considered over the field of complex numbers $\mathbf{C}$. We can assume that the Z is element of space $\mathbf{C}^{nm^2}$. In this set of vectors we consider the matrix scalar multiplication:

$$\langle Z, W \rangle = Z_1 W_1^* + Z_2 W_2^* + \cdots + Z_n W_n^*,$$

where the matrix W_j^*, conjugate and transposed for the matrices W_j.
    The domain (see [1])

$$B_{m,n}^{(1)} = \{ Z = (Z_1, \ldots, Z_n) \in \mathbf{C}^n[m \times m] : I^{(m)} - \langle Z, Z \rangle > 0 \},$$

is called the matrix ball (of the first type), where $\langle Z, Z \rangle = Z_1 Z_1^* + Z_2 Z_2^* + \cdots + Z_n Z_n^*$ the scalar multiplication, I - is the unit $[m \times m]$ matrix, $Z_\nu^* = \bar{Z}_\nu'$ matrix, conjugate and transposed to Z_ν, $\nu = 1, 2, \ldots, n$,. Here $I^{(m)} - \langle Z, Z \rangle > 0$ means that the Hermitian matrix is positively definite, i. e. all eigenvalues are positive.
    The domain $B_{m,n}^{(3)}$ of a space $\mathbf{C}^{nm^2}$:

G. Khudayberganov (✉)
National University of Uzbekistan named after M. Ulugbek, Tashkent, Uzbekistan
e-mail: gkhudaiberg@mail.ru

U. S. Rakhmonov
Tashkent State Technical University named after I. Karimov, Tashkent, Uzbekistan
e-mail: uktam_rakmonov@mail.ru

© Springer Nature Switzerland AG 2018
Z. Ibragimov et al. (eds.), *Algebra, Complex Analysis, and Pluripotential Theory*,
Springer Proceedings in Mathematics & Statistics 264,
https://doi.org/10.1007/978-3-030-01144-4_9

$$B_{m,n}^{(3)} = \left\{ Z \in \mathbf{C}^n \left[m \times m \right] : I^{(m)} + \langle Z, Z \rangle > 0, \ Z_\nu' = -Z_\nu, \ \nu = 1, \dots, n \right\} \quad (1)$$

is called a matrix ball of the third type, where I, as usual, the unit matrix of order m (see [1]).

The skeleton of this domain is a manifold of the form:

$$X_{m,n}^{(3)} = \left\{ Z = (Z_1, \dots, Z_n) \in \mathbf{C}^n \left[m \times m \right] : I^{(m)} + \langle Z, Z \rangle = 0, \ Z_\nu' = -Z_\nu, \ \nu = 1, \dots, n \right\}.$$

Note that, $B_{1,1}^{(1)}$ and $B_{2,1}^{(1)}$ - unit disks, $X_{1,1}^{(1)}$ and $X_{2,1}^{(3)}$ - unit circles in a complex plane.

If $n = 1, m > 1$, than $B_{m,1}^{(1)}$ and $B_{m,1}^{(3)}$ - classical domains of the first and third types (on E. Cartan's classification) and skeletons, $X_{m,1}^{(1)}$ and $X_{m,1}^{(3)}$ - unitary and skew-symmetric unitary matrices, respectively.

If we arrange the elements of a matrix $Z \in \mathbf{C}[m \times m]$ in form of a vector, namely

$$Z = (z_{11}, \dots, z_{1m}, z_{21}, \dots, z_{2m}, \dots, z_{m1}, \dots, z_{mm}) \in \mathbf{C}^{m \times m},$$

then we arrive at an isomorphism $\mathbf{C}[m \times m] \simeq \mathbf{C}^{m^2}$.

We denote by D_3 a classical domain of the third type, and by S_3 its distinguished boundary (Shilov boundary)(see [2, p.10]). Let $L^2(S_3, d\mu)$ denote the space of functions f square integrable with respect to Haar measure $d\mu$, and by $H^2(S_3, d\mu)$ the space of functions in $L^2(S_3, d\mu)$ admitting a holomorphic extension into D_3, i.e. a holomorphic function $f \in H^2(S_3, d\mu)$ if

$$\sup_{0 < r < 1} \int_{S_3} |f(r\zeta)|^2 \, d\mu < C.$$

We denote $G \subset \mathbf{C}[m \times m]$ the bounded set. By Gershgorin's theorem [3], there exists a bounded simply connected domain $D \subset \mathbf{C}^1$ with piecewise smooth boundary containing all eigenvalues of all matrices $W \in G$. Then for every function $f \in \mathcal{O}(D) \cap C(\overline{D})$ it is defined $f(W)$, using the Cauchy formula

$$f(W) = \frac{1}{2\pi i} \int_{\partial D} f(z)(zI - W)^{-1} dz, \quad (2)$$

where $W \in G$ [3, §5.8]. Formula (2) easily extends to a class $H^1(D)$. We give a matrix analogue of the classical Carleman formula.

Theorem 1 ([4]) *If $f \in H^1(D)$ and a set $M \subset \partial D$ of positive Lebesgue measure, then for any matrix $W \in G$ we have following the Carleman formula*

$$f(W) = \lim_{m \to \infty} \frac{1}{2\pi i} \int_M f(\zeta)(\zeta I - W)^{-1} \{\varphi(\zeta)[\varphi(W)]^{-1}\}^m d\zeta,$$

where the function φ is the same as in the one-dimensional Carleman formula.

For the matrix unit disc

$$\tau = \{Z \in \mathbf{C}^n[m \times m] : I - ZZ* > 0\}$$

we can take the usual unit circle $U(0, 1)$ as D [2].

2 Main Results

The aim of this paper is to prove the analogue of Theorem 1 in a matrix ball of the third type.

For a function $f \in L^1(X_{m,n}^{(3)})$, we define Poisson transformation $P[f]$ as follows

$$P[f](Z) = \int\limits_{X_{m,n}^{(3)}} f(W)P(Z, W)d\sigma(W), \, Z \in B_{m,n}^{(3)}.$$

where $d\sigma$ - Lebesgue measure, the $P(Z, W)$ is Poisson kernel for a matrix ball of the third type [5] that have the following form for even m:

$$P(Z, W) = \left[\frac{\det(I^{(m)} + \langle Z, Z \rangle)}{\left| \det(I^{(m)} + \langle Z, W \rangle) \right|^2} \right]^{\frac{(m-1)n}{2}}, \, Z \in B_{m,n}^{(3)}, \, W \in X_{m,n}^{(3)},$$

for odd m:

$$P(Z, W) = \left[\frac{\det(I^{(m)} + \langle Z, Z \rangle)}{\left| \det(I^{(m)} + \langle Z, W \rangle) \right|^2} \right]^{\frac{mn}{2}}, \, Z \in B_{m,n}^{(3)}, \, W \in X_{m,n}^{(3)}.$$

Lemma 1 ([5]) *If ψ an arbitrary automorphism of the matrix ball $B_{m,n}^{(3)}$ such that $\psi^{-1}(0) = Z$, then*

$$P(Z, W)d\sigma(W) = d\sigma(\psi(W)). \tag{3}$$

We define the space $H^1(B_{m,n}^{(3)})$ in the following way: a function f belongs $H^1(B_{m,n}^{(3)})$ if it is holomorphic in $B_{m,n}^{(3)}$ and

$$\sup_{0<r<1} \int\limits_{X_{m,n}^{(3)}} |f(rZ)| \, d\sigma(Z) < +\infty.$$

Since $B_{m,n}^{(3)}$ is bounded complete circular domain, any function $f \in H^1(B_{m,n}^{(3)})$ has the following properties [6, ch.4], [7], :

(1) for almost all (in measure σ) $Z \in X_{m,n}^{(3)}$ the shear-functions $f_Z(\lambda) = f(\lambda Z)$ belong to the space H^1 in the unit disk $\Delta = \{\lambda \in \mathbf{C}^1 : |\lambda| < 1\}$;
(2) the function f has radial boundary values

$$\lim_{r \to 1} f(rZ) = f^*(Z), Z \in X_{m,n}^{(3)}$$

and they belong to the class $L^1(X_{m,n}^{(3)})$;
(3) the following formula holds

$$\lim_{r \to 1} \int\limits_{X_{m,n}^{(3)}} |f(rZ)|d\sigma = \int\limits_{X_{m,n}^{(3)}} |f^*(Z)|d\sigma;$$

Let M be a set of positive measure σ on the skeleton $X_{m,n}^{(3)}$. Now we reconstruction the values of function $f \in H^1(B_{m,n}^{(3)})$ in the domain $B_{m,n}^{(3)}$ using the radial boundary values of f on M.

First of all, we parametrize the set $X_{m,n}^{(3)}$ as follows: for $Z \in X_{m,n}^{(3)}$ we put $Z = e^{i\theta}U$, where $0 \le \theta \le 2\pi$, and in the matrix U_1 the element $u_{11}^{(1)}$ in the upper left corner is a positive number. We denote a manifold of such matrices by X_1^+. Let σ^+ be a normalized Lebesgue measure on the X_1^+. Note that not the whole set $X_{m,n}^{(3)}$ is parametrized in this way, but a set smaller than $X_{m,n}^{(3)}$, differing by a set of measure 0.

Applying the Fubini theorem and the invariance of the Lebesgue measure with respect to the rotations, we the following

Lemma 2 ([8]) *Measure*

$$d\sigma = h(U)d\theta d\sigma^+(U), \quad U \in X_1^+, \tag{4}$$

where the smooth positive function $h(U)$ does not depend on θ.

Thus from (4) we have

$$d\sigma = \frac{1}{2\pi i}\frac{d\lambda}{\lambda}d\sigma_1(U), \tag{5}$$

where $\lambda = e^{i\theta}$ and the measure σ_1, is positive on X_1^+.

We denote by

$$M_{0,U} = \left\{W \in M : \ W = \lambda U, \ \lambda = e^{i\theta}, \ 0 \le \theta \le 2\pi\right\}, \ U \in X_1^+,$$

$$M_0' = \left\{U \in X_1^+ : \ m_1(M_{0,U}) > 0\right\},$$

where m_1- the one-dimensional normalized Lebesgue measure on the unit disc. It follows Fubini's theorem and formula (5) that $\sigma_1(M_0') > 0$.

We introduce the one-dimensional Carleman function:

$$g_0(W) = \frac{1}{2\pi i} \int\limits_{M_{0,U}} \frac{\eta + \lambda}{\eta - \lambda} \frac{d\eta}{\eta}, \quad h_0 = \exp g_0.$$

Theorem 2 *For point 0 and any function $f \in H^1(B_{m,n}^{(3)})$ the Carleman formula holds*

$$f(0) = \frac{1}{\int\limits_{M_0'} d\sigma_1} \lim_{j \to \infty} \lim \int\limits_M f(W) \left[\frac{h_0(W)}{h_0(0)}\right]^j d\sigma(W). \tag{6}$$

Proof In the section of the ball $B_{m,n}^{(3)}$ of the complex line

$$\{Z : Z = \lambda U\}, U \in M_0',$$

we use the properties of the shear-function and the classical Carleman formula (see [8])

$$f(0) = \lim_{j \to \infty} \lim \frac{1}{2\pi i} \int\limits_{M_{0,U}} f(\lambda U) \left[\frac{h_0(\lambda)}{h_0(0)}\right]^j \frac{d\lambda}{\lambda}.$$

Multiplying both sides of this equation by $d\sigma_1(U)$ and integrating by M_0', we obtain (6). Since Lebesgue's theorem on bounded convergence, the limit can be entried into the sign of integral. Indeed (Δ the unit circle)

$$\left| \int\limits_{M_{0,U}} f(\lambda U)\left[\frac{h_0(\lambda)}{h_0(0)}\right]^j \frac{d\lambda}{\lambda} \right| \leq \left| f(0) \right| + \int\limits_{\Delta \backslash M_{0,U}} \left| f(\lambda U) \right| \left| \frac{h_0(\lambda)}{h_0(0)} \right|^j d\varphi \leq \left| f(0) \right|$$

$$+ \int\limits_{\Delta} \left| f(\lambda U) \right| d\varphi,$$

as

$$\left| \frac{h_0(\lambda)}{h_0(0)} \right| \leq 1$$

on $\Delta \setminus M_{0,U}$.

The theorem is proved.

Let, $\varphi_P(Z)$ be an automorphism in the matrix ball $B_{m,n}^{(3)}$, that $\varphi_P(P) = 0$ for $P \in B_{m,n}^{(3)}$ (see [1]). We define

$$\sigma_P(K) = \sigma_1(\varphi_P^{-1}(K)), \quad K \in X_P^+,$$

$$M_{P,U} = \left\{ W \in M : W = \varphi_P^{-1}(\lambda \varphi_P^{-1}(U)), |\lambda| = 1 \right\},$$

$$U \in X_P^+ = \varphi_P(X_1^+), \quad M_P' = \left\{ U \in X_P^+ : \ m_1(M_{P,U}) > 0 \right\},$$

$$g_P(W) = \frac{1}{2\pi i} \int\limits_{M_{P,U}} \frac{\eta + \lambda}{\eta - \lambda} \frac{d\eta}{\eta}, \quad h_P = \exp g_P.$$

Since λ and U the functions of W, it is clear that the function g_P depends on W.

Theorem 3 *For any function $f \in H^1(B_{m,n}^{(3)})$, for even m and every point $P \in B_{m,n}^{(3)}$ the following Carleman formula holds*

$$f(P) = \frac{1}{\int\limits_{M_P'} d\sigma_P} \lim_{j \to \infty} \lim \int\limits_M f(W) \left[\frac{h_P(W)}{h_P(P)} \right]^j C(P, W) d\sigma(W). \qquad (7)$$

Proof For $P = 0$ formula (7) coincides with formula (6). Assume $f_P(W) = f(\varphi_p^{-1}(W))$, then $f_P \in H^1(B_{m,n}^{(3)})$ and $f_P(0) = f(P)$. We apply Theorem 2 to the function f_P and make transformation φ_P under the sign of integral, i.e.

$$f_P(0) = \frac{1}{\int\limits_{M_0'} d\sigma_1(\varphi_P(W))} \lim_{j \to \infty} \lim \int\limits_M f_P(\varphi_P(W)) \left[\frac{h_0(\varphi_P(W))}{h_0(\varphi_P(0))} \right]^j d\sigma(\varphi_P(W)).$$

Then we get

$$f(P) = \frac{1}{\int\limits_{M_P'} d\sigma_P} \lim_{j \to \infty} \lim \int\limits_M f(W) \left[\frac{h_P(W)}{h_P(P)} \right]^j d\sigma(\varphi_P(W)). \qquad (8)$$

By the formula (4)

$$d\sigma(\varphi_P(W)) = P(P, W) d\sigma(W),$$

where (see. [9])

$$P(P, W) = \frac{C(P, W)C(W, P)}{C(P, P)}.$$

Applying formula (8) to the function we obtain

$$f(W) = C^{-1}(W, P) \in H^1(B_{m,n}^{(3)}),$$

where $C(P, W)$ is Cauchy–Szego kernel for a matrix ball of the third type [10].

Since

$$f(P)C^{-1}(P, P) = \frac{1}{\int\limits_{M'_P} d\sigma_P} \lim_{j\to\infty} \lim \int\limits_M f(W)C^{-1}(W, P)\left[\frac{h_P(W)}{h_P(P)}\right]^j \frac{C(P, W)C(W, P)}{C(P, P)} d\sigma(W),$$

we get

$$f(P) = \frac{1}{\int\limits_{M'_P} d\sigma_P} \lim_{j\to\infty} \lim \int\limits_M f(W)\left[\frac{h_P(W)}{h_P(P)}\right]^j C(P, W)d\sigma(W).$$

The theorem is proved.

Theorem 4 *For any function $f \in H^1(B_{m,n}^{(3)})$, for odd m and every point $P_1 \in B_{m,n}^{(3)}$, the Carleman formula holds*

$$f(P_1) = \frac{m+1}{\mu_P\left(\widetilde{M}'_A\right)} \lim_{j\to\infty} \lim \int\limits_M f(W)\left[\frac{h_P\begin{pmatrix} 0 & 0 \\ 0 & W_1 \end{pmatrix}}{h_P(P)}\right]^j C(P, W)d\sigma(W),$$

where $C(Z, W)$ is Cauchy–Szego kernel for a matrix ball of the third type [5], a $P = \begin{pmatrix} 0 & 0 \\ 0 & P_1 \end{pmatrix}$ is matrix of order $m + 1$,

$$\widetilde{M} = \left\{ \begin{pmatrix} 0 & \xi \\ -\xi' & W_1 \end{pmatrix} : W_1 \in M_{P,U}, \ \xi \in E \right\},$$

E consists of k-vectors satisfying conditions

$$\xi\xi' = 1, \quad \xi W_1 = 0, \quad W_1'\overline{W_1} + \xi\xi' = I^{(m)}.$$

The proof of the theorem is similar to the proof of Theorem 3. For this it is necessary to extend $f(W_1)$ [11] to a function $f \in \mathscr{O}((B_{m,n}^{(3)})) \cap C(\overline{(B_{m,n}^{(3)})})$, where $W = \begin{pmatrix} 0 & 0 \\ 0 & W_1 \end{pmatrix}$ and calculate the integral.

References

1. Khudayberganov, G., Khidirov, B.B., Rakhmonov, U.S.: Automorphisms of matrix balls. Vestnik NUUz **4**, 205–209 (2010). (in Russian)
2. Hua, L.-K.: Harmonic analysis of functions of several complex variables, in classical domains. Moscow, IL (1959) (in Russian)

3. Lancaster, P.: The Theory of Matrices. Nauka, Moscow (1982). (in Russian)
4. Khudayberganov, G.: The Carleman formula for functions of matrices. Sib. Math. J. **28**(1), 207–208 (1988). (in Russian)
5. Rakhmonov, U.S.: Poisson kernel for matrix ball of third type. Uzbek Math. J. **3**, 123–125 (2012). (in Russian)
6. Vladimirov, V.S., Sergeev, A.G.: Complex analysis in the future tube. Encycl. Math. Sci. Fundam. Dir. Itogi nauki i tekhniki **8**, 191–266 (1985). (in Russian)
7. Khudayberganov, G., Kytmanov, A.M., Shaimkulov B.A.: Analysis in matrix domains. Monograph. Krasnoyarsk, Siberian Federal University. 296 p. (in Russian) (2017)
8. Aizenberg, L.A.: Carleman Formulas in complex analysis. Novosibirsk, Science. 248 p. (in Russian) (1990)
9. Koranyi, A.: The Poisson integral for generalized half-planes and bounded symmetric domains. Ann. Math. **82**(2), 332–350 (1965)
10. Khudayberganov, G., Rakhmonov, U.S.: Berghman and Cauchy–Szego kernels for matrix ball of third type. Uzbek Math. J. **2**, 152–157 (2012). (in Russian)
11. Kytmanov, A.M., Nikitina, T.N.: Analogs of Carleman's formula for classical domains. Math. Notes **45**(3), 87–93 (1989)

A Multidimensional Boundary Analogue of Hartogs's Theorem on n-Circular Domains for Integrable Functions

Bairambay Otemuratov

Abstract In the present paper we consider integrable functions given on the boundary of n-circular domain $D \subset \mathbb{C}^n$, $n > 1$ and having one-dimensional property of holomorphic extension along the families of complex lines, passing through finite number of points of D. We prove the existence of holomorphic extension of such functions in D.

Keywords Holomorphic extension · Szego kernel · Poisson kernel · Complex lines

1 Introduction

One of the important problems of complex analysis is holomorphic extension of integrable functions defined on the boundary of a domain $D \subset \mathbb{C}^n$, $n > 1$. In particular, it is interesting to study than problem for integrable functions with one-dimensional holomorphic extension property along complex lines. On the complex plane $\mathbb{C}$ problem of the one-dimensional holomorphic extension is trivial. The result in this area are essentially multidimensional.

The first result related to our subject was obtained M. L. Agranovsky and R.E. Valsky [1], who studied functions with the one-dimensional holomorphic extension property into a ball. The proof was based on the properties of the group of automorphisms of a sphere.

E. L. Stout in [27], using the complex Radon transformation to generalize the Agranovsky and Val'sky Theorem for an arbitrary bounded domain with a smooth boundary. An alternative proof of Stout's Theorem was obtained by A.M. Kytmanov in [5], by using the Bochner–Martinelli integral. The idea of using the integral representations (Bochner–Martinelli, Cauchy–Fantappie, logarithmic residue) has been

B. Otemuratov (✉)
Ch. Abdirov 1, Department of Mathematics, Karakalpak State University,
Nukus 230113, Uzbekistan
e-mail: bayram_utemurato@mail.ru

© Springer Nature Switzerland AG 2018
Z. Ibragimov et al. (eds.), *Algebra, Complex Analysis, and Pluripotential Theory*,
Springer Proceedings in Mathematics & Statistics 264,
https://doi.org/10.1007/978-3-030-01144-4_10

useful in the study of functions with one- dimensional holomorphic continuation property (see review [15]).

The question of finding different familes of complex lines sufficient for holomorphic extension was put in [12]. Clearly, the family of complex lines passing through one point is not enough. As shown in [16], the family of complex lines passing through a finite number of points also, generally speaking, is not sufficient.

In [16] it was proved that the family of complex lines crossing the germ of a generic manifold γ, is sufficient for the holomorphic extension. In [17] considered continuous functions given on the boundary of a bounded domain D in $\mathbb{C}^n$, $n > 1$, with the one-dimensional holomorphic extension property along families of complex lines. Also it was studied the existence of holomorphic extensions of these functions to D depending on the dimension and location of the families of complex lines. Various other families and related problems were studied by many authours [2, 3, 6, 7, 11]. We note that it was shown in the papers [3, 11] were shown that a family of complex lines passing through finite numbers of points is sufficient for holomorphic extension. But it was proved only for real-analytic or infinitely differentiable functions defined on the boundary. In [8, 10] were shown that for holomorphic extension of continuous functions is sufficient a family of complex lines passing through $n + 1$ points lying at the interior of ball. Another proof of this result based on applications of integral representation was given in [18, 21]. In [22–24] sufficient conditions for holomorphic extension of integrable functions for a family of complex lines passing through open subset lying in a domain D, were considered.

The example of Globevnik [11] shows that for continuous functions on the boundary of ball in $\mathbb{C}^2$ two points is not enough for holomorphic extension. In paper [19] were considered continuous functions given on the boundary of a ball B of $\mathbb{C}^n$, $n > 1$, having one-dimensional property of holomorphic extension along the families of complex lines, passing through finite number of points of D.

In this paper we generalize this result for integrable functions. In Sect. 2 we consider the Szego kernel in n-circular domains. In Sects. 3 and 4 we consider the Poisson kernel and modified Poisson kernel on n-circular domains. In Sect. 5 we will consider integrable functions given on the boundary n-circular domain $D \subset \mathbb{C}^n$, $n > 1$ and having one-dimensional property of holomorphic extension along the families of complex lines, passing through finite number of points of D. We prove the existence of holomorphic extension of such functions in the domain D (see Theorem 7).

2 The Szego Kernel on n-Circular Domains

Let D be a bounded complete n-circular domain in $\mathbb{C}^n$ with the center at the origin, that is, together with a point $z^0 = (z_1^0, \ldots, z_n^0) \in D$, it contains

$$\{z \in \mathbb{C}^n : |z_k| \le |z_k^0|, \ k = 1, \ldots, n\}.$$

Denote by $D^+ = \{(|z_1|, \ldots, |z_n|) : z \in D\}$ the image of a domain D in the absolute octant

$$\mathbb{R}_n^+ = \{(x_1, \ldots, x_n) : |x_k| \geq 0, \ k = 1, \ldots, n\}.$$

Let $\partial D^+ = \{(|z_1|, \ldots, |z_n|) : z \in \partial D\}$.

Consider a finite measure μ on ∂D^+. A measure μ is said to be massive on the Shilov boundary [5, P. 76], if for any subset $E \subset \partial D^+$ of measure zero measure μ satisfied a condition $\overline{\partial D^+ \setminus E} \supset S(D^+)$, where $S(D^+)$ is the image of the Shilov boundary $S(D)$ in the absolute octant.

Further we need the following result from [13, Sect. 3.1].

Proposition 1 *If D is a strongly pseudoconvex n-circular domain (i.e., strictly logarithmically convex) then the Shilov boundary $S(D)$ coincides with the boundary of the domain.*

Proposition 1 implies that the Lebesgue measure μ on the boundary of such domain is massive. From now on we shall assume that μ is a massive measure.

Define the Szego kernel of domain D :

$$h(\bar{\zeta}, z) = \sum_{\alpha \geq 0} a_\alpha \bar{\zeta}^\alpha z^\alpha, \tag{1}$$

where

$$a_\alpha = \frac{1}{\int\limits_{\partial D^+} |\zeta|^{2\alpha} d\mu} = \frac{1}{\int\limits_{\partial D^+} |\zeta_1|^{2\alpha_1} \cdot \ldots \cdot |\zeta_n|^{2\alpha_n} d\mu}$$

and $\alpha = \{\alpha_1, \ldots, \alpha_n\}$ is a multi-index such that $\alpha \geq 0$ (i.e., $\alpha_k \geq 0$, $k = 1, \ldots, n$) and $z^\alpha = z_1^{\alpha_1} \cdot \ldots \cdot z_n^{\alpha_n}$, $\|\alpha\| = \alpha_1 + \ldots + \alpha_n$.

Recall a definition of a class $\mathcal{H}^p(D)$. A holomorphic function $f \in \mathcal{H}^p(D)$ ($p > 0$), if

$$\sup_{\varepsilon > 0} \int_{\partial D} |f(\zeta - \varepsilon \nu(\zeta))|^p d\sigma < +\infty,$$

where $d\sigma$ is an element of the surface ∂D and $\nu(\zeta)$ is the outer unit normal vector to the surface ∂D at the point ζ. It is well-known that normal boundary values of $f \in \mathcal{H}^p(D)$ belong to the class $\mathcal{L}^p(\partial D)$ (with respect to the measure $d\sigma$).

The following result gives us the existence of the Szego kernels on the n-circular domains.

Theorem 1 *Let μ be a finite measure on ∂D^+. For any function $f \in \mathcal{H}^p(D)$, ($p \geq 1$) there exists a the Szego representation*

$$f(z) = \lim_{r \to 1} \frac{1}{(2\pi i)^n} \int_{\partial D^+} d\mu \int_{\Delta_{|\zeta|}} f(\zeta) h(\bar{\zeta}, rz) \frac{d\zeta}{\zeta}, \quad z \in D, \tag{2}$$

where

$$\Delta_{|\zeta|} = \{\zeta : \zeta_1 = |\zeta_1|e^{i\theta_1}, \ldots, \zeta_n = |\zeta_n|e^{i\theta_n}, \ 0 \leq \theta_k \leq 2\pi, \ k = 1, \ldots, n, \ |\zeta| \in \partial D^+\},$$

$$\frac{d\zeta}{\zeta} = \frac{d\zeta_1}{\zeta_1} \wedge \cdot \frac{d\zeta_n}{\zeta_n},$$

and for any fixed $z \in D$ the Szego kernel $h(\bar{\zeta}, z) = h(\bar{\zeta}_1 z_1, \ldots, \bar{\zeta}_n z_n)$ belongs to $\mathscr{O}(\overline{D})$ as a function of $\bar{\zeta}$, and for any fixed $\zeta \in \partial D$ the function $h(\bar{\zeta}, z)$ belongs to $\mathscr{O}(D)$ as a function of z.

This Theorem was proved for continuous functions in [5] and for functions from the class $\mathscr{H}^p$, it is obtained by approximation of $f(z)$ by functions $f(r\zeta)$ when $r \to 1 - 0, r < 1$, with respect to the metric of $\mathscr{H}^p$.

So, by Theorem 1 the series (1) converges absolutely for $\zeta \in \overline{D}$ and $z \in D$ and uniformly for $\zeta \in \overline{D}$ and $z \in K$, where K is an arbitrary compact subset in D.

Clearly $\partial D = \bigcup\limits_{|\zeta| \in \partial D^+} \Delta_{|\zeta|}$. The following property of the Szego kernel is evidently:

$$h(\bar{\zeta}, z) = \overline{h(\zeta, \bar{z})} = h(z, \bar{\zeta}).$$

3   The Poisson Kernel on n-Circular Domains

Let us recall the Poisson kernel

$$P(\zeta, z) = \frac{h(\bar{\zeta}, z)h(\zeta, \bar{z})}{h(\bar{z}, z)} = \frac{|h(\bar{\zeta}, z)|^2}{h(\bar{z}, z)}.$$

Note that the kernel $P(\zeta, z)$ is defined for $(\zeta, z) \in D \times D$, because $h(\bar{z}, z) > 0$.

Proposition 2 *If $f \in \mathscr{H}^p(D)$ $(p \geq 1)$, the following formula is true*

$$f(z) = \lim_{r \to 1} \frac{1}{(2\pi i)^n} \int\limits_{\partial D^+} d\mu \int\limits_{\Delta_{|\zeta|}} f(\zeta)P(\zeta, rz)\frac{d\zeta}{\zeta}, \quad z \in D.$$

The proof follows from the bottom of the Poisson kernel and Theorem 1 domain D satisfies.

Lemma 1 *Consider the Szego kernel at $\zeta = z$*

$$h(\bar{z}, z) = \sum_{\alpha \geq 0} a_\alpha |z|^{2\alpha} > 0$$

in D. Then $h(\bar{z}, z) \to \infty$, when $z \to \partial D$.

Suppose that a domain D satisfied the following property (A):

$h(\bar{\zeta}, rz)$ is uniformly bounded in z outside any neighborhood of ζ for $\zeta, z \in \partial D$ and $\zeta \neq z, r \to 1$.

Theorem 2 *Let D be a domain with the property (A) and $f \in \mathscr{L}^p(\partial D)$. Then the Poisson integral*

$$F(z) = P[f](z) = \lim_{r \to 1} \frac{1}{(2\pi i)^n} \int_{\partial D^+} d\mu \int_{\Delta_{|\zeta|}} f(\zeta) P(\zeta, rz) \frac{d\zeta}{\zeta}$$

is a real-analytically function on D and its values on the boundary with respect to the metric $\mathscr{L}^p$ coincides with f on ∂D.

Proof Real-analytically of $F(z)$ follows from the real-analyticity of the Szego and Poisson kernels. By condition (A) and Lemma 1 we obtain that $P(\zeta, rz)$ uniformly converges to zero outside any neighborhood's at the point ζ for $\zeta, z \in \partial D$, $\zeta \neq z$ and $r \to 1$. Besides $P(\zeta, z) > 0$ and $P[1](\zeta) = 1$. Hence the Poisson kernel $P(\zeta, z)$ is an approximation to identity [26, P. 49], where $d\zeta = d\zeta_1 \wedge \ldots \wedge d\zeta_n$, $d\bar{\zeta}[k] = d\bar{\zeta}_1 \wedge \ldots \wedge d\bar{\zeta}_{k-1} \wedge d\bar{\zeta}_{k+1} \wedge \ldots \wedge d\bar{\zeta}_n$. The proof is complete. $\qquad\square$

4 The Modified Poisson Kernel

Consider the following differential form

$$\omega = c \sum_{k=1}^{n} (-1)^{k-1} \bar{\zeta}_k \, d\bar{\zeta}[k] \, d\zeta,$$

where $c = \frac{(n-1)!}{(2\pi i)^n}$.

Let us find the restriction of this form the boundary ∂D of the domain

$$D = \{z \in \mathbb{C}^n : \rho\left(|z_1|^2, \ldots, |z_n|^2\right) < 0\},$$

where $\rho(z)$ is a twice smooth function and $\mathrm{grad}\rho = \left(\frac{\partial \rho}{\partial z_1}, \ldots, \frac{\partial \rho}{\partial z_n}\right) \neq 0$ on ∂D.

Denote $|z_k|^2 = t_k$, $k = 1, \ldots, n$. Then

$$\mathrm{grad}\rho = \left(\frac{\partial \rho}{\partial t_1} \bar{z}_1, \ldots, \frac{\partial \rho}{\partial t_n} \bar{z}_n\right) \neq 0.$$

The function ρ can be choosen so that $|\mathrm{grad}\rho|\big|_{\partial D} = 1$. Let $\nu = \omega\big|_{\partial D}$, and in the case it is not hard to check that (see for example [14, Lemma 3.5]),

$$v = c \sum_{k=1}^{n} \bar{\zeta}_k \frac{\partial \rho}{\partial \bar{\zeta}_k} d\sigma = c \sum_{k=1}^{n} t_k \frac{\partial \rho}{\partial t_k} d\sigma,$$

where $d\sigma$ is the Lebesgue measure on ∂D. In a case of n-circular domain we have $d\sigma = d\sigma_+ \cdot d\sigma'$, where $d\sigma'$ is a measure defined by the form

$$\frac{1}{(2\pi i)^n} \frac{d\zeta_1}{\zeta_1} \wedge \cdots \wedge \frac{d\zeta_n}{\zeta_n},$$

and $d\sigma_+$ is the Lebesque measure on ∂D^+. Hence

$$v = c \sum_{k=1}^{n} t_k \frac{\partial \rho}{\partial t_k} d\sigma_+ \cdot d\sigma'.$$

Set

$$\mu = c \sum_{k=1}^{n} t_k \frac{\partial \rho}{\partial t_k} d\sigma_+. \qquad (3)$$

Lemma 2 *If D is a complete n-circular domain, then μ is a measure on ∂D^+.*

The proof can be found in [20, P. 292].

Corollary 1 *If D is a complete strongly pseudoconvex n-circular domain, then μ is a massive measure on ∂D^+.*

Consider *the modified Poisson kernel*

$$Q(\zeta, z, w) = \frac{h(\bar{\zeta}, z)h(\zeta, w)}{h(w, z)}.$$

Then for $w = \bar{\zeta}$ we get $Q(\zeta, z, \bar{z}) = P(\zeta, z)$ and $h(\bar{z}, z) > 0$. Therefore there exists a neighbored U of the diagonal $w = \bar{z}$ in $D_z \times D_w$ such that $h(w, z) \neq 0$.

Consider a function

$$\Phi(z, w) = c \int_{\partial D} f(\zeta) Q(\zeta, z, w) dv =$$

$$= c \int_{\partial D^+} d\mu \int_{\Delta_{|\zeta|}} f(\zeta) Q(\zeta, z, w) \frac{d\zeta}{\zeta}, \qquad (z, w) \in D \times D.$$

This function is holomorphic in the variables $(z, w) \in U$, and if $w = \bar{z}$ then $\Phi(z, w) = F(z)$ and

$$\frac{\partial^{\delta+\gamma} \Phi(z, w)}{\partial z^\delta \partial w^\gamma}\bigg|_{w=\bar{z}} = \frac{\partial^{\delta+\gamma} F(z)}{\partial z^\delta \partial \bar{z}^\gamma}, \qquad (4)$$

where

$$\frac{\partial^{\delta+\gamma} \Phi(z,w)}{\partial z^\delta \partial w^\gamma} = \frac{\partial^{\delta_1+\dots+\delta_n+\gamma_1+\dots\gamma_n} \Phi(z,w)}{\partial z_1^{\delta_1} \cdots \partial z_n^{\delta_n} \partial w_1^{\gamma_1} \cdots \partial w_n^{\gamma_n}},$$

$$\frac{\partial^{\delta+\gamma} F(z)}{\partial z^\delta \partial \bar{z}^\gamma} = \frac{\partial^{\delta_1+\dots+\delta_n+\gamma_1+\dots\gamma_n} F(z)}{\partial z_1^{\delta_1} \cdots \partial z_n^{\delta_n} \partial \bar{z}_1^{\gamma_1} \cdots \partial \bar{z}_n^{\gamma_n}},$$

and $\delta = (\delta_1, \dots, \delta_n)$, $\gamma = (\gamma_1, \dots, \gamma_n)$.

Let $\zeta = bt$, $b \in \mathbb{CP}^{n-1}$. As it was proved in [15] (see below also [14, Sect. 15])

$$\omega = c\frac{dt}{t} \wedge \lambda(b), \tag{5}$$

where $\lambda(b)$ is a differential form of type $(n-1, n-1)$ independent at t.

From now on we shall assume the existence of a direction $b^0 \neq 0$ such that

$$\langle b^0, \bar{\zeta} \rangle \neq 0 \quad \text{for all} \quad \zeta \in \overline{D}. \tag{6}$$

Denote by $\mathcal{L}_\Gamma$ the set of all complex lines of the form

$$\ell_{z,b} = \{\zeta \in \mathbb{C}^n : \zeta_j = z_j + b_j t, \ j = 1, \dots, n, \ t \in \mathbb{C}\}, \tag{7}$$

passing through a point $z \in \Gamma$ in the direction of vector $b \in \mathbb{CP}^{n-1}$ (the direction b is defined up to multiplication to a complex number $\lambda \neq 0$).

By Sard's Theorem for almost all points $z \in \mathbb{C}^n$ and for a fixed point $b \in \mathbb{CP}^{n-1}$ the intersection $\ell_{z,b} \cap \partial D$ consists of finite number of piecewise-smooth curves (except for the degenerate case $\partial D \cap \ell_{z,b} = \emptyset$).

It is known that if $f \in \mathcal{L}^p(\partial D)$, $p \geq 1$, then for almost all $z \in D$ and almost all $b \in \mathbb{CP}^{n-1}$ the function $f \in \mathcal{L}^p(\partial D \cap \ell_{z,b})$ (see [22]).

We will say that a function $f \in \mathcal{L}^p(\partial D)$ has the *one-dimensional holomorphic extension property along the family* $\ell_{z,b} \in \mathcal{L}_\Gamma$ of the form (7), if for almost all lines $\ell_{z,b}$ such that $\partial D \cap \ell_{z,b} \neq \emptyset$ there exists a function f_ℓ with properties

(1) $f_\ell \in \mathcal{H}^p(D \cap \ell_{z,b})$,
(2) normal boundary values by the metric $\mathcal{H}^p$ of function f_ℓ coincides with f on $\partial D \cap \ell_{z,b}$ almost everywhere.

Consider the Bochner–Martinelli kernel

$$U(\zeta, z) = \frac{(n-1)!}{(2\pi i)^n} \sum_{k=1}^n (-1)^{k-1} \frac{\bar{\zeta}_k - \bar{z}_k}{|\zeta - z|^{2n}} d\bar{\zeta}[k] \wedge d\zeta,$$

where $d\zeta = d\zeta_1 \wedge \dots \wedge d\zeta_n$, and $d\bar{\zeta}[k]$ is obtained from $d\bar{\zeta}$ by deleting the differential $d\bar{\zeta}_k$.

For a function $f \in \mathscr{L}^p(\partial D)$ define the Bochner–Martinelli integral by the following way

$$F(z) = \int\limits_{\partial D_\zeta} f(\zeta)U(\zeta, z), \qquad z \notin \partial D. \tag{8}$$

A function $F(z)$ is harmonic outside of the boundary of domain and tends to zero when $|z| \to \infty$.

A subset $\mathfrak{L}_\Gamma$ is said to be *sufficient for holomorphic extension*, if any function $f \in \mathscr{L}^p(\partial D)$ having the one-dimensional property along almost all complex lines from a family $\mathfrak{L}_\Gamma$, and then the function f extends holomorphically to D as a function of the class $\mathscr{H}^p$.

From now on without loss of generality we assume that $0 \in D$.

Theorem 3 *Let D be a bounded strongly convex n-circular domain and a function $f \in \mathscr{L}^p(\partial D)$ has the one-dimensional holomorphic extension property along complex lines passing through the origin. Then $\Phi(0, w) = const$ and*

$$\left.\frac{\partial^\delta \Phi(z, w)}{\partial z^\delta}\right|_{z=0}$$

is a polynomial in w of degree not higher than $\|\delta\|$.

Proof Let $\ell_{0,b}$ be the line passing through the origin in the direction of a vector $b \in \mathbb{CP}^{n-1}$. Consider

$$Q(bt, z, w) = \frac{h(\bar{b}\bar{t}, z)h(bt, w)}{h(z, w)}.$$

Then

$$h(\bar{b}\bar{t}, z) = \sum_{\alpha \geq 0} a_\alpha (\bar{b}z)^\alpha \bar{t}^{\|\alpha\|} \tag{9}$$

and $h(0, 0) = h(\zeta, 0) = a_0$. Thus

$$\Phi(0, 0) = \int\limits_{\partial D} f(\zeta) \frac{h(\bar{\zeta}, 0)h(\zeta, 0)}{h(0, 0)} dv =$$

$$= \frac{1}{h(0, 0)} \int\limits_{\partial D} f(\zeta)h(\bar{\zeta}, 0)h(\zeta, 0) dv =$$

$$= \frac{c}{h(0, 0)} \int\limits_{\partial D \cap l_{0,b}} \lambda(b) \int\limits_{l_{0,b}} h(\bar{b}\bar{t}, 0)h(bt, 0)\frac{f(bt)}{t} dt =$$

$$= \frac{ca_0^2}{a_0} \int\limits_{\partial D \cap l_{0,b}} \lambda(b) \int\limits_{l_{0,b}} \frac{f(bt)}{t} \, dt = ca_0 \int\limits_{\partial D \cap l_{0,b}} \lambda(b) \int\limits_{l_{0,b}} \frac{f(bt)}{t} \, dt.$$

Let us consider derivatives

$$\frac{\partial^{\delta+\gamma} \Phi(z,w)}{\partial z^\delta \partial w^\gamma} = \frac{\partial^{\delta_1+\ldots+\delta_n+\gamma_1+\ldots\gamma_n} \Phi(z,w)}{\partial z_1^{\delta_1} \cdots \partial z_n^{\delta_n} \partial w_1^{\gamma_1} \cdots \partial w_n^{\gamma_n}},$$

where $\delta = (\delta_1, \ldots, \delta_n)$, $\gamma = (\gamma_1, \ldots, \gamma_n)$. We have

$$\frac{\partial^\gamma \Phi(0,w)}{\partial w^\gamma} = \int\limits_{\partial D} f(\zeta) \frac{\partial^\gamma Q(\zeta,0,w)}{\partial w^\gamma} \, dv =$$

$$= \frac{a_0}{a_0} \int\limits_{\partial D} f(\zeta) \frac{\partial^\gamma h(\zeta,w)}{\partial w^\gamma} \, dv = \int\limits_{\partial D} f(\zeta) \frac{\partial^\gamma h(\zeta,w)}{\partial w^\gamma} \, dv.$$

Now compute

$$\frac{\partial^\gamma h(\zeta,w)}{\partial w^\gamma} = \sum_{\alpha-\gamma \geq 0} a_\alpha d_{\alpha,\gamma} \zeta^\alpha w^{\alpha-\gamma},$$

where $d_{\alpha,\gamma}$ are constants. Then for $w = 0$ we get

$$\left. \frac{\partial^\gamma h(\zeta,w)}{\partial w^\gamma} \right|_{w=0} = a_\alpha d_{\gamma,\gamma} \zeta^\gamma.$$

Thus we obtain that

$$\left. \frac{\partial^\gamma \Phi(0,w)}{\partial w^\gamma} \right|_{w=0} = c \int\limits_{\partial D \cap l_{0,b}} \lambda(b) \int\limits_{l_{0,b}} f(bt) a_\gamma d_{\gamma,\gamma} b^\gamma t^{\|\gamma\|} \frac{dt}{t} = 0$$

for $\|\gamma\| > 0$. This means that $\Phi(0,w) = \text{const}$.
Compute

$$\left. \frac{\partial^{\delta+\gamma} \left(h(\bar\zeta,z) \cdot h(\zeta,w) \right)}{\partial z^\delta \partial w^\gamma} \right|_{\substack{z=0 \\ w=0}} = \left. \frac{\partial^\delta h(\bar\zeta,z)}{\partial z^\delta} \cdot \frac{\partial^\gamma h(\zeta,w)}{\partial w^\gamma} \right|_{\substack{z=0 \\ w=0}} =$$

$$= \delta! a_\delta \bar\zeta^\delta \gamma! a_\gamma \zeta^\gamma = \delta! \gamma! a_\delta a_\gamma \bar\zeta^\delta \zeta^\gamma,$$

where $\delta! = \delta_1! \cdot \ldots \cdot \delta_n!$, $\gamma! = \gamma_1! \cdot \ldots \cdot \gamma_n!$. Then

$$\frac{\partial^{\delta+\gamma} Q(\zeta, z, w)}{\partial z^\delta \partial w^\gamma}\bigg|_{\substack{z=0 \\ w=0}} = \frac{\partial^{\delta+\gamma}}{\partial z^\delta \partial w^\gamma}\left(\frac{h(\bar{\zeta}, z)h(\zeta, w)}{h(z, w)}\right)\bigg|_{\substack{z=0 \\ w=0}} = \sum_{k=1}^{N} C_k \bar{\zeta}^{\delta-\varepsilon^k} \zeta^{\gamma-\varepsilon^k},$$

(10)

because by substituting $z = 0$ and $w = 0$ in derivatives, we can see that derivatives in z and w with different order are equal to zero.

Let $\|\delta\| < \|\gamma\|$. Then from (5) and (10) we get that

$$\frac{\partial^{\delta+\gamma} \Phi(z, w)}{\partial z^\delta \partial w^\gamma}\bigg|_{\substack{z=0 \\ w=0}} = c \int_{\partial D} f(\zeta) \frac{\partial^{\delta+\gamma} Q(\zeta, z, w)}{\partial z^\delta \partial w^\gamma}\bigg|_{\substack{z=0 \\ w=0}} dv =$$

$$= c \sum_{k=1}^{N} C_k \int_{\partial D} f(\zeta) \bar{\zeta}^{\delta-\varepsilon^k} \zeta^{\gamma-\varepsilon^k} dv =$$

$$= c \sum_{k=1}^{N} C_k \int_{\partial D \cap l_{0,b}} \lambda(b) \int_{l_{0,b}} f(bt) \bar{t}^{\|\delta-\varepsilon^k\|} t^{\|\gamma-\varepsilon^k\|-1} dt =$$

$$= c \sum_{k=1}^{N} C_k \int_{\partial D \cap l_{0,b}} \lambda(b) \int_{l_{0,b}} f(bt) |t|^{\|\delta-\varepsilon^k\|} t^{\|\gamma-\delta\|-1} dt =$$

$$= C_\delta \sum_{k=1}^{N} C_k \int_{\partial D \cap l_{0,b}} \lambda(b) \int_{l_{0,b}} f(bt) t^{\|\gamma-\delta\|-1} dt = 0,$$

because the intersection of D and $l_{0,b}$ is a disc. So, derivatives $\frac{\partial^\delta \Phi(z,w)}{\partial z^\delta}\big|_{z=0}$ are polynomials in w of degree not higher than $\|\delta\|$. The proof is complete. $\qquad\square$

For continuous functions Theorem 3 was proved in [19].

5 Main Results

Consider a mapping $\zeta = \chi(\eta) : \overline{B} \to \overline{D}$, where B is the unit ball in $\mathbb{C}^n$ centered at zero taking zero to a $a \in D$. The mapping χ will be constructed as follows:

Consider the complex lines $\lambda_b = \{\eta \in \mathbb{C}^n : \eta = b\tau, \ \tau \in \mathbb{C}\}$ and $\ell_{a,b} = \{\zeta \in \mathbb{C}^n : \zeta = a + bt, \ t \in \mathbb{C}\}$, where $b \in \mathbb{CP}^{n-1}$. The intersection $D_{a,b} = \ell_{a,b} \cap D$ is a strongly convex domain in $\mathbb{C}$, and therefore, there exists a conformal mapping $\chi_b(\tau)$ of the unit ball in $\mathbb{C}$ into $D_{a,b}$, taking $\tau = 0$ to $t = 0$. By Caratheodory's Theorem [25, P. 228] this mapping extends to a homeomorphism of the closed domains. Then to a point $\eta = b\tau \in D \cap \lambda_b$ there is assigned the point $\chi(\eta) = a + b\chi_b(\tau) \in D_{a,b}$.

The proofs of the following two Lemmata can be find in [20].

Lemma 3 *Let D be a bounded strongly convex n-circular domain. Then the mapping $\chi(\eta)$ is a well defined and is diffeomorphism from $\overline{B}$ onto $\overline{D}$ in the class $\mathscr{C}^1$.*

Henceforth, we assume that D is a bounded strongly convex n-circular domain with twice smooth boundary.

Lemma 4 *The derivatives of $\chi(\eta)$ are holomorphic functions in τ for b fixed.*

Lemma 5 *Let the function $f \in \mathscr{L}^p(\partial D)$ have the one-dimensional holomorphic extension property along almost all complex lines passing through the point $a \in D$. Then the function $f^*(\eta) = f(\chi(\eta)) \in \mathscr{L}^p(\partial B)$ and has the one-dimensional holomorphic extension property along almost all complex lines passing through zero.*

Proof Consider a holomorphic extension $f_{a,b}(\zeta)$ of the function f on $D_{a,b}$. Then the function $f_b^*(\eta) = f_{a,b}(\chi_b(\tau))$ is holomorphic in τ in $B \cap \lambda_b$ by the construction of $\chi(\eta)$. The proof is complete.

Performing a change of variables in integral for a function Φ, we obtain

$$\Phi(z, w) = \int_{\partial D} f(\zeta) Q(\zeta, z, w) \, dv(\zeta) =$$

$$= \int_{\partial B} f(\chi(\eta)) Q(\chi(\eta), z, w) \, dv(\chi(\eta)) = \int_{\partial B} f^*(\eta) Q^*(\eta, z, w) \, dv^*(\eta).$$

Consider the form

$$\omega^*(\eta) = \omega(\chi(\eta)) = \sum_{k=1}^{n} (-1)^{k-1} \bar{\chi}_k(\eta) \, d\bar{\chi}(\eta)[k] \wedge d\chi(\eta).$$

By Lemma 4, the form $d\chi(b\tau)$ is holomorphic in τ for fixed b, while the form $d\bar{\chi}(b\tau)[k]$ is antiholomorphic in τ for fixed b.

Lemma 6 *The forms $d\bar{\chi}(b\tau)\big|_{|\tau|=1}$, $k = 1, \ldots, n$ are forms with holomorphic coefficients in τ.*

Theorem 4 *Let $f \in \mathscr{L}^p(\partial D)$ be a function with the one-dimensional holomorphic extension property along complex lines passing through $a \in D$. Then*

$$\frac{\partial^\gamma \Phi(z, w)}{\partial w^\gamma}\bigg|_{\substack{z=a \\ w=\bar{a}}} = 0$$

for $\|\gamma\| > 0$.

Proof Consider the derivative

$$\frac{\partial^\gamma \Phi(z, w)}{\partial w^\gamma}\bigg|_{w=\bar{a}} = \int_{\partial B} f^*(\eta)\frac{\partial^\gamma Q^*(\eta, z, w)}{\partial w^\gamma} \times$$

$$\times \sum_{k=1}^{n}(-1)^{k-1}\bar{\chi}_k(\eta)\, d\bar{\chi}(\eta)[k] \wedge d\chi(\eta)\bigg|_{w=\bar{a}} =$$

$$= \int_{\partial B \cap \lambda_b} \int_{\lambda_b} f^*(bt)\frac{\partial^\gamma Q^*(bt, z, w)}{\partial w^\gamma} \times$$

$$\times \sum_{k=1}^{n}(-1)^{k-1}t^{-1}\psi_k\left(\bar{b}\frac{1}{t}\right) d\chi\left(\bar{b}t^{-1}\right)[k] \wedge d\chi(bt)\bigg|_{w=\bar{a}}.$$

We have

$$\frac{\partial^\beta h(\chi(\eta), w)}{\partial w^\beta}\bigg|_{w=\bar{a}} = \sum_{\substack{\alpha-\beta\geq 0 \\ \|\alpha-\beta\|\neq 0}} a_\alpha d_\beta \chi^\alpha(\eta)\bar{a}^{\alpha-\beta},$$

where d_β are constants. Then

$$\frac{\partial^\gamma Q^*(\eta, z, w)}{\partial w^\gamma}\bigg|_{\substack{z=a \\ w=\bar{a}}} = h(\bar{\eta}, a)\frac{\partial^\gamma}{\partial w^\gamma}\left(\frac{h(\zeta, w)}{h(w, z)}\right)\bigg|_{\substack{z=a \\ w=\bar{a}}} =$$

$$= h(\bar{\eta}, a)\sum_{0\leq\beta\leq\gamma} c_\beta \frac{\frac{\partial^\beta h(\zeta, w)}{\partial w^\beta} \cdot \frac{\partial^{\gamma-\beta} h(w, z)}{\partial w^{\gamma-\beta}}}{h^{\|\gamma\|+1}(w, z)}\bigg|_{\substack{z=a \\ w=\bar{a}}} =$$

$$= \frac{h(\bar{\eta}, a)}{h^{\|\gamma\|+1}(\bar{a}, a)}\sum_{0\leq\beta\leq\gamma} c_\beta \frac{\partial^\beta h((\zeta, w))}{\partial w^\beta} \cdot \frac{\partial^{\gamma-\beta} h(w, a)}{\partial w^{\gamma-\beta}}\bigg|_{w=\bar{a}}.$$

Therefore the right side of $\frac{\partial^\gamma \Phi(z, w)}{\partial w^\gamma}\bigg|_{w=\bar{a}}$ is a linear combination of forms

$$\int_{\partial B \cap \lambda_b} \int_{\lambda_b} f^*(bt)h\left(\frac{\bar{b}}{t}, a\right)\frac{\partial^\beta h(bt, w)}{\partial w^\beta} \cdot \frac{\partial^{\gamma-\beta} h(w, a)}{\partial w^{\gamma-\beta}}\bigg|_{w=\bar{a}}.$$

$$\cdot \sum_{k=1}^{n}(-1)^{k-1}\frac{1}{t}\psi_k\left(\frac{\bar{b}}{t}\right) d\chi\left(\frac{\bar{b}}{t}\right)[k] \wedge d\chi(bt).$$

By Lemma 6 and by the assertion of Theorem these components are equal

$$
\int\limits_{\partial B \cap \lambda_b} \int\limits_{\lambda_b} f^*(bt) \frac{1}{t} h\left(\frac{\bar{b}}{t}, a\right) \cdot \sum_{\substack{\alpha-\beta \geq 0 \\ \|\alpha-\beta\| \neq 0}} a_\alpha d_\beta b^\alpha t^\alpha \bar{a}^{\alpha-\beta} \cdot \frac{\partial^{\gamma-\beta} h(w, a)}{\partial w^{\gamma-\beta}}\bigg|_{w=\bar{a}} \cdot
$$

$$
\cdot \sum_{k=1}^{n} (-1)^{k-1} \psi_k\left(\frac{\bar{b}}{t}\right) d\chi\left(\frac{\bar{b}}{t}\right)[k] \wedge d\chi(bt) = 0
$$

for $\|\gamma\| > 0$. The proof is complete.

Corollary 2 *Under the conditions of Theorem 4 the function* $\Phi(a, w) = const$.

Theorem 5 *Let D be a complete strongly convex n-circular domain with the twice smooth boundary and let* $f(\zeta) \in \mathscr{L}^p(\partial D)$, $a, c \in D$. *Suppose that the function* $\Phi(z, w)$ *satisfied the conditions:*

$$
\Phi(a, w) = const, \quad \Phi(c, w) = const,
$$

and

$$
\frac{\partial^\alpha \Phi(a, w)}{\partial z^\alpha}, \quad \frac{\partial^\alpha \Phi(c, w)}{\partial z^\alpha}
$$

are polynomials in w of degree not higher than $\|\alpha\|$. *Then for any fixed point z on the complex plane*

$$
\ell_{a,c} = \{(z, w) : z = at + c(1 - t), \ w = \bar{a}t + \bar{c}(1 - t), \ t \in \mathbb{C}\}
$$

the equality $\Phi(z, w) = const$ *in w, i.e.,* $\frac{\partial^\gamma \Phi(z,w)}{\partial w^\gamma} = 0$ *at* $\|\gamma\| > 0$.

The proof of this Theorem repeated the proof of Theorem 3 from [20].

Corollary 3 *Under the conditions of Theorem 5 the equality* $\frac{\partial^\gamma F(z)}{\partial \bar{z}^\gamma}\bigg|_{z=at+(1-t)c} = 0$ *holds if* $\|\gamma\| > 0$.

Theorem 6 *Let* $n = 2$ *and the function* $f \in \mathscr{L}^p(\partial D)$ *have the one-dimensional holomorphic extension property along the family* $\mathfrak{L}_{a,c,d}$ *and the points* $a, c, d \in D$ *do not lie on one complex line in* $\mathbb{C}^2$. *Then* $\frac{\partial^\gamma \Phi(z,w)}{\partial w^\gamma} = 0$ *for any* $z \in D$ *and* $\|\gamma\| > 0$, *and the function* f *extends holomorphically into D.*

Proof Let $\tilde{z}$ be an arbitrary point on $\ell_{a,c}$. By Theorem 5 we have

$$
\frac{\partial^\gamma \Phi(\tilde{z}, w)}{\partial w^\gamma} = 0 \tag{11}
$$

for $\|\gamma\| > 0$. Joining $\tilde{z}$ with the point d by the line $\ell_{\tilde{z},d}$ and again applying Theorem 5 with $\tilde{\tilde{z}} \in \ell_{\tilde{z},d}$, we conclude that $\frac{\partial^\gamma \Phi(\tilde{\tilde{z}},w)}{\partial w^\gamma} = 0$ for $\|\gamma\| > 0$. Therefore for all points $\tilde{z}$ in some open subset for condition (11) holds.

Inserting $w = \bar{z}$ in equality (11) and using equality (4), we have $\frac{\partial^{\gamma} F(z)}{\partial \bar{z}^{\gamma}} = 0$ in some open set in D. The real analyticity of the function $F(z)$ implies that $\frac{\partial F(z)}{\partial \bar{z}_j} = 0$ for any $z \in D$ and $j \in \{1, \ldots, n\}$. Since by Theorem 2 we have $F(\zeta)\big|_{\partial D} = f(\zeta)$, then the function $f(\zeta)$ extends holomorphically into D. The proof is complete.

Denote by $\mathfrak{A}$ the set of points $a_k \in D \subset \mathbb{C}^n$, $k = 1, \ldots, n+1$, do not lie on complex hyperplane in $\mathbb{C}^n$.

The following is the main result of this paper.

Theorem 7 *Let the function $f \in \mathscr{L}^p(\partial D)$ have the one-dimensional holomorphic extension property along the family $\mathfrak{L}_{\mathfrak{A}}$. Then $\frac{\partial^{\gamma} \Phi(z,w)}{\partial w^{\gamma}} = 0$ for any $z \in D$ and $\|\gamma\| > 0$, and the function f extends holomorphically into D.*

Proof Proceed by induction on n. The inductions base in Theorem 6 ($n = 2$). Suppose that the theorem holds for all $k < n$. Consider the complex plane Γ passing through points $a_1, \ldots, a_n$, the dimension is by hypothesis equal to $n - 1$ and $a_{n+1} \notin \Gamma$. The intersection $\Gamma \cap D$ is a strongly convex domain in $\mathbb{C}^{n-1}$. The restriction $f\big|_{\Gamma \cap \partial D}$ of f is integrable and has the property of one-dimensional holomorphic extension along the family $\mathfrak{L}_{\mathfrak{A}_1}$, where $\mathfrak{A}_1 = \{a_1, \ldots, a_n\}$. By the induction assumption, $\frac{\partial^{\gamma} \Phi(z',w)}{\partial w^{\gamma}} = 0$ for $\|\gamma\| > 0$ and for all $z' \in \Gamma \cap D$.

Now connecting points $z' \in \Gamma$ with the point a_{n+1}, we find by Theorem 6, that $\frac{\partial^{\gamma} \Phi(z,w)}{\partial w^{\gamma}} = 0$ for some open set in $D \times D$ for all $\|\gamma\| > 0$. Thus, analogously to Theorem 6, we have that $F(z)$ is holomorphic in D, and therefore the function f extends holomorphically into D. The proof is complete.

References

1. Agranovskii, M.L., Val'skii, R.E.: Maxiniality of invariant algebras of functions. Sibirsk. Mat. Zh. **12**(1), 3–12 (1971); English translation in Siberian Math. J. **12** (1971)
2. Agranovsky, M.L.: Propagation of boundary CR-foliations and Morera type theorems for manifolds with attached analytic discs. Adv. Math. **211**, 284–326 (2007)
3. Agranovsky, M.L.: Analog of a theorem of Forelli for boundary values of holomorphic functions on the unit ball of $\mathbb{C}^n$. J. d'Analyse Mathematique **113**, 293–304 (2011)
4. Aizenberg, L.A.: Integral representations of functions holomorphic in n-circular domains (Continuation of Szego kernels). Mat. Sb. **65**, 104–143 (1964)
5. Aizenberg, L.A., Yuzhakov, A.P.: Integral Representations and Residues in Multidimensional Complex Analysis. American Mathematical Society, Providence (1983). (translated from the Russian)
6. Aytuna, A., Sadullaev, A.: S^*-parabolic manifolds. TWMS J. Pure Appl. Math. **2**, 6–9 (2011)
7. Baracco, L.: Holomorphic extension from the sphere to the ball. J. Math. Anal. Appl. **388**, 760–762 (2012)
8. Baracco, L.: Separate holomorphic extension along lines and holomorphic extension from the sphere to the ball. Amer. J. Math. **135**, 493–497 (2013)
9. Egorychev, G.P.: Integral Representation and the Computation of Combinatorial Sums, Nauka, Novosibirsk, 1977 (in Russian). English transl. Transl. Math. Monographs 59, Amer. Math. Soc., Providence. RI 1984; 2nd ed. (1989)

10. Globevnik, J.: Meromorphic extension from small families of circles and holomorphic textension from spheres Trans. Amer. Math. Soc. **364**, 5867–5880 (2012)
11. Globevnik, J.: Small families of complex lines for testing holomorphic extendibility. Amer. J. Math. **134**, 1473–1490 (2012)
12. Globevnik, J., Stout, E.L.: Boundary Morera theorems for holomorphic functions of several complex variables. Duke Math. J. **64**, 571–615 (1991)
13. Henkin, G.M.: The method of integral representations in complex analysis, Complex analysis-several variables **1**, Itogi Nauki i Tekhniki. Ser. Sovrem. Probl. Mat. Fund. Napr., 7, VINITI, Moscow, pp. 23–124 (1985)
14. Kytmanov, A.M.: The Bochner-Martinelli Integral and Its Applications. Birkhauser, Basel (1995)
15. Kytmanov, A.M., Myslivets, S.G.: Higher-dimensional boundary analogs of the Morera theorem in problems of analytic continuation of functions. J. Math. Sci. **120**, 1842–1867 (2004)
16. Kytmanov, A.M., Myslivets, G.: On families of complex lines sufficient for holomorphic extension. Math. Notes **83**, 500–505 (2008)
17. Kytmanov, A.M., Myslivets, S.G., Kuzovatov, V.I.: Minimal dimension families of complex lines sufficient for holomorphic extension of functions. Siberian Math. J. **52**, 256–266 (2011)
18. Kytmanov, A.M., Myslivets, S.G.: Holomorphic continuation of functions along finite families of complex lines in the ball. J. Sib. Fed. Univ. Math. Phys. **5**(4), 547–557 (2012)
19. Kytmanov, A.M., Myslivets, S.G.: On the families of complex lines which are sufficient for holomorphic continuation of functions given on the boundary of the domain. J. Sib. Fed. Univ. Math. Phys. **5**(2), 213–222 (2012)
20. Kytmanov, A.M., Myslivets, S.G.: Holomorphic extension of continuous functions along finite families of complex lines in a ball. J. Sib. Fed. Univ. Math. Phys. **8**(3), 291–302 (2015)
21. Kytmanov, A.M., Myslivets, S.G.: Holomorphic extension of functions along the finite families of complex lines in a ball of $\mathbb{C}^n$. Math. Nahr **288**(2–3), 224–234 (2015)
22. Otemuratov, B.P.: Functions of class L^p with the one-dimensional holomorphic extension property. Vestnik KrasGU **9**, 95–100 (2006)
23. Otemuratov, B.P.: A multidimensional Morera's theorem for integrable functions. Uzbek Math. J. **2**, 112–119 (2009)
24. Otemuratov, B.P.: Some families of complex lines of minimal dimension which are sufficient for holomorphic continuation of integrable functions. J. Sib. Fed. Univ. Math. Phys. **5**(1), 97–105 (2012)
25. Shabat, B.V.: Introduction to Complex Analysis, Part I, 2nd edn. Nauka, Moscow (1976). (in Russian)
26. Stein, E.M., Weiss, G.: Introduction to Fourier Analysis on Euclidean Spaces. Princeton University Press, Princeton (1975)
27. Stout, E.L.: The boundary values of holomorphic functions of several complex variables. Duke Math. J. **44**(1), 105–108 (1977)

On the Number of Real Zeros of Entire Functions of Finite Order of Grows

B. B. Prenov

Abstract Analogues of the Descartes rule and the Budan–Fourier theorem for entire functions of not higher than the first order of the minimal type are obtained.

Keywords Entire function · Real zero · Polynomial · Coefficient

To determine the number of real roots of a real polynomial on a segment of the real axis, the classical Sturm method is frequently used (see, for example, [1]). But this method is rather cumbersome and uncomfortable. Usually the Hermite theorem is used in combination with the rule of the signs of Descartes and the theorem of the Budan–Fourier (see, for example, [1]).

Theorem 1 (Descartes) *Let P(x) be a real polynomial of the form*

$$P(x) = a_0 x^n + a_1 x^{n-1} + \cdots + a_n.$$

Denote by $W(a_0, a_1, \ldots, a_n)$- the number of sign changes in the sequence of coefficients of the polynomial $a_0, a_1, \ldots, a_n$ (zero coefficients are not taken). The number of positive roots $P(x)$ (calculated with their multiplicities) is equal to the number of sign changes in the sequence of non-zero coefficients P (ie $W(a_0, a_1, \ldots, a_n)$), or less than an even number. If all the zeros of $P(x)$ are real, then the number of positive roots is equal to $W(a_0, a_1, \ldots, a_n)$.

Theorem 2 (Budan–Fourier) *Let $P(x)$ be a real polynomial of degree n. The number of real roots of P on the interval $[a, b]$ is equal to the difference*

$$W(P(a), P'(a), \ldots, P^{(n)}(a)) - W(P(b), P'(b), \ldots, P^{(n)}(b)),$$

or less than this value by an even number. (Here it is assumed that $P(a) \neq 0$, $P(b) \neq 0$. If a part of the derivatives is 0 at a or b, then they are $W(P(a), P'(a), \ldots, P^{(n)}(a))$ and $W(P(b), P'(b), \ldots, P^{(n)}(b))$ not taken into account.)

B. B. Prenov (✉)
Nukus State Pedagogical Institute named after Ajiniyaz, Nukus, Uzbekistan
e-mail: prenov@mail.ru

© Springer Nature Switzerland AG 2018   125
Z. Ibragimov et al. (eds.), *Algebra, Complex Analysis, and Pluripotential Theory*,
Springer Proceedings in Mathematics & Statistics 264,
https://doi.org/10.1007/978-3-030-01144-4_11

If all the roots of P are real, then the number of roots of the polynomial P is equal to the number $W(P(a), P'(a), \ldots, P^{(n)}(a)) - W(P(b), P'(b), \ldots, P^{(n)}(b))$.

The proof of these theorems is based on Rolle's theorem.

In the problems of chemical kinetics, systems of non-algebraic equations arise (see, for example, [1–3]). By eliminating unknowns from such systems, we obtain an entire transcendental function (see [4]). Therefore, the problem naturally arises of an analogue of Descartes' rule of signs (Theorem 1) on the number of positive zeros of an entire function, and on the analogue of the Budan–Fourier theorem for entire functions (Theorem 2).

Suppose that an entire function $f(z)$ with real coefficients has the form

$$f(z) = 1 + b_1 z + \cdots + b_n z^n + \ldots \tag{1}$$

Let us denote the number of sign changes in the sequence

$$1, b_1, \ldots b_n, \ldots \tag{2}$$

by W_-, and the number of sign constants in (2) by W_+. If there are zero elements in the sequence (2), then we simply omit them. Thus in (2) we leave only non-zero elements. For an entire transcendental function f, the numbers W_-, W_+ can be finite or infinite.

Proposition 1 *If W is finite, then the function f has finitely many positive zeros.*

Proof Let W_- be finite and equal to m and let the sequence of coefficients has no sign changes after some b_s. Then the sequence of coefficients of the derivative $f^{(s)}$ has no sign changes. Either they are all negative, or all are positive. We assume that they are all positive (recall that the zero coefficients are removed). Then the function $f^{(s)}$ is positive on the positive part of the real axis.

If the function f has an infinite number of positive zeros, then by Rolle's theorem its derivative f' also has an infinite number of positive zeros, and so on. Therefore, the function $f^{(s)}$ must also have an infinite number of positive zeros, which is impossible.

Rolle's theorem also shows that the number of positive zeros (with multiplicities) does not exceed $m + 1$.

Corollary 1 *If the number of positive zeros of f is infinite, then the number of sign changes W_- is also infinite.*

If the number of sign changes W_- is infinite, then the number of positive zeros can be finite.

Example 1 Let $f(z) = 2 - \sin z$. Then the number of sign changes W_- is infinite, but the function $f(z)$ has no real zeros.

For entire functions, there is a refinement of the previous assertions, namely, an analogue of the Descartes rule for polynomials (Theorem 1). It can be found in [5], division [4], Chap. 1, [3].

Theorem 3 *Suppose that for an entire function $f(z)$ of the form (1) with real coefficients the number of positive zeros (counting multiplicities) is equal to N_+, and the number of sign changes in the sequence of Taylor coefficients f is equal to W_-. If the number W_- is finite, then the number N_+ is also finite and the difference $W_- - N_+$ is a nonnegative even number.*

We now discuss the question for which functions f the number of sign changes coincides with the number of positive zeros.

Let all zeros of f are real and that the Hadamard expansion of a function $f(z)$ has the form

$$f(z) = \prod_{j=1}^{\infty} \left(1 - \frac{z}{\beta_j}\right) \tag{3}$$

where β_j are real.

It is satisfied if, for example, $f(z)$ is an entire function of not higher than the first order of the minimal type (see [6], Chap. 7).

Theorem 4 *If a function of the form (3) has a number of changes of sign W_- in the series of Taylor coefficients equal to m, then the number of positive zeros (with multiplicity) is also equal to m. If the Taylor coefficients of the function have an infinite number of sign changes, then the number of positive zeros of $f(z)$ is infinite.*

Proof Let the number W_- be finite and equal to m. The number of positive roots of the function is equal to k (k can be zero) and by Theorem 3 $k \le m$. Let them are equal to $\alpha_1, \ldots, \alpha_k$. If the number of negative roots is also finite, then f is a polynomial and the theorem follows from the rule of signs Descartes (Theorem 1). Let the number of negative roots $f(z)$ are infinite and they are equal to $-\gamma_1, \ldots, -\gamma_s, \ldots$ Consider the polynomial

$$P_s(z) = \prod_{j=1}^{s} \left(1 + \frac{z}{\gamma_j}\right) \prod_{j=1}^{k} \left(1 - \frac{z}{\alpha_j}\right)$$

Then the number of sign changes in the series of coefficients of the polynomial $P_s(z)$ is equal to k by the Descartes Theorem 1 for each s. We denote its coefficients by $b_0^{(s)}, \ldots, b_s^{(s)}, b_{s+1}^{(s)} = 0, \ldots$ ($b_0^{(s)} = 1$). Since the polynomials $P_s(z)$ converge uniformly to the function $f(z)$ for $s \to \infty$, then for each fixed j the sequence $b_j^{(s)} \to b_j$ for m. Therefore, for each sign change in the function $f(z)$, the corresponding coefficients of the polynomial $P_s(z)$ also form a change of sign for sufficiently large s. Therefore, $m \le k$, and $m = k$. Let the number of changes in the sign of W_- be infinite. Then two cases are possible: the number of negative zeros is finite or infinite. Let the number of negative zeros of $f(z)$ be finite and equal to p. If the number of positive roots is finite and equal to k, so that, by Theorem 1 of Descartes, the number of sign changes is k, which contradicts the assumption, then

$$f(z) = \prod_{j=1}^{p} \left(1 + \frac{z}{\gamma_j}\right) \prod_{j=1}^{k} \left(1 - \frac{z}{\alpha_j}\right)$$

Let the number of negative zeros be infinite, and the number of positive roots is finite and equal to k. Then from the preceding argument for sufficiently large s the number of sign changes for the polynomial $P_s(z)$ is equal to k and it is not less than the number of sign changes for the function f. This contradicts the assumption.

The condition that all zeros of an entire function $f(z)$ are real are studied in [7]. For polynomials they are transformed into the classical Hermite theorem (see, for example, [8]).

If the function f is not represented in the form (3), then the previous Theorem 4 is not true.

Example 2 Let $f(z)$ is

$$f(z) = (1 - z)e^{-z}.$$

Its order is equal to 1. In the series of its coefficients there is an infinite number of sign changes. A real root is only one. Therefore Theorem 4 does not hold for the function $f(z)$.

A situation is possible where the canonical product of a function can be a function of not higher than the first order, of minimal type, and the function itself is not. In this case, the question arises as to the factorization of the function, i.e. on the allocation in it of the canonical product. This question was considered in [9].

As a corollary, we obtain a theorem on the number of negative zeros of $f(z)$.

Corollary 2 *Suppose that for an entire function $f(z)$ of the form (3) with real coefficients, the number of negative zeros (with multiplicities) is equal to N_-, and the number of sign changes in the sequence of numbers*

$$1, -b_1, \ldots, (-1)^k b_k, \ldots \tag{4}$$

is equal to W_+. If the number W_+ is finite, then the number N is also finite, and $N = N_-$. If the sequence (4) has an infinite number of sign changes, then the number of negative zeros of $f(z)$ is infinite.

The proof is obtained in an obvious way from Theorem 4 when z is replaced by $-z$.

Note that the number W_+ does not need to coincide with the number of sign constants in the series of Taylor coefficients of the function f, since some place Taylor coefficients may vanish. However, if all the Taylor coefficients of a function are nonzero, then the number of negative roots of $f(z)$ coincides with the number of sign constants in the sequence of Taylor coefficients of the function.

Let us now consider the analogue of the Budan–Fourier theorem for entire functions.

Corollary 3 *Suppose that the function has the form (3), a is a real number and $f(a) \neq 0$. Then the number of real zeros functions f (with their multiplicities) is equal to the number of sign changes in the sequence*

$$f(a), f'(a), \ldots, f^{(k)}(a), \ldots \tag{5}$$

(zero numbers are not taken). This means that if the number of sign changes in (5) is finite and equal to s, then the number of zeros functions f lying to the right of a is also equal to (5). If the number of sign changes in (5) is infinite, then the number of zeros of the function lying to the right of a is also infinite.

The proof is obtained in an obvious way from Theorem (4) when $w = z - a$. We consider the case when $a, b \in \mathscr{R}$, $a < b$ and $f(a) \neq 0$, $f(b) \neq 0$. Note that if for a function of the form (3) the number of sign changes in the sequence S is finite, then the number of sign changes in sequence

$$f(b), f'(b), \ldots, f^{(k)}(b), \ldots \tag{6}$$

also of course, since to the right of a there is only a finite number of zeros of f by Corollary 3, so the number of zeros lying to the right of b is also finite.

Since the entire function can not have an infinite number of zeros on the finite interval $[a, b]$, the converse is also true: if the number of sign changes in the sequence (6) is finite for a function of the form (3), then the number of sign changes in the sequence (5) also of course.

We obtain an obvious result.

Corollary 4 *Suppose that for a function of the form (3) the number of sign changes in the sequence (5) is finite and equal to p, and the number of sign changes in the sequence (6) is finite and equal to q, then the number of zeros of the function on $[a, b]$ (with allowance for their multiplicities) is equal to $p - q$.*

If though the number of sign changes in (5) or in (6) is infinite, then the second one is also infinite. while the number of sign changes on the segment $[a, b]$ can not be said. It can be any finite number.

References

1. Kurosh, A.G.: A Course in Higher Algebra. M., Nauka (1971) (Russian)
2. Bykov, V.I.: Modeling of Critical Phenemenon in Chemical Kinetics. Komkniga, Moscow (2006). (Russian)
3. Bykov, V.I., Tsibenova, S.B.: Nonlinear Models in Chemical Kinetics. KRA-CAND, Moscow (2011)
4. Khodos, O.V.: On some systems of non-algebraic equations in C^n. J. Seberian Fed. Univ., Ser. "Mathematics and Physics" **7**(4), 455–465 (2014)
5. Polya, G., Szego, G.: Aufgaben und lehrsatze aus der analysis. Springer, Berlin (1964)
6. Markushevich, A.I.: Theory of Analytical Functions. V. 2, M., Nauka (1968) (Russian)
7. Kytmanov, A.M., Khodos, O.V.: On localization of zeros of an entire function of finite order of growth. Complex Anal. Oper. Theory **11** (2017). https://doi.org/10.1007/S11785-016-0606-8
8. Gantmaher, F.R.: Theory of matrices. Second supplemented edition. "Nauka", Moscow (1967)
9. Kytmanov, A.M., Naprienko, Ya.M.: One approach to finding the resultant of two entire functions. In: Complex Variables and Elliptic Equations, vol. 62 (2017). https://doi.org/10.1080/17476933.20161218855

On Extensions of Some Classes
of Algebras

Isamiddin Rakhimov

Abstract The paper consists of three parts. In the first part we discuss on extensions of Lie algebras and their importance in Physics. Then we deal with the extensions of some classes of algebras with one binary operation. The third part is devoted to the study of extensions of two classes of algebras, possessing two algebraic operations, called dialgebras. In all the cases we propose 2-cocycles and respective extensions. The motivation to study the extensions is to use them further for the classification problem of the classes algebras considered in low dimensional cases.

Keywords Lie algebra · Associative algebra · Pre-Lie algebra · Leibniz algebra
Zinbiel algebra · Dialgebra

1 Extensions of Lie Algebras

The central extensions of groups and algebras mostly are extensively studied in Lie algebras case and the results are successfully applied in various branches of physics. In the theory of Lie groups, Lie algebras and their representation theory, a Lie algebra extension M is an enlargement of a given Lie algebra L by another Lie algebra K. Extensions arise in several ways. There is the trivial extension obtained by taking a direct sum of two Lie algebras. Other types are the split extension and the central extension. Extensions may arise naturally, for instance, when forming a Lie algebra from projective group representations. It is proven that a finite-dimensional simple Lie algebra has only trivial central extensions. Central extensions are needed in physics, because the symmetry group of a quantized system usually is a central

I. Rakhimov (✉)
Institute for Mathematical Research (INSPEM), Universiti Putra Malaysia,
Serdang, Malaysia
e-mail: risamiddin@gmail.com

© Springer Nature Switzerland AG 2018   131
Z. Ibragimov et al. (eds.), *Algebra, Complex Analysis, and Pluripotential Theory*,
Springer Proceedings in Mathematics & Statistics 264,
https://doi.org/10.1007/978-3-030-01144-4_12

extension of the classical symmetry group, and in the same way the corresponding symmetry Lie algebra of the quantum system is, in general, a central extension of the classical symmetry algebra. Kac-Moody algebras have been conjectured to be a symmetry groups of a unified superstring theory. The centrally extended Lie algebras play a dominant role in quantum field theory, particularly in conformal field theory, string theory and in M-theory.

The central extensions also are applied in the process of pre-quantization, namely in the construction of prequantum bundles in geometric quantization.

Definition 1 A Lie algebra, $\mathfrak{g}$ is a vector space over a field $\mathbb{K}$ with a bilinear map $\lambda : \mathfrak{g} \times \mathfrak{g} \longrightarrow \mathfrak{g}$ (usually, denoted by $[\cdot, \cdot]$) possessing the following properties:

$$[x, x] = 0, \text{ for all } x \in \mathfrak{g},$$
$$[x, [y, z]] + [y, [z, x]] + [z, [x, y]] = 0, \text{ for all } x, y \in \mathfrak{g}.$$

The center $Z(\mathfrak{g})$ of a Lie algebra $\mathfrak{g}$ is defined as follows:

$$Z(\mathfrak{g}) = \{x \in \mathfrak{g} | [x, y] = 0, \text{ for all } y \in \mathfrak{g}\}.$$

Let us start with some introductory considerations on Lie algebra extensions. Loosely speaking an extension of a Lie algebra M is an enlargement of M by another Lie algebra. Starting with two Lie algebras K and M over the same field $\mathbb{K}$, we consider the Cartesian product $L := M \times K$. Elements of this set are ordered pairs (m, k) with $m \in M$ and $k \in K$. Defining addition of such pairs by $(m, k) + (m', k') := (m + m', k + k')$ and multiplication by scalars $\alpha \in \mathbb{K}$ as $\alpha(m, k) := (\alpha m, \alpha k)$ the set L becomes a vector space. Using the Lie brackets on M and K we define bracket on L as follows:

$$[(m, k), (m', k')] := ([m, m'], [k, k']).$$

One easily sees that this is a Lie bracket on L. Defining the maps $\alpha : k \in K \longmapsto (0, k) \in L$ and $\beta : (m, k) \in L \longmapsto m \in M$ one readily verifies that α is an injective, while β is a surjective Lie algebra homomorphisms. Moreover, im $\alpha = \ker \beta$. The Lie algebra L with these properties is called *a trivial extension* of M by K (or of K by M). Instead of denoting the elements of L by ordered pairs we will frequently use the notation $(m, k) = m + k$. The vector space L is then written as $L = M \oplus K$ and the Lie bracket is in this notation given by

$$[m + k, m' + k']_L = [m, m']_M + [k, k']_K.$$

This example of a trivial extension is generalized in the following concept. Let K, L and M be Lie algebras and let these algebras be related in the following way.

- There exists an injective Lie algebra homomorphism

$$\alpha : K \longrightarrow L.$$

- There exists a surjective Lie algebra homomorphism

$$\beta : L \longrightarrow M.$$

- The Lie algebra homomorphisms α and β are related by

$$\text{im } \alpha = \ker \beta.$$

Then the Lie algebra L is called *an extension* of M by K. The relationship is summarized by the sequence

$$K \xrightarrow{\alpha} L \xrightarrow{\beta} M.$$

One obtains that $\ker \beta$ is an ideal in L and since β is surjective we have a Lie algebra isomorphism between the quotient algebra $L/\ker \beta$ and M :

$$L/\ker \beta \cong M.$$

Using $\text{im } \alpha = \ker \beta$ this relation can be written as

$$L/\text{im } \alpha \cong M.$$

Since α is injective the Lie algebras K and $\text{im } \alpha$ are Lie isomorphic. From these properties one sees that it makes sense to call the Lie algebra L is an extension of M by K. The discussion above leads to the following

Proposition 1 *The sequence $K \longrightarrow L \longrightarrow M$ of Lie algebras is an extension of M by K if and only if the sequence*

$$0 \longrightarrow K \xrightarrow{\alpha} L \xrightarrow{\beta} M \longrightarrow 0$$

is exact.

Definition 2 An extension $K \rightarrow L \rightarrow M$ is called:

- **trivial** if there exists an ideal $I \subset L$ complementary to $\ker \beta$, i.e.,

$$L = \ker \beta \oplus I \quad \text{(Lie algebra direct sum)},$$

- **split** if there exists a Lie subalgebra $S \subset L$ complementary to $\ker \beta$, i.e.,

$$L = \ker \beta \oplus S \quad \text{(vector space direct sum)},$$

- **central** if the $\ker \beta$ is contained in the center $Z(L)$ of L, i.e., $\ker \beta \subset Z(L)$.

Here is an example of the split extension of Lie algebras. Let $\mathfrak{g}$ be a Lie algebra over a field $\mathbb{K}$ and $\delta : \mathfrak{g} \longrightarrow \mathfrak{g}$ be a derivation of $\mathfrak{g}$. Define a Lie bracket on the vector space direct sum $\mathfrak{g} \oplus \mathbb{K}\delta$ by

$$[x + \lambda\delta, y + \mu\delta] := [x, y]_{\mathfrak{g}} + \lambda\delta(y) - \mu\delta(x),$$

where $x, y \in \mathfrak{g}$, $\lambda, \mu \in \mathbb{K}$ and $[\cdot, \cdot]_{\mathfrak{g}}$ is the Lie bracket on $\mathfrak{g}$. The vector space $\mathfrak{g} \oplus \mathbb{K}\delta$ equipped with this Lie bracket is a Lie algebra which is denoted by $\mathfrak{g}_\delta$. Notice that the right-hand side of the equation above is always an element of $\mathfrak{g}$, and consequently $\mathfrak{g}$ is an ideal in $\mathfrak{g}_\delta$. Furthermore, $\mathbb{K}\delta$ is a (one-dimensional abelian) subalgebra of $\mathfrak{g}_\delta$ and the vector space $\mathbb{K}\delta$ is complementary to $\mathfrak{g}$. Elements in $\mathfrak{g}_\delta = \mathfrak{g} \oplus \mathbb{K}\delta$ have the form $(x, \mu\delta) = x + \mu\delta$ $(x \in \mathfrak{g}, \mu \in \mathbb{K})$. Defining α and β by $\alpha : x \in \mathfrak{g} \longmapsto (x, 0) \in \mathfrak{g}_\delta$ and $\beta : (x, \mu\delta) \in \mathfrak{g}_\delta \longmapsto \mu\delta \in \mathbb{K}\delta$ one sees that α is injective while β is surjective. Furthermore, im $\alpha = \ker \beta$ and hence the sequence

$$\mathbb{K}\delta \longrightarrow \mathfrak{g}_\delta \longrightarrow \mathfrak{g}$$

is an extension of $\mathfrak{g}$ by $K = \mathbb{K}\delta$. Since $\mathfrak{g}_\delta = \ker \beta \oplus \mathbb{K}\delta$ with $\ker \beta$ an ideal and $\mathbb{K}\delta$ a subalgebra, we have a split extension

$$\mathfrak{g}_\delta = \mathfrak{g} \oplus \mathbb{K}\delta,$$

which is called an *extension by a derivation*.

Let us consider a few examples of complex Lie algebras which have important applications in Physics and are related to each other by central extensions. We start with so-called Loop algebras. *The polynomial loop algebra associated to a finite-dimensional Lie algebra* is given as follows. Let $\mathfrak{g}$ be a Lie algebra. Elements of the polynomial loop algebra $\tilde{\mathfrak{g}}$ are linear combinations of elements of the form $P(\lambda)x$ with $x \in \mathfrak{g}$ and $P(t)$ is a Laurent polynomial in the indeterminate t. The commutation relations are given by

$$[\lambda^m x, \lambda^n y]_{\tilde{\mathfrak{g}}} = \lambda^{m+n}[x, y]_{\mathfrak{g}} \quad m, n \in \mathbb{Z}. \tag{1}$$

The affine Kac-Moody algebra $\hat{\mathfrak{g}}$ can be represented by the following generators: $\{e_i\}_{0 \le i \le n}$, $\{f_i\}_{0 \le i \le n}$, $\{h_i\}_{0 \le i \le n}$ and relations

$$[h_i, h_j] = 0, \quad [e_i, f_j] = \delta_{ij}h_i, \quad [h_i, e_j] = c_{ij}e_j, \quad [h_i, f_j] = -c_{ij}f_j,$$

$$(\text{ad}_{e_i})^{1-c_{ij}}(e_j) = 0, \text{ for } i \ne j, \quad (\text{ad}_{f_i})^{1-c_{ij}}(f_j) = 0, \text{ for } i \ne j,$$

where δ_{ij} denotes the Kronecker delta symbol, $C = (c_{ij})_{0 \le i, j \le n}$ is a singular matrix over $\mathbb{Z}$ with the conditions: $c_{ii} = 2$, $i \ne j$ implies $c_{ij} \le 0$, ($c_{ij} = 0$ if and only if $c_{ji} = 0$) and all proper principal minors are strictly positive, i.e.,

$$\det((c_{ij})_{0 \le i, j \le m}) > 0 \text{ for all } m \le n - 1.$$

The matrix $C = (c_{ij})_{0 \le i, j \le n}$ is called the Cartan matrix and

$$\{\{e_i\}_{0 \le i \le n}, \{f_i\}_{0 \le i \le n}, \{h_i\}_{0 \le i \le n}\}$$

are said to be Chevalley generators.

Starting with a polynomial loop algebra over finite-dimensional simple Lie algebra and performing two extensions, a central extension and an extension by a derivation, one obtains a Lie algebra which is isomorphic with an untwisted affine Kac-Moody algebra. Using the centrally extended loop algebra one may construct a current algebra in two space time dimensions.

The Witt algebra is a complex Lie algebra constructed as follows. Let $\mathbb{C}[t, t^{-1}]$ be the algebra of Laurent polynomials over $\mathbb{C}$ and $\partial := d/dt$ be its derivation. Consider the vector space $V = \text{Span}_{\mathbb{C}}\{e_m = t^{m+1}\partial \mid m \in \mathbb{Z}\}$ with the bracket

$$[e_i, e_j] = t^{i+1}\partial(t^{j+1})\partial - t^{j+1}\partial(t^{i+1})\partial = (i - j)e_{i+j}.$$

It is readily to verify that $\mathscr{W} = (V, [\cdot, \cdot])$ is a Lie algebra called *Witt algebra*. The Witt algebra $\mathscr{W}$ is the Lie algebra of vector fields on the unit cycle S^1.

The Virasoro algebra $\mathscr{V}$ over a field $\mathbb{K}$ is defined as follows: it is a vector space V spanned by a basis $\{e_i, \ i \in \mathbb{Z}, \ c\}$, i.e., $V = \text{Span}_{\mathbb{K}}\{e_i, \ i \in \mathbb{Z}, \ c\}$ such that

$$V = \bigoplus_{i \in \mathbb{Z}} \mathbb{K}e_i \oplus \mathbb{K}c$$

with the following commutation relations:

$$[e_i, e_j] = (j - i)e_{j+i} + \frac{1}{12}(j^3 - j)\delta_{j+i,0}c, \quad i, j \in \mathbb{Z}, \text{ and } [V, c] = 0,$$

where δ_{ij} is the Kronecker delta symbol. One easily can check that $\mathscr{V} = (V, [\cdot, \cdot])$ is a Lie algebra.

The Virasoro algebra $\mathscr{V}$ is the central extension of the Witt algebra $\mathscr{W}$ and is used to establish the boson-fermion correspondence in Physics.

2 Extensions of Some Classes of Algebras with One Operation

In the paper for some classes of algebras we fix the following notations.

- Associative algebra $(A, \cdot)$;
- Lie algebra $(\mathfrak{g}, [\cdot, \cdot])$;
- Pre-Lie algebra $(\mathbf{g}, [\cdot, \cdot])$;
- Leibniz algebra $(L, [\cdot, \cdot])$;
- Zinbiel algebra $(R, \circ)$;
- Associative dialgebra $(D, \dashv, \vdash)$;

• Dendriform algebra $(E, \prec, \succ)$.

Definition 3 An algebra $\mathfrak{A}$ over a field $\mathbb{K}$ is a vector space V over $\mathbb{K}$ equipped by a bilinear binary function $\lambda : V \times V \longrightarrow V$.

The vector space V is called underlying vector space of the algebra $\mathfrak{A}$, as sets they are the same.

Definition 4 An algebra A is said to be associative if its bilinear map $\lambda : V \times V \to V$ satisfies the associative law, i.e.,

$$\lambda(\lambda(x, y), z) = \lambda(x, \lambda(y, z)) \text{ for all } x, y, z \in V.$$

Further the notation $x \cdot y$ (even just xy) will be used for $\lambda(x, y)$.

Definition 5 A pre-Lie algebra $\mathbf{g}$ is a vector space V (over $\mathbb{K}$) with a binary operation $V \times V \longrightarrow V$ $((x, y) \longmapsto [x, y])$ satisfying

$$[[x, y], z] - [x, [y, z]] = [[y, x]z] - [y, [x, z]], \text{ for all } x, y, z \in \mathbf{g}.$$

Remark 1 • The term of "pre-Lie algebra" is due to its close relations with Lie algebras.
• Pre-Lie algebras have several other names. For example, left-symmetric algebra. right-symmetric algebra, quasi-associative algebra (due to identity that they satisfy), Vinberg algebra or Koszul algebra or Koszul-Vinberg algebra. The later are due to a pioneer work of Koszul [10] in the study of affine manifolds and a paper on affine structures on Lie groups by Vinberg [20].

Definition 6 A Leibniz algebra, L is a vector space over a field $\mathbb{K}$ equipped with a bilinear map $[\cdot, \cdot] : L \times L \longrightarrow L$ $(\lambda = [\cdot, \cdot])$ satisfying the Leibniz identity:

$$[x, [y, z]] = [[x, y], z] - [[x, z], y] \text{ for all } x, y, z \in L.$$

Note that under the condition of antisymmetricity $[x, y] = -[y, x]$ the Leibniz identity is transferred to the Jacobi identity, hence a Leibniz algebra is a generalization of Lie algebra.

Definition 7 Zinbiel algebra R is an algebra with a binary operation $\circ : R \times R \to R$, satisfying the condition:

$$(x \circ y) \circ z = x \circ (y \circ z) + x \circ (z \circ y), \text{ for all } x, y, z \in R.$$

Zinbiel algebras have been studied in [6], where the authors proved that any finite-dimensional Zinbiel algebra is nilpotent and the lists of isomorphism classes of Zinbiel algebras in dimension three have been given.

3 Cocycles

Let $\mathfrak{A}$ be an algebra and V be a vector space (both over the same field $\mathbb{K}$). A bilinear function $\theta : \mathfrak{A} \times \mathfrak{A} \longrightarrow V$ is said to be

- an associative 2-cocycle of $\mathfrak{A} = A$ with values in V if

$$\theta(x \cdot y, z) = \theta(x, y \cdot z) \text{ for all } x, y, z \in A.$$

- a Lie 2-cocycle of $\mathfrak{A} = \mathfrak{g}$ with values in V if

$$\theta(x, y) = -\theta(y, x)$$
$$\theta([x, y], z) + \theta([y, z], x) + \theta([z, x], y) = 0, \text{ for all } x, y, z \in \mathfrak{g}.$$

- a Pre-Lie 2-cocycle of $\mathfrak{A} = \mathbf{g}$ with values in V if

$$\theta([x, y], z) - \theta(x, [y, z]) = \theta([x, z], y) - \theta(x, [z, y]),$$

for all $x, y, z \in \mathbf{g}$.

- a Leibniz 2-cocycle of $\mathfrak{A} = L$ with values in V if

$$\theta([x, y], z) = \theta([x, z], y) + \theta(x, [y, z]),$$

for all $x, y, z \in L$.
- a Zinbiel 2-cocycle of $\mathfrak{A} = R$ with values in V if

$$\theta(x \circ y, z) = \theta(x, z \circ y) + \theta(x, y \circ z),$$

for all $x, y, z \in R$.

The set of all 2-cocycles on an algebra $\mathfrak{A}$ with values in V is denoted by $Z^2(\mathfrak{A}, V)$.

Let $\nu : \mathfrak{A} \longrightarrow \mathfrak{A}$ be a linear map. Define $\eta(x, y) = \nu(\lambda(x, y))$. Then η is a 2-cocycle called coboundary. The set of all 2-coboundaries on $\mathfrak{A}$ with values in V is denoted by $B^2(\mathfrak{A}, V)$ and $H^2(\mathfrak{A}, V) = Z^2(\mathfrak{A}, V)/B^2(\mathfrak{A}, V)$ is called the second group of cohomologies with values in V.

Theorem 1 *Let $(\mathfrak{A}, \lambda)$ be an algebra, (where $\mathfrak{A} = A$, $\mathfrak{g}$, $\mathbf{g}$, L or R,) and V a vector space, $\theta : \mathfrak{A} \times \mathfrak{A} \longrightarrow V$ be bilinear map. Put $\mathfrak{A}_\theta = \mathfrak{A} \oplus V$. For $x, y \in \mathfrak{A}$ and $v, w \in V$ we define*

$$\lambda_\theta((x + v), (y + w)) = \lambda(x, y) + \theta(x, y).$$

Then $(\mathfrak{A}, \lambda_\theta)$ is an associative, Lie, Pre-Lie, Leibniz or Zinbiel algebra if and only if θ is the corresponding associative, Lie, Leibniz, Pre-Lie or Zinbiel 2-cocycle, respectively.

Proof The proof can be carried out by a straightforward verification.

The main purpose to consider the extensions is the construction of algebras in higher dimensions having the list of algebras in fixed dimension. Some specific classes of algebras can be obtained by using the extensions. This method of classification of nilpotent algebras for associative algebras is given in [5], for Lie algebras it is an objective of [4, 7, 19] in various cases. Application of the extension method for subclasses of Pre-Lie algebras and Zinbiel algebras cases are under consideration. For applications of the method to classify some subclasses of Leibniz algebras the reader is refered to [15–17]. We also remind some applications of the method to Jordan and Malcev algebras [8, 9].

4 Dialgebras

Definition 8 A dialgebra $\mathfrak{A} = (V, \lambda_1, \lambda_2)$ is a vector space V over a field $\mathbb{K}$ equipped with two bilinear maps $\lambda_1 : V \times V \to V$ and $\lambda_2 : V \times V \to V$ (sometimes, λ_1 and λ_2 also are called left and right products in $\mathfrak{A}$, respectively).

Let $(\mathfrak{A}_1, \lambda_1, \lambda_2)$ and $(\mathfrak{A}_2, \mu_1, \mu_2)$ be two dialgebras with binary operations λ_1, λ_2 and μ_1, μ_2, respectively. A function $f : \mathfrak{A}_1 \longrightarrow \mathfrak{A}_2$ is a *homomorphism* if

$$f(\lambda_1(x, y)) = \mu_1(f(x), f(y)) \text{ and } f(\lambda_2(x, y)) = \mu_2(f(x), f(y)),$$

for all $x, y \in \mathfrak{A}_1$.

The kernel and the image of a homomorphism is defined naturally. A bijective homomorphism is called an *isomorphism*. Main problem in structural theory of algebras is the problem of classification. The classification means the description of the orbits under base change linear transformations and list representatives of the orbits. An isomorphism of an algebra to itself is said to be an *automorphism*. The set of all automorphisms of an algebra $\mathfrak{A}$ forms a group with respect to the composition operation. The automorphism group is denoted by $\operatorname{Aut}\mathfrak{A}$.

The subset of $(\mathfrak{A}, \lambda_1, \lambda_2)$ defined by

$$K(\mathfrak{A}) = \{x \in \mathfrak{A} \mid \lambda_i(x, y) = \lambda_i(y, x) = 0 \text{ for all } y \in \mathfrak{A} \text{ and } i = 1, 2\}$$

is said to be the *annihilator* of $\mathfrak{A}$. It is an ideal of $\mathfrak{A}$.

Definition 9 A subspace S of a dialgebra $(\mathfrak{A}, \lambda_1, \lambda_2)$ is said to be a subalgebra, if

$$\lambda_i(S, S) \subset S \text{ i.e., } \lambda_i(x, y) \in S \text{ for all } x, y \in S.$$

Definition 10 A subspace I of a dialgebra $(\mathfrak{A}, \lambda_1, \lambda_2)$ is called a two-sided ideal of $\mathfrak{A}$ if

$$\lambda_i(I, \mathfrak{A}) \subset I \text{ and } \lambda_i(\mathfrak{A}, I) \subset I \text{ for } i = 1, 2.$$

In this section we deal with two classes of dialgebras introduced by Loday. They were called Dendriform and Diassociative algebras. Depending on properties of λ_1 and λ_2 we specify two classes of algebras. In [11] Loday has given motivations to introduce and applications of them in difference areas of algebra (also see [12]).

Definition 11 An associative dialgebra, (or diassociative algebra), D over a field $\mathbb{K}$ is a $\mathbb{K}$-module D equipped with two bilinear binary operation maps

$$\lambda_1 : D \times D \longrightarrow D, \qquad \lambda_2 : D \times D \longrightarrow D,$$

satisfying the following axioms:

$$\lambda_1(\lambda_1(x, y), z) = \lambda_1(x, \lambda_1(y, z)),$$
$$\lambda_2(\lambda_2(x, y), z) = \lambda_2(x, \lambda_2(y, z)),$$
$$\lambda_1(\lambda_2(x, y), z) = \lambda_2(x, \lambda_1(y, z)),$$
$$\lambda_1(\lambda_1(x, y), z) = \lambda_1(x, \lambda_2(y, z)),$$
$$\lambda_2(x, \lambda_2(y, z)) = \lambda_2(\lambda_1(x, y), z),$$

for all $x, y, z \in D$.

For a convenience we write $\lambda_1(x, y) = x \dashv y$ and $\lambda_2(x, y) = x \vdash y$. In these notations the axioms above are rewritten as follows:

$$
\begin{aligned}
(x \dashv y) \dashv z &= x \dashv (y \dashv z), \\
(x \vdash y) \vdash z &= x \vdash (y \vdash z), \\
(x \vdash y) \dashv z &= x \vdash (y \dashv z), \\
(x \dashv y) \dashv z &= x \dashv (y \vdash z), \\
x \vdash (y \vdash z) &= (x \dashv y) \vdash z.
\end{aligned}
\tag{2}
$$

Remark 2 Observe that an analogue of the formula $(x \vdash y) \dashv z = x \vdash (y \dashv z)$, for the product symbols pointing outward, is not valid in general, i.e., : $(x \dashv y) \vdash z \neq x \dashv (y \vdash z)$.

J.-L. Loday, his colleagues have constructed and studied the (co)homology theory for diassociative algebras. Since an associative algebra is a particular case of the diassociative algebra, one gets a new (co)homology theory for associative algebras as well. The natural extensions of the concepts of nilpotency and solvability of algebras to diassociative algebras case have been given in [2].

Definition 12 A dendriform algebra, E over $\mathbb{K}$ is a $\mathbb{K}$-module E equipped with two binary operations $\lambda_1 = \prec$, $\lambda_2 = \succ$

$$\prec : E \times E \longrightarrow E, \qquad \succ : E \times E \longrightarrow E,$$

satisfying the following axioms:

$$(x \prec y) \prec z = x \prec (y \prec z) + x \prec (y \succ z),$$
$$(x \succ y) \prec z = x \succ (y \prec z),$$
$$(x \prec y) \succ z + (x \succ y) \succ z = x \succ (y \succ z)$$

for all $x, y, z \in E$.

The concept of a dendriform algebra was introduced by Loday [13] with motivation from algebraic $K-$theory, and has been further studied with connections to several areas in Mathematics and Physics, including operads, homology, Hopf algebras, Lie and Leibniz algebras, combinatorics and quantum field theory. In [13], Loday showed relationships of Dendriform algebras with other classes, such as Zinbiel, Associative, Diassociative, Lie and Leibniz algebras. Besides of Loday's motivations, the key point is the intimate relation between the so-called Rota-Baxter algebras and such Dendriform algebras. Rota-Baxter algebra is an algebra $\mathfrak{A}$ over a field $\mathbb{K}$ with a linear endomorphism β satisfying the Rota-Baxter relation:

$$\beta(x)\beta(y) = \beta(\beta(x)y + x\beta(y)) + a\beta(xy), \forall x, y \in \mathfrak{A}. \tag{3}$$

where a is a fixed element of $\mathbb{K}$. The map β is called a Rota-Baxter operator of weight a. Associative Rota-Baxter algebras arise in many mathematical contexts, i.e., in integral and finite differences calculus, but also in perturbative renormalization in quantum field theory. The Rota-Baxter algebra was introduced by Baxter [3] and was popularized mainly by Rota [18] and his colleagues.

In [13] Loday has noted that the operad of Dendriform algebras is the Koszul dual of the one of Associative dialgebras. The Dendriform algebra characterizes an associative multiplication, i.e., the sum of two products, $x * y := x \prec y + x \succ y$, is associative. According to [1], one may associate a Dendriform algebra to an Associative algebra equipped with a Rota-Baxter operator, i.e., it was shown that if $\beta : \mathfrak{A} \to \mathfrak{A}$ is a Rota-Baxter operator on an associative algebra $\mathfrak{A}$ then the operations $x \prec y := \beta(x)y$ and $x \succ y := x\beta(y)$ equip $\mathfrak{A}$ with a Dendriform algebra structure [3].

5   Extensions and Cocycles

5.1   Extensions of Dialgebras

Now we introduce the concept of annihilator extension of associative dialgebras.

Definition 13 Let $\mathfrak{A}_1$, $\mathfrak{A}_2$ and $\mathfrak{A}_3$ be dialgebras over a field $\mathbb{K}$. The dialgebra $\mathfrak{A}_2$ is called an **extension** of $\mathfrak{A}_3$ by $\mathfrak{A}_1$ if there are dialgebra homomorphisms $\alpha : \mathfrak{A}_1 \longrightarrow \mathfrak{A}_2$ and $\beta : \mathfrak{A}_2 \longrightarrow \mathfrak{A}_3$ such that the following sequence

$$0 \longrightarrow \mathfrak{A}_1 \xrightarrow{\alpha} \mathfrak{A}_2 \xrightarrow{\beta} \mathfrak{A}_3 \longrightarrow 0$$

is exact.

Definition 14 An extension is called **trivial** if there exists an ideal I of $\mathfrak{A}_2$ complementary to $\ker \beta$, i.e.,

$$\mathfrak{A}_2 = \ker \beta \oplus I \quad \text{(the direct sum of algebras)}.$$

Definition 15 Two sequences

$$0 \longrightarrow \mathfrak{A}_1 \xrightarrow{\alpha} \mathfrak{A}_2 \xrightarrow{\beta} \mathfrak{A}_3 \longrightarrow 0$$

and
$$\| \qquad \downarrow f \qquad \|$$

$$0 \longrightarrow \mathfrak{A}_1 \xrightarrow{\alpha'} \mathfrak{A}'_2 \xrightarrow{\beta'} \mathfrak{A}_3 \longrightarrow 0$$

are called **equivalent extensions** if there exists a dialgebra isomorphism $f : \mathfrak{A}_2 \longrightarrow \mathfrak{A}'_2$ such that

$$f \circ \alpha = \alpha' \text{ and } \beta' \circ f = \beta.$$

The equivalence of extensions is an equivalent relation.

Definition 16 An extension

$$0 \longrightarrow \mathfrak{A}_1 \xrightarrow{\alpha} \mathfrak{A}_2 \xrightarrow{\beta} \mathfrak{A}_3 \longrightarrow 0 \tag{4}$$

is called an **annihilator** (the term **central** also is used) extension if the kernel of β is contained in the annihilator $K(\mathfrak{A}_2)$ of $\mathfrak{A}_2$, i.e., $\ker \beta \subset K(\mathfrak{A}_2)$.

From the definition above it is easy to see that in the extension the algebra $\mathfrak{A}_1$ must be abelian. A central extension of a dialgebra $\mathfrak{A}_2$ by an abelian dialgebra $\mathfrak{A}_1$ can be obtained with the help of 2-cocycles on $\mathfrak{A}_3$. (For extensions of associative dialgebras also see [14]).

5.2 Cocycles on Dialgebras

This section introduces the concept of 2-cocycle for dialgebras and gives some simple but important properties of the 2-cocycles.

Definition 17 Let E be a dendriform algebra over a field $\mathbb{K}$ and V be a vector space over the same field. A pair $\Theta = (\theta_1, \theta_2)$ of bilinear maps $\theta_1 : E \times E \longrightarrow V$ and $\theta_2 : E \times E \longrightarrow V$ is called a **2-cocycle** on E with values in V if θ_1 and θ_2 satisfy the conditions

$$\theta_1(x \prec y, z) = \theta_1(x, y \prec z) + \theta_1(x, y \succ z),$$
$$\theta_2(x \succ y, z) + \theta_2(x \prec y, z) = \theta_2(x, y \succ z), \qquad (5)$$
$$\theta_1(x \succ y, z) = \theta_2(x, y \prec z),$$

for all $x, y, z \in E$.

Definition 18 Let D be an associative dialgebra over a field $\mathbb{K}$ and V be a vector space over the same field. A pair $\Theta = (\theta_1, \theta_2)$ of bilinear maps $\theta_1 : D \times D \longrightarrow V$ and $\theta_2 : D \times D \longrightarrow V$ is called a 2-**cocycle** on D with values in V if θ_1 and θ_2 satisfy the conditions

$$\theta_1(x \dashv y, z) = \theta_1(x, y \dashv z) = \theta_1(x, y \vdash z),$$
$$\theta_2(x \vdash y, z) = \theta_2(x, y \vdash z) = \theta_2(x \dashv y, z), \qquad (6)$$
$$\theta_1(x \vdash y, z) = \theta_2(x, y \dashv z),$$

for all $x, y, z \in D$.

Let $(\mathfrak{A}, \lambda_1, \lambda_2)$ be a dialgebra, where $\lambda_1 = \dashv$, $\lambda_2 = \vdash$ for $\mathfrak{A} = D$ and $\lambda_1 = \prec$, $\lambda_2 = \succ$ for $\mathfrak{A} = E$. The set of all 2-cocycles on $\mathfrak{A}$ with values in V is denoted by $Z^2(\mathfrak{A}, V)$. One easily sees that $Z^2(\mathfrak{A}, V)$ is a vector space if one defines the vector space operations as follows:

$$(\Theta_1 \oplus \Theta_2)(x, y) := \Theta_1(x, y) + \Theta_2(x, y) \text{ and } (\lambda \odot \Theta)(x) := \lambda\Theta(x).$$

It is easy to see that a linear combination of 2-cocycles again is a 2-cocycle.

Further all statements we write for dialgebra $\mathfrak{A}$ but the proofs of them are given for associative dialgebras D case only since for dendriform algebras E the proofs can be easily obtained by small modifications of that given for associative dialgebras applying dendriform algebra and 2-cocycle axioms.

A special type of 2-cocycles given by the following lemma is called 2-**coboundaries**.

Lemma 1 *Let* $v : \mathfrak{A} \longrightarrow V$ *be a linear map, define* $\varphi_1(x, y) = v(\lambda_1(x, y))$ *and* $\varphi_2(x, y) = v(\lambda_2(x, y))$. *Then* $\Phi = (\varphi_1, \varphi_2)$ *is a cocycle.*

Proof Let us check the axioms (6) one by one.

$$\varphi_1(x \dashv y, z) = v((x \dashv y) \dashv z)$$
$$= v(x \dashv (y \dashv z)) = \varphi_1(x, y \dashv z).$$

$$\varphi_1(x, y \dashv z) = v(x \dashv (y \dashv z))$$
$$= v(x \dashv (y \vdash z)) = \varphi_1(x, y \vdash z).$$

$$\varphi_2(x \vdash y, z) = v((x \vdash y) \vdash z)$$
$$= v(x \vdash (y \vdash z)) = \varphi_2(x, y \vdash z).$$

$$\varphi_2(x, y \vdash z) = \nu(x \vdash (y \vdash z))$$
$$= \nu((x \vdash y) \vdash z)$$
$$= \nu((x \dashv y) \vdash z) = \varphi_2(x \dashv y, z).$$

$$\varphi_1(x \vdash y, z) = \nu((x \vdash y) \dashv z)$$
$$= \nu(x \vdash (y \dashv z)) = \varphi_2(x, y \dashv z).$$

The set of all coboundaries is denoted by $B^2(\mathfrak{A}, V)$. Clearly, $B^2(\mathfrak{A}, V)$ is a subgroup of $Z^2(\mathfrak{A}, V)$. The group $H^2(\mathfrak{A}, V) = Z^2(\mathfrak{A}, V)/B^2(\mathfrak{A}, V)$ is said to be a second cohomology group of $\mathfrak{A}$ with values in V. Two cocycles Θ_1 and Θ_2 are said to be **cohomologous** cocycles if $\Theta_1 \ominus \Theta_2$ is a coboundary. If we view V as a trivial $\mathfrak{A}$-bimodule, then $H^2(\mathfrak{A}, V)$ is an analogue of the second Hochschild-cohomology space.

Theorem 2 *Let $\mathfrak{A}$ be a dialgebra over a field $\mathbb{K}$, V a vector space over $\mathbb{K}$,*

$$\theta_1 : \mathfrak{A} \times \mathfrak{A} \longrightarrow V \ and \ \theta_2 : \mathfrak{A} \times \mathfrak{A} \longrightarrow V$$

be bilinear maps. Set $\mathfrak{A}_\Theta = \mathfrak{A} \oplus V$, where $\Theta = (\theta_1, \theta_2)$. For $x, y \in \mathfrak{A}$, $v, w \in V$ we define

$$\lambda_{1,\Theta}((x + v), (y + w)) = \lambda_1(x, y) + \theta_1(x, y)$$

and

$$\lambda_{2,\Theta}((x + v), (y + w)) = \lambda_2(x, y) + \theta_2(x, y).$$

Then the dialgebra $(\mathfrak{A}_\Theta, \lambda_{1,\Theta}, \lambda_{2,\Theta})$ is an associative dialgebra (dendriform algebra) if and only if $\Theta = (\lambda_{1,\Theta}, \lambda_{2,\Theta})$ is an associative dialgebra 2-cocycle (dendriform algebra 2-cocycle).

Proof Let $\mathfrak{A}_\Theta$ be the diassociative algebra defined above, we denote $\lambda_{1,\Theta} = \dashv_\Theta$, $\lambda_{2,\Theta} = \vdash_\Theta$ and show that the functions θ_1, θ_2 are 2-cocycles. Indeed, according to the axioms (2) we have

$$((x + v) \dashv_\Theta (y + w)) \dashv_\Theta (z + t) = (x + v) \dashv_\Theta ((y + w) \dashv_\Theta (z + t))$$
$$= (x + v) \dashv_\Theta ((y + w) \vdash_\Theta (z + t)).$$
$$((x + v) \vdash_\Theta (y + w)) \vdash_\Theta (z + t) = (x + v) \vdash_\Theta ((y + w) \vdash_\Theta (z + t))$$
$$= ((x + v) \dashv_\Theta (y + w)) \vdash_\Theta (z + t).$$
$$((x + v) \vdash_\Theta (y + w)) \dashv_\Theta (z + t)) = (x + v) \vdash_\Theta ((y + w) \dashv_\Theta (z + t)).$$

Since

$$((x + v) \dashv_\Theta (y + w)) \dashv_\Theta (z + t) = ((x \dashv y) + \theta_1(x, y)) \dashv_\Theta (z + t)$$
$$= ((x \dashv y) \dashv z) + \theta_1(x \dashv y, z).$$

$$(x + v) \dashv_\Theta ((y + w)) \dashv_\Theta (z + t)) = (x + v) \dashv_\Theta ((y \dashv z) + \theta_1(y, z))$$
$$= (x \dashv (y \dashv z)) + \theta_1(x, y \dashv z).$$

$$(x + v) \dashv_\Theta ((y + w) \vdash_\Theta (z + t)) = (x + v) \dashv_\Theta ((y \vdash z) + \theta_2(y, z))$$
$$= (x \dashv (y \vdash z)) + \theta_1(x, y \vdash z).$$

Comparing these equations we get

$$\boxed{\theta_1(x \dashv y, z) = \theta_1(x, y \dashv z) = \theta_1(x, y \vdash z).}$$

$$((x + v) \vdash_\Theta (y + w)) \vdash_\Theta (z + t) = ((x \vdash y) + \theta_2(x, y)) \vdash_\Theta (z + t)$$
$$= ((x \vdash y) \vdash z) + \theta_2(x \vdash y, z).$$
$$(x + v) \vdash_\Theta ((y + w)) \vdash_\Theta (z + t)) = (x + v) \vdash_\Theta ((y \vdash z) + \theta_2(y, z)) \quad (7)$$
$$= (x \vdash (y \vdash z)) + \theta_2(x, y \vdash z).$$
$$((x + v) \dashv_\Theta (y + w)) \vdash_\Theta (z + t) = (x \dashv y + \theta_1(x, y)) \vdash_\Theta (z + t) \quad (8)$$
$$= (x \dashv y) \vdash z) + \theta_2(x \dashv y, z).$$

Comparing these equations we obtain

$$\boxed{\theta_2(x \vdash y, z) = \theta_2(x, y \vdash z) = \theta_2(x \dashv y, z).}$$

$$((x + v) \vdash_\Theta (y + w)) \dashv_\Theta (z + t)) = (x \vdash y + \theta_2(x, y) \dashv_\Theta (z + t)$$
$$= ((x \vdash y) \dashv z)) + \theta_1(x \vdash y, z).$$

$$(x + v) \vdash_\Theta ((y + w)) \dashv_\Theta (z + t)) = (x + v) \vdash_\Theta ((y \dashv z) + \theta_1(y, z))$$
$$= (x \vdash (y \dashv z)) + \theta_2(x, y \dashv z).$$

Comparing the later two we derive

$$\boxed{\theta_1(x \vdash y, z) = \theta_2(x, y \dashv z).}$$

Lemma 2 *Let Θ be a cocycle and Φ be a coboundary. Then $D_\Theta \cong D_{\Theta \oplus \Phi}$.*

Proof The isomorphism

$$f : D_\Theta \longrightarrow D_{\Theta \oplus \Phi},$$

is given by $f(x + w) = x + v(x) + w$, where $\Phi = (\varphi_1, \varphi_2)$, and $\varphi_1(x, y) = v(x \dashv y)$, $\varphi_2(x, y) = v(x \vdash y)$. First of all we note that f is a bijective linear transformation. The following shows that it obeys the algebraic operations $\dashv_\Theta$ and $\vdash_\Theta$ as well.

Indeed,

$$
\begin{aligned}
f((x + w_1) \dashv_\Theta (y + w_2)) &= f(x \dashv y + \theta_1(x, y)) \\
&= f(x \dashv y) + f(\theta_1(x, y)) \\
&= x \dashv y + v(x \dashv y) + \theta_1(x, y) \\
&= x \dashv y + \varphi_1(x, y) + \theta_1(x, y) \\
&= x \dashv y + (\varphi_1 + \theta_1)(x, y) \\
&= ((x + (\varphi_1 + \theta_1)(x) + w_1)) \dashv_\Theta (y + (\varphi_1 + \theta_1)(y) + w_2) \\
&= f(x + w_1) \dashv_\Theta f(y + w_2).
\end{aligned}
$$

For $\vdash_\Theta$ the proof is carried out similarly.

The proof of the following corollary is straightforward.

Corollary 1 *Let Θ_1, Θ_2 be cohomologous 2-cocycles on a dialgebra $\mathfrak{A}$ and $\mathfrak{A}_1$, $\mathfrak{A}_2$ be extensions constructed with these 2-cocycles, respectively. Then the extensions $\mathfrak{A}_1$ and $\mathfrak{A}_2$ are equivalent. Particularly, a central extension defined by a coboundary is equivalent with a trivial central extension.*

Let $\Theta \in Z^2(\mathfrak{A}, V)$. The subset

$$
\mathrm{Rad}(\Theta) = \{x \in \mathfrak{A} \mid \Theta(x, y) = 0, \text{ for all } y \in \mathfrak{A}\}
$$

is said to be the *radical* of the cocycle Θ. Note that $K(\mathfrak{A}_\Theta) = (\mathrm{Rad}(\Theta) \cap K(\mathfrak{A}) + V$. One has the following simple lemma.

Lemma 3 *Let $\Theta \in Z^2(\mathfrak{A}, V)$, then $\mathrm{Rad}(\Theta) \cap K(\mathfrak{A}) = 0$ if and only if $K(\mathfrak{A}_\Theta) = V$.*

Let V be a k-dimensional vector space with a basis $e_1, e_2, ..., e_k$ and $\Theta = (\theta_1, \theta_2) \in Z^2(\mathfrak{A}, V)$. Then we have $Z^2(\mathfrak{A}, V) \cong Z^2(\mathfrak{A}, \mathbb{K}^k)$, $B^2(\mathfrak{A}, V) \cong B^2(\mathfrak{A}, \mathbb{K}^k)$ and

$$
\Theta(x, y) = (\theta_1(x, y), \theta_2(x, y)) = \left(\sum_{i=1}^{k} \theta_1^i(x, y)e_i, \sum_{i=1}^{k} \theta_2^i(x, y)e_i \right).
$$

Lemma 4 $\Theta = (\theta_1, \theta_2) \in Z^2(\mathfrak{A}, V)$ *if and only if* $\Theta^i = (\theta_1^i, \theta_2^i) \in Z^2(\mathfrak{A}, \mathbb{K})$.

Proof The proof is a straightforward verification of the axioms (6).

5.3 Annihilator Extensions and Cocycles

There is a close relationship between the annihilator extensions and cocycles on dialgebras that is established in this section.

Theorem 3 *There exists one to one correspondence between elements of $H^2(\mathfrak{A}, V)$ and nonequivalents annihilator extensions of the dialgebra $\mathfrak{A}$ by V.*

Proof The fact that for a 2-cocycle Θ the algebra D_Θ is an annihilator extension of D has been proven earlier.

Let us prove the converse, i.e., suppose that D_2 is an annihilator extension of D_3:

$$0 \longrightarrow D_1 \xrightarrow{\alpha} D_2 \xrightarrow{\beta} D_3 \longrightarrow 0$$

such that im $\alpha = \ker \beta \subset K(D_2)$. We construct a 2-cocycle $\Theta \in Z^2(D, V)$ ($D_2 = D_\Theta$, $D_1 = K(D_\Theta)$ and $D_3 = D$) such that generates this extension. Let us consider bilinear maps

$$\theta_1 : D_3 \times D_3 \longrightarrow D_1 \text{ and } \theta_2 : D_3 \times D_3 \longrightarrow D_1$$

which satisfy (6) for all $x, y, z \in D_3$ ($V = D_1$).

To obtain maps satisfying equations (6) we consider a linear map $s : D_3 \longrightarrow D_2$ satisfying $\beta \circ s = id_{D_3}$. A map with this property is called a *section* of D_2.

With the help of a section one can define a bilinear maps $\vartheta_i : D_3 \times D_3 \longrightarrow D_2$ ($i = 1, 2$) by taking for all $x, y \in D_2$

$$\vartheta_1(x, y) := s(x \dashv y) - s(x) \dashv s(y), \tag{9}$$

$$\vartheta_2(x, y) := s(x \vdash y) - s(x) \vdash s(y). \tag{10}$$

Note that ϑ_i, $i = 1, 2$ is identically zero if s is an associative dialgebra homomorphism. Moreover, since β is an associative dialgebra homomorphism one sees that $\beta \circ \vartheta_i = 0$, $i = 1, 2$. Hence, for all $x, y \in D_3$ we have

$$\vartheta_i(x, y) \in \ker \beta \subset K(D_2). \tag{11}$$

Using this property and the associative dialgebra axioms one readily verifies that $\Omega = (\vartheta_1, \vartheta_2)$ satisfies

$$\begin{aligned}
\vartheta_1(x \dashv y, z) &= \vartheta_1(x, y \dashv z) = \vartheta_1(x, y \vdash z), \\
\vartheta_2(x \vdash y, z) &= \vartheta_2(x, y \vdash z) = \vartheta_2(x \dashv y, z), \\
\vartheta_1(x \vdash y, z) &= \vartheta_2(x, y \dashv z).
\end{aligned} \tag{12}$$

In the final step to obtain the 2-cocycle $\Theta = (\theta_1, \theta_2)$ we use the injectivity of the map $\alpha : D_1 \longrightarrow D_2$. The map $\Theta = (\theta_1, \theta_2)$ is the required cocycle, where $\theta_i : D_3 \times D_3 \longrightarrow D_1$, $i = 1, 2$ are defined by $\theta_i := \alpha^{-1} \circ \vartheta_i$, $i = 1, 2$.

6 Action of Automorphisms

In identifying the central extensions of $\mathfrak{A}$ the action of the $\mathrm{Aut}(\mathfrak{A})$ on the Grassmanian of $H^2(\mathfrak{A}, \mathbb{K})$ is very helpful. The action of $\sigma \in \mathrm{Aut}(\mathfrak{A})$ on $\Theta \in Z^2(\mathfrak{A}, V)$ is defined as follows:

$$\sigma(\Theta(x, y)) = (\sigma(\theta_1(x, y)), \sigma(\theta_2(x, y))),$$

where $\sigma(\theta_i(x, y)) = \theta_i(\sigma(x), \sigma(y))$, $i = 1, 2$, for $x, y \in \mathfrak{A}$. Thus $\mathrm{Aut}(\mathfrak{A})$ operates on $Z^2(\mathfrak{A}, V)$. It is easy to see that $B^2(\mathfrak{A}, V)$ is stable by this action, so that there is an induced action of $\mathrm{Aut}(\mathfrak{A})$ on $H^2(\mathfrak{A}, V)$.

Theorem 4 *Let*

$$\Theta(x, y) = (\theta_1(x, y), \theta_2(x, y)) = \left(\sum_{i=1}^{k} \theta_1^i(x, y)e_i, \sum_{i=1}^{k} \theta_2^i(x, y)e_i \right), \tag{13}$$

$$\Omega(x, y) = (\omega_1(x, y), \omega_2(x, y)) = \left(\sum_{i=1}^{k} \omega_1^i(x, y)e_i, \sum_{i=1}^{k} \omega_2^i(x, y)e_i \right) \tag{14}$$

and $\mathrm{Rad}(\Theta) \cap K(\mathfrak{A}) = \mathrm{Rad}(\Theta) \cap K(\mathfrak{A}) = 0$. *Then* $\mathfrak{A}_\Theta \cong \mathfrak{A}_\Omega$ *if and only if there exists* $\varphi \in \mathrm{Aut}(\mathfrak{A})$ *such that* $\varphi(\omega_j^i)$ *span the same subspace of* $H^2(\mathfrak{A}, \mathbb{K})$ *as the* θ_j^i, $j = 1, 2$ *and* $i = 1, 2, ..., k$.

Proof As vector spaces $D_\Theta = D \oplus V$ and $D_\Omega = D \oplus V$. Let $\sigma : D_\Theta \longrightarrow D_\Omega$ be an isomorphism. Since V is the kernel of both dialgebras, we have $\sigma(V) = V$. So σ induces an isomorphism of $D_\Theta/V = D$ to $D_\Omega/V = D$, i.e., it generates an automorphism φ of D. Let $D = \mathrm{Span}\{x_1, ..., x_n\}$. Then we write $\sigma(x_i) = \varphi(x_i) + v_i$, where $v_i \in V$, and $\sigma(e_i) = \sum_{j=1}^{s} a_{ji}e_j$. Also write

$$x_i \dashv x_j = \sum_{k=1}^{n} \gamma_{ij}^k x_k, \quad x_i \vdash x_j = \sum_{k=1}^{n} \delta_{ij}^k x_k,$$

and $v_i = \sum_{l=1}^{s} \beta_{il}^l e_l$.

Then the relations

$$\sigma(x_i \dashv_\Theta x_j) = \sigma(x_i) \dashv_\Omega \sigma(x_j) \text{ and } \sigma(x_i \vdash_\Theta x_j) = \sigma(x_i) \vdash_\Omega \sigma(x_j)$$

amount to

$$\omega_1^l(\varphi(x_i), \varphi(x_j)) = \sum_{k=1}^{s} a_{lk}\theta_r^k(x_i, x_j) + \sum_{k=1}^{n} \gamma_{ij}^k \beta_{kl}, \text{ for } 1 \leq l \leq s \tag{15}$$

and

$$\omega_2^l(\varphi(x_i), \varphi(x_j)) = \sum_{k=1}^{s} a_{lk}\theta_r^k(x_i, x_j) + \sum_{k=1}^{n} \delta_{ij}^k \beta_{kl}, \text{ for } 1 \le l \le s. \qquad (16)$$

Now define the linear function $f_l : D \longrightarrow \mathbb{K}$ by $f_l(x_k) = \beta_{kl}$. Then

$$f_l^1(x_i \dashv x_j) = \sum_{k=1}^{n} \gamma_{ij}^k \beta_{kl} \text{ and } f_l^2(x_i \vdash x_j) = \sum_{k=1}^{n} \delta_{ij}^k \beta_{kl}.$$

From (15) and (16) it is obvious that modulo $B^2(D, \mathbb{K})$, the cocycles $\varphi(\omega_r^i)$ and θ_r^i span the same space.

Let now suppose that there exists an automorphism $\varphi \in \mathrm{Aut}(D)$ such that the cocycles $\varphi(\omega_j^i)$ and θ_j^i, $j = 1, 2$ and $i = 1, 2, ..., k$ span the same space in $Z^2(D, \mathbb{K})$, modulo $B^2(D, \mathbb{K})$. Then there are linear functions $f_l : D \longrightarrow \mathbb{K}$ and $a_{lk} \in \mathbb{K}$ so that

$$\omega_1^l(\varphi(x_i), \varphi(x_j)) = \sum_{k=1}^{s} a_{lk}\theta_r^k(x_i, x_j) + f_l(x_i \dashv x_j) \text{ for } 1 \le l \le s,$$

$$\omega_2^l(\varphi(x_i), \varphi(x_j)) = \sum_{k=1}^{s} a_{lk}\theta_r^k(x_i, x_j) + f_l(x_i \vdash x_j) \text{ for } 1 \le l \le s.$$

If we take $\beta_{kl} = f(x_k)$ then (15) and (16) hold. Now define $\sigma : D_\Theta \longrightarrow D_\Omega$ as follows

$$\sigma(x_i) = \varphi(x_i) + \sum_{l=1}^{s} \beta_{il}^l e_l, \quad \sigma(e_i) = \sum_{j=1}^{s} a_{ji} e_j.$$

Then the σ is the required isomorphism.

7 Application

Let $\mathfrak{A}'$ be an algebra and $K(\mathfrak{A}')$ be its annihilator (center in Lie, PreLie and Leibniz algebras cases) which we suppose to be nonzero. Set $V = K(\mathfrak{A}')$ and $\mathfrak{A} = \mathfrak{A}'/K(\mathfrak{A}')$. Then there is a $\Theta \in H^2(\mathfrak{A}, V)$ such that $\mathfrak{A}' = \mathfrak{A}_\Theta$. We conclude that any algebra with a nontrivial annihilator can be obtained as an annihilator extension of an algebra of smaller dimension. So in particular, all nilpotent algebras can be constructed by this way.

Procedure: Let $\mathfrak{A}$ be an algebra of dimension $n - s$. The procedure outputs all nilpotent algebras $\mathfrak{A}'$ of dimension n such that $\mathfrak{A}'/K(\mathfrak{A}') = \mathfrak{A}$. It runs as follows:

(1) Determine $Z^2(\mathfrak{A}, \mathbb{K})$, $B^2(\mathfrak{A}, \mathbb{K})$ and $H^2(\mathfrak{A}, \mathbb{K})$.
(2) Determine the orbits of $\mathrm{Aut}(\mathfrak{A})$ on s-dimensional subspaces of $H^2(\mathfrak{A}, \mathbb{K})$.

(3) For each of the orbits let Θ be the cocycle corresponding to a representative of it, and construct $\mathfrak{A}_\Theta$.

Acknowledgements The author would like to thank the referee for the comments made. The research was supported by FRGS Grant 01-01-16-1869FR, MOHE, Malaysia.

References

1. Aguiar, M.: Pre-poisson algebras. Lett. Math. Phys. **54**(4), 263–277 (2000)
2. Basri, W., Rakhimov, I.S., Rikhsiboev, I.M.: On nilpotent diassociative algebras. J. Gen. Lie Theor. Appl. **9**(1), 1 (2015)
3. Baxter, G.: An analytic problem whose solution follows from a simple algebraic identity. Pacific. J. Math. **10**, 731–742 (1960)
4. De Graaf, W.A.: Classification of 6-dimensional nilpotent Lie algebras over fields of characteristic not 2. J. Algebra **309**(2), 640–653 (2007)
5. De Graaf, W.A.: Classification of nilpotent associative algebras of small dimension. Int. J. Algebr. Comput. **28**(01), 133–161 (2017)
6. Dzhumadildaev, A.S., Tulenbaev, K.M.: Nilpotency of Zinbiel algebras. J. Dyn. Control Syst. **11**(2), 195–213 (2005)
7. Gong, M.-P.: Classification of nilpotent Lie algebras of dimension 7 (over algebraic closed fields and $\mathbb{R}$). Ph.D. Thesis, Waterloo, Ontario, Canada (1992)
8. Hegazi, A., Abdelwahab, H.: The classification of n-dimensional non-associative Jordan algebras with $(n - 3)$-dimensional annihilator. Commun. Algebra **46**(2), 629–643 (2018)
9. Hegazi, A., Abdelwahab, H., Calderon Martin, A.: The classification of n-dimensional non-Lie Malcev algebras with $(n - 4)$- dimensional annihilator. Linear Algebra Appl. **505**, 32–56 (2016)
10. Koszul, J.-L.: Domaines bornes homogenes et orbites de groupes de transformation affines. Bull. Soc. Math. France **89**, 515–533 (1961)
11. Loday, J.-L.: Une version non commutative des algèbres de Lie: les algèbres de Leibniz. L'Ens. Math. **39**, 269–293 (1993). (in French)
12. Loday, J.-L.: Overview on Leibniz algebras, dialgebras and their homology. Fields Inst. Commun. **17**, 91–102 (1997)
13. Loday, J.-L., Frabetti, A., Chapoton, F., Goichot, F.: Dialgebras and Related Operads. Lecture Notes in Math., 1763. Springer, Berlin (2001)
14. Rakhimov, I.S.: On central extensions of associative dialgebras. J. Phys. Conf. Ser. **697**(1), 012009 (2016)
15. Rakhimov, I.S., Langari, S.J.: A cohomological approach for classifying Nilpotent Leibniz algebras. Int. J. Algebra **4**(4), 153–163 (2010)
16. Rakhimov, I.S., Hassan, M.A.: On one-dimensional central extension of a filiform Lie algebra. Bull. Aust. Math. Soc. **84**, 205–224 (2011)
17. Rakhimov, I.S., Hassan, M.A.: On isomorphism criteria for Leibniz central extensions of a linear deformation of μ_n. Int. J. Algebra Comput. **21**(5), 715–729 (2011)
18. Rota, G.: Baxter algebras and combinatorial identities I-II. Bull. Amer. Math. Soc. **75**(2), 325–334 (1969)
19. Skjelbred, T., Sund, T.: Sur la classification des algèbres de Lie nilpotentes. C. R. Acad. Sci. Paris Sér. A-B **286**(5), A241–A242 (1978)
20. Vinberg, E.B.: Convex homogeneous cones. Transl. Moscow Math. Soc. **12**, 340–403 (1963)

Holliday Junctions for the Potts Model of DNA

Utkir Rozikov

Abstract In this paper a DNA is considered as a configuration of 4-state Potts model and it is embed on a path of a Cayley tree. Then we give a Hamiltonian of the set of DNAs by an analogue of Potts model with 5-spin values $0, \pm 1, \pm 2$ (considered as DNA base pairs and 0 means vacant) on a set of admissible configurations. To study thermodynamic properties of the model of DNAs we describe corresponding translation invariant Gibbs measures (TIGM) of the model on the Cayley tree of order two. We show that there are two critical temperatures $T_{i,c}$, $i = 1, 2$ such that (i) If the temperature $T > T_{1,c}$ then there is at last one (TIGM), (ii) If $T = T_{1,c}$ then there are at least 4 TIGMs, (iii) If $T_{2,c} < T < T_{1,c}$ then there are at least 7 TIGMs, (iv) If $T = T_{2,c}$ then there are at least 10 TIGMs, and (v) If $T < T_{2,c}$ then there are at least 13 TIGMs. Each such measure describes a phase of the set of DNAs. We use these results to study distributions of Holliday junctions of DNAs. In case of very low temperatures we give stationary distributions and typical configurations of the Holliday junctions.

Keywords Cayley tree · Potts model · Gibbs measure · Holliday junction

1 Introduction and Definitions

It is known that each molecule of DNA is a double helix formed from two complementary strands of nucleotides held together by hydrogen bonds between $G - C$ and $A - T$ base pairs [1], where cytosine (C), guanine (G), adenine (A), and thymine (T).

U. Rozikov (✉)
Institute of Mathematics, 81, Mirzo Ulug'bek str., 100170 Tashkent, Uzbekistan
e-mail: rozikovu@yandex.ru

© Springer Nature Switzerland AG 2018

Z. Ibragimov et al. (eds.), *Algebra, Complex Analysis, and Pluripotential Theory*,
Springer Proceedings in Mathematics & Statistics 264,
https://doi.org/10.1007/978-3-030-01144-4_13

A Holliday junction [5] is a branched nucleic acid structure that contains four double-stranded arms joined together. These arms may adopt one of several conformations depending on buffer salt concentrations and the sequence of nucleobases closest to the junction.

The structure of DNA, at the microscopic level, can be described using ideas from statistical physics [13], where by a single DNA strand is modelled as a stochastic system of interacting bases with long-range correlations. This approach makes an important connection between the structure of DNA sequence and *temperature*; e.g., phase transitions in such a system may be interpreted as a conformational (topological) restructuring.

In the recent paper [9] we gave an Ising model of DNAs, to study its thermodynamics. Translation invariant Gibbs measures (TIGMs) of the set of DNAs on the Cayley tree of order two are studied. It is shown that, depending on temperature, number of TIGMs can be up to three. Note that non-uniqueness of Gibbs measure corresponds to phase coexistence in the system of DNAs. By properties of Markov chains (corresponding to TIGMs) Holliday junction and branches of DNAs are studied. In case of very high and very low temperatures, stationary distributions and typical configurations of the Holliday junctions are described.

This paper is continuation of [9], here we consider the set of DNAs as configurations of the Potts model on Cayley tree.

In our model we consider a set of DNAs which 'live' on a Cayley tree. Let l be an edge of this tree we have a function $\sigma(l)$ with five possible values $0, \pm 1, \pm 2$ (an analogue of spin values in physical systems), in case $\sigma(l) = 0$ we say the edge l does not belong to a DNA (i.e. a vacant), we denote

$$-1 = A - T, \quad 1 = T - A, \quad -2 = C - G, \quad 2 = G - C. \tag{1}$$

If this l separates two DNA then the value $\sigma(l) \neq 0$ means that these two DNA have a Holliday junction.

Now following [9] we recall some definitions.

Cayley tree. The Cayley tree Γ^k of order $k \geq 1$ is an infinite tree, i.e., a graph without cycles, such that exactly $k + 1$ edges originate from each vertex. Let $\Gamma^k = (V, L, i)$, where V is the set of vertices Γ^k, L the set of edges and i is the incidence function setting each edge $l \in L$ into correspondence with its endpoints $x, y \in V$. If $i(l) = \{x, y\}$, then the vertices x and y are called the *nearest neighbors*, denoted by $l = \langle x, y \rangle$. The distance $d(x, y), x, y \in V$ on the Cayley tree is the number of edges of the shortest path from x to y:

$$d(x, y) = \min\{d \mid \exists x = x_0, x_1, \ldots, x_{d-1}, x_d = y \in V \text{ such that } \langle x_0, x_1 \rangle, \ldots, \langle x_{d-1}, x_d \rangle\}.$$

For a fixed $x^0 \in V$ we set $W_n = \{x \in V \mid d(x, x^0) = n\}$,

$$V_n = \{x \in V \mid d(x, x^0) \leq n\}, \quad L_n = \{l = \langle x, y \rangle \in L \mid x, y \in V_n\}. \tag{2}$$

For any $x \in V$ denote

$$W_m(x) = \{y \in V : d(x, y) = m\}, \quad m \geq 1.$$

Group representation of the tree. Let G_k be a free product of $k + 1$ cyclic groups of the second order with generators $a_1, a_2, \ldots, a_{k+1}$, respectively, i.e. $a_i^2 = e$, where e is the unit element.

It is known that there exists a one-to-one correspondence between the set of vertices V of the Cayley tree Γ^k and the group G_k (see Chap. 1 of [8] for properties of the group G_k).

We consider a normal subgroup $\mathcal{H}_0 \subset G_k$ of infinite index constructed as follows. Let the mapping $\pi_0 : \{a_1, \ldots, a_{k+1}\} \longrightarrow \{e, a_1, a_2\}$ be defined by

$$\pi_0(a_i) = \begin{cases} a_i, & \text{if } i = 1, 2 \\ e, & \text{if } i \neq 1, 2. \end{cases}$$

Denote by G_1 the free product of cyclic groups $\{e, a_1\}$, $\{e, a_2\}$. Consider

$$f_0(x) = f_0(a_{i_1} a_{i_2} \ldots a_{i_m}) = \pi_0(a_{i_1}) \pi_0(a_{i_2}) \ldots \pi_0(a_{i_m}).$$

Then it is easy to see that f_0 is a homomorphism and hence $\mathcal{H}_0 = \{x \in G_k : f_0(x) = e\}$ is a normal subgroup of infinite index. Consider the factor group

$$G_k / \mathcal{H}_0 = \{\mathcal{H}_0, \mathcal{H}_0(a_1), \mathcal{H}_0(a_2), \mathcal{H}_0(a_1 a_2), \ldots\},$$

where $\mathcal{H}_0(y) = \{x \in G_k : f_0(x) = y\}$. Denote

$$\mathcal{H}_n = \mathcal{H}_0(\underbrace{a_1 a_2 \ldots}_{n}), \quad \mathcal{H}_{-n} = \mathcal{H}_0(\underbrace{a_2 a_1 \ldots}_{n}).$$

In this notation, the factor group can be represented as

$$G_k / \mathcal{H}_0 = \{\ldots, \mathcal{H}_{-2}, \mathcal{H}_{-1}, \mathcal{H}_0, \mathcal{H}_1, \mathcal{H}_2, \ldots\}.$$

We introduce the following equivalence relation on the set G_k: $x \sim y$ if $xy^{-1} \in \mathcal{H}_0$. Then G_k can be partitioned to countably many classes $\mathcal{H}_i$ of equivalent elements. The partition of the Cayley tree Γ^2 w.r.t. $\mathcal{H}_0$ is shown in Fig. 1 (the elements of the class $\mathcal{H}_i$, $i \in \mathbf{Z}$, are merely denoted by i).

Z-*path.* Denote

$$q_i(x) = |W_1(x) \cap \mathcal{H}_i|, \quad x \in G_k,$$

where $|\cdot|$ is the counting measure of a set. We note that (see [10]) if $x \in \mathcal{H}_m$, then

$$q_{m-1}(x) = 1, \quad q_m(x) = k - 1, \quad q_{m+1}(x) = 1.$$

From this fact it follows that for any $x \in V$, if $x \in \mathcal{H}_m$ then there is a unique two-side-path (containing x) such that the sequence of numbers of equivalence classes

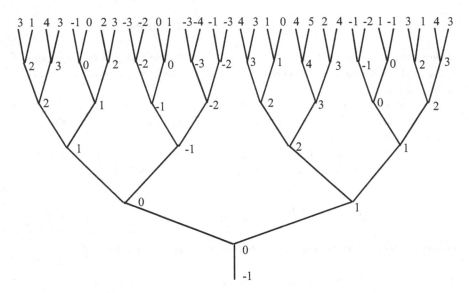

Fig. 1 The partition of the Cayley tree Γ^2 w.r.t. $\mathcal{H}_0$, the elements of the class $\mathcal{H}_i$, $i \in \mathbf{Z}$, are denoted by i

for vertices of this path in one side are $m, m+1, m+2, \ldots$ in the second side the sequence is $m, m-1, m-2, \ldots$. Thus the two-side-path has the sequence of numbers of equivalent classes as $\mathbf{Z} = \{\ldots, -2, -1, 0, 1, 2, \ldots\}$. Such a path is called $\mathbf{Z}$-path (In Fig. 1 one can see the unique $\mathbf{Z}$-paths of each vertex of the tree.)

Since each vertex x has its own $\mathbf{Z}$-path one can see that the Cayley tree considered with respect to normal subgroup $\mathcal{H}_0$ contains infinitely many (countable) set of $\mathbf{Z}$-paths.

The model. Let L be the set of edges of a Cayley tree. Consider function σ which assigns to each edge $l \in L$, values $\sigma(l) \in \{-2, -1, 0, 1, 2\}$.

A configuration $\sigma = \{\sigma(l), \ l \in L\}$ on edges of the Cayley tree is given by a function from L to $\{-2, -1, 0, 1, 2\}$. The set of all configurations in L is denoted by Ω. Configurations in L_n are defined analogously and the set of all configurations in L_n is denoted by Ω_n.

A configuration $\sigma = \{\sigma(l), \ l \in L\}$ is called *admissible* if $\sigma(l) \neq 0$ for any $l \in \mathbf{Z}$-path.

The restriction of an admissible configuration on a $\mathbf{Z}$-path is called *a DNA*.

We note that for an admissible configuration σ we may have $\sigma(l) = 0$ only for edges $l = \langle x, y \rangle$ with $x \sim y$, i.e. $f_0(x) = f_0(y)$. Comparing this with Holliday junction one can see that DNAs corresponding to two $\mathbf{Z}$-paths can have a junction only through the edge $l = \langle x, y \rangle$ with $x \sim y$.

The most common discrete models treat DNA as a collection of rigid subunits representing the base pairs (see [12]). This description has long been used by chemists to characterize DNA crystal structures. We consider the following Potts model of the energy of the configuration σ of a set of DNAs:

$$H(\sigma) = J \sum_{\langle l,t \rangle \in L \times L} \delta_{\sigma(l)\sigma(t)}, \tag{3}$$

where $J > 0$ is a coupling constant, δ is the Kronecker delta, $\sigma(l) \in \{-2, -1, 0, 1, 2\}$ and $\langle l, t \rangle$ stands for nearest neighbor edges, i.e. edges which have a common endpoint. See [6, 7] for very recent development of the theory of Gibbs measures for the Potts model on Cayley trees.

Tree-hierarchy of the set of DNAs. We shall give a Cayley tree hierarchy of the set of DNAs, as a DNA crystal.

Given an admissible configuration σ on a Cayley tree, since there are countably many **Z**-paths we have a countable many distinct DNAs. We say that two DNA *are neighbors* if there is an edge (of the Cayley tree) such that one its endpoint belongs to the first DNA and another endpoint of the edge belongs to the second DNA. By our construction it is clear (see Fig. 1) that such an edge is unique for each neighboring pair of DNAs. This edge has equivalent endpoints, i.e. both endpoints belong to the same class $\mathcal{H}_m$ for some $m \in \mathbf{Z}$.

Moreover these countably infinite set of DNAs have a hierarchy that: (i) each DNA has its own countably many set of neighboring DNAs; (ii) for any two neighboring DNAs, say D_1 and D_2, there exists a unique edge $l = l(D_1, D_2) = \langle x, y \rangle$ with $x \sim y$ which connects DNAs; (iii) two DNAs, D_1, D_2 (corresponding to two **Z**-paths) can have a Holliday junction if and only if they are neighbors, i.e. through the unique edge $l(D_1, D_2)$; (iv) for any finite $n \geq 1$ the ball V_n has intersection only with finitely many DNAs.

2 System of Functional Equations of Finite Dimensional Distributions

Let Ω_n^a (resp. Ω^a) be the set of all admissible configurations on L_n (resp. L). Denote

$$E_n = \{\langle x, y \rangle \in L : x \in W_{n-1}, \ y \in W_n\},$$

$$\Omega_n^{ba} = \text{the set of admissible configurations on } E_n.$$

For $l \in E_{n-1}$ denote

$$S(l) = \{t \in E_n : \langle l, t \rangle\}.$$

We have

$$S(l) \cap \mathbf{Z} - \text{path} = \begin{cases} \{l_0, l_1\} \subset L, & \text{if } l \notin \mathbf{Z} - \text{path} \\ \{l_1\} \subset L, & \text{if } l \in \mathbf{Z} - \text{path} \end{cases}.$$

Denote

$$S_0(l) = S(l) \setminus \{l_0, l_1\}, \quad l \notin \mathbf{Z} - \text{path},$$

$$S_1(l) = S(l) \setminus \{l_1\}, \quad l \in \mathbf{Z} - \text{path.}$$

Define a finite-dimensional distribution of a probability measure μ on Ω_n^a as

$$\mu_n(\sigma_n) = Z_n^{-1} \exp \left\{ \beta H_n(\sigma_n) + \sum_{l \in E_n} h_{\sigma(l),l} \right\}, \tag{4}$$

where $\beta = 1/T$, $T > 0$ is temperature, Z_n^{-1} is the normalizing factor, $\{h_{i,l} \in \mathbf{R}, i = -2, -1, 0, 1, 2, l \in L\}$ is a collection of real numbers and

$$H_n(\sigma_n) = J \sum_{\substack{l,t \in L_n: \\ \langle l,t \rangle}} \delta_{\sigma(l)\sigma(t)}.$$

We say that the probability distributions (4) are compatible if for all $n \geq 1$ and $\sigma_{n-1} \in \Omega_{n-1}^a$:

$$\sum_{\omega_n \in \Omega_n^{ba}} \mu_n(\sigma_{n-1} \vee \omega_n) = \mu_{n-1}(\sigma_{n-1}). \tag{5}$$

Here $\sigma_{n-1} \vee \omega_n$ is the concatenation of the configurations. In this case there exists a unique measure μ on Ω^a such that, for all n and $\sigma_n \in \Omega_n^a$,

$$\mu(\{\sigma|_{L_n} = \sigma_n\}) = \mu_n(\sigma_n).$$

Such a measure is called a *Gibbs measure* corresponding to the Hamiltonian (3) and functions $h_{i,l}, l \in L$.

Remark 1 The above construction of a Gibbs measure uses the facts that Ω^a is a compact set (which can be shown as Proposition 6.20 of [2]) and Kolmogorov's extension theorem (which is Theorem 6.6 of [2] suitable for our setting.)

The following statement describes conditions on $h_{i,l}$ guaranteeing compatibility of $\mu_n(\sigma_n)$.

Theorem 1 *Probability distributions $\mu_n(\sigma_n)$, $n = 1, 2, \ldots$, in (4) are compatible iff for any $l \in L$ the following equations hold:*
if $l \notin \mathbf{Z} - \text{path}$ then

$$u_{-1,l} = \prod_{i=0}^{1} \frac{1 + \alpha v_{-1,l_i} + v_{1,l_i} + v_{2,l_i}}{\alpha + v_{-1,l_i} + v_{1,l_i} + v_{2,l_i}} \prod_{t \in S_0(l)} \frac{1 + \alpha u_{-1,t} + u_{0,t} + u_{1,t} + u_{2,t}}{\alpha + u_{-1,t} + u_{0,t} + u_{1,t} + u_{2,t}},$$

$$u_{0,l} = \prod_{i=0}^{1} \frac{1 + v_{-1,l_i} + v_{1,l_i} + v_{2,l_i}}{\alpha + v_{-1,l_i} + v_{1,l_i} + v_{2,l_i}} \prod_{t \in S_0(l)} \frac{1 + u_{-1,t} + \alpha u_{0,t} + u_{1,t} + u_{2,t}}{\alpha + u_{-1,t} + u_{0,t} + u_{1,t} + u_{2,t}},$$

$$u_{1,l} = \prod_{i=0}^{1} \frac{1 + v_{-1,l_i} + \alpha v_{1,l_i} + v_{2,l_i}}{\alpha + v_{-1,l_i} + v_{1,l_i} + v_{2,l_i}} \prod_{t \in S_0(l)} \frac{1 + u_{-1,t} + u_{0,t} + \alpha u_{1,t} + u_{2,t}}{\alpha + u_{-1,t} + u_{0,t} + u_{1,t} + u_{2,t}},$$

$$u_{2,l} = \prod_{i=0}^{1} \frac{1 + v_{-1,l_i} + v_{1,l_i} + \alpha v_{2,l_i}}{\alpha + v_{-1,l_i} + v_{1,l_i} + v_{2,l_i}} \prod_{t \in S_0(l)} \frac{1 + u_{-1,t} + u_{0,t} + u_{1,t} + \alpha u_{2,t}}{\alpha + u_{-1,t} + u_{0,t} + u_{1,t} + u_{2,t}}, \qquad (6)$$

if $l \in \mathbf{Z} -$ path then

$$v_{-1,l} = \frac{1 + \alpha v_{-1,l_1} + v_{1,l_1} + v_{2,l_1}}{\alpha + v_{-1,l_1} + v_{1,l_1} + v_{2,l_1}} \prod_{t \in S_1(l)} \frac{1 + \alpha u_{-1,t} + u_{0,t} + u_{1,t} + u_{2,t}}{\alpha + u_{-1,t} + u_{0,t} + u_{1,t} + u_{2,t}},$$

$$v_{1,l} = \frac{1 + v_{-1,l_1} + \alpha v_{1,l_1} + v_{2,l_1}}{\alpha + v_{-1,l_1} + v_{1,l_1} + v_{2,l_1}} \prod_{t \in S_1(l)} \frac{1 + u_{-1,t} + u_{0,t} + \alpha u_{1,t} + u_{2,t}}{\alpha + u_{-1,t} + u_{0,t} + u_{1,t} + u_{2,t}},$$

$$v_{2,l} = \frac{1 + v_{-1,l_1} + v_{1,l_1} + \alpha v_{2,l_1}}{\alpha + v_{-1,l_1} + v_{1,l_1} + v_{2,l_1}} \prod_{t \in S_1(l)} \frac{1 + u_{-1,t} + u_{0,t} + u_{1,t} + \alpha u_{2,t}}{\alpha + u_{-1,t} + u_{0,t} + u_{1,t} + u_{2,t}},$$

Here, $\alpha = \exp(J\beta)$,

$$\begin{aligned}
u_{i,l} &= \exp\left(h_{i,l} - h_{-2,l}\right), i = -1, 0, 1, 2, \ l \notin \mathbf{Z} - \text{path}; \\
v_{j,l} &= \exp\left(h_{j,l} - h_{-2,l}\right), j = -1, 1, 2, \ l \in \mathbf{Z} - \text{path}.
\end{aligned} \qquad (7)$$

Proof Similar to the proof of Theorem 5.1 in [8], see also the proof of Theorem 1 in [11].

From Theorem 1 it follows that for any set of vectors

$$\mathbf{z} = \{(u_{i,l}, i = -1, 0, 1, 2, \ l \notin \mathbf{Z} - \text{path}), (v_{j,t}, j = -1, 1, 2, \ t \in \mathbf{Z} - \text{path})\}$$

satisfying the system of functional equations (6) there exists a unique Gibbs measure μ and vice versa. However, the analysis of solutions to (6) is not easy.

In next section we shall give several solutions to (6).

One says that a "phase" transition occurs for the model of DNAs, if Eq. (6) has more than one solution [4].

The number of the solutions of Eq. (6) depends on the temperature-parameter $\alpha = \exp(\frac{J}{T})$. In this paper without loss of generality we take $J = 1$.

3 Translation Invariant Gibbs Measures of the Set of DNAs

In this section, in case $J = 1$, we find solutions $\mathbf{z}_l$ to the system of functional equations (6), which does not depend on l, i.e.,

$$u_{i,l} = u_i, \quad \forall l \notin \mathbf{Z} - \text{path}; \quad v_{j,l} = v_j, \quad \forall l \in \mathbf{Z} - \text{path}. \tag{8}$$

The Gibbs measure corresponding to such a solution is called *translation invariant*.
For u_i, v_j from (6) we get

$$u_{-1} = \left(\frac{1 + \alpha v_{-1} + v_1 + v_2}{\alpha + v_{-1} + v_1 + v_2}\right)^2 \left(\frac{1 + \alpha u_{-1} + u_0 + u_1 + u_2}{\alpha + u_{-1} + u_0 + u_1 + u_2}\right)^{k-2},$$

$$u_0 = \left(\frac{1 + v_{-1} + v_1 + v_2}{\alpha + v_{-1} + v_1 + v_2}\right)^2 \left(\frac{1 + u_{-1} + \alpha u_0 + u_1 + u_2}{\alpha + u_{-1} + u_0 + u_1 + u_2}\right)^{k-2},$$

$$u_1 = \left(\frac{1 + v_{-1} + \alpha v_1 + v_2}{\alpha + v_{-1} + v_1 + v_2}\right)^2 \left(\frac{1 + u_{-1} + u_0 + \alpha u_1 + u_2}{\alpha + u_{-1} + u_0 + u_1 + u_2}\right)^{k-2},$$

$$u_2 = \left(\frac{1 + v_{-1} + v_1 + \alpha v_2}{\alpha + v_{-1} + v_1 + v_2}\right)^2 \left(\frac{1 + u_{-1} + u_0 + u_1 + \alpha u_2}{\alpha + u_{-1} + u_0 + u_1 + u_2}\right)^{k-2}, \tag{9}$$

$$v_{-1} = \frac{1 + \alpha v_{-1} + v_1 + v_2}{\alpha + v_{-1} + v_1 + v_2} \left(\frac{1 + \alpha u_{-1} + u_0 + u_1 + u_2}{\alpha + u_{-1} + u_0 + u_1 + u_2}\right)^{k-1},$$

$$v_1 = \frac{1 + v_{-1} + \alpha v_1 + v_2}{\alpha + v_{-1} + v_1 + v_2} \left(\frac{1 + u_{-1} + u_0 + \alpha u_1 + u_2}{\alpha + u_{-1} + u_0 + u_1 + u_2}\right)^{k-1},$$

$$v_2 = \frac{1 + v_{-1} + v_1 + \alpha v_2}{\alpha + v_{-1} + v_1 + v_2} \left(\frac{1 + u_{-1} + u_0 + u_1 + \alpha u_2}{\alpha + u_{-1} + u_0 + u_1 + u_2}\right)^{k-1},$$

where $u_i, v_j > 0$. By assumption $J = 1$ we have $\alpha > 1$.

For simplicity we consider the case $k = 2$. Then from (9) denoting $x = v_{-1}$, $y = v_1$, $z = v_2$ we get

$$\begin{cases} x = f(x, y, z, \alpha) \\ y = f(y, x, z, \alpha) \\ z = f(z, y, x, \alpha) \end{cases} \tag{10}$$

where

$$f(x, y, z, \alpha) \equiv \frac{1 + \alpha x + y + z}{\alpha + x + y + z} \times$$

$$\frac{(\alpha + x + y + z)^2 + \alpha(1 + \alpha x + y + z)^2 + (1 + x + y + z)^2 + (1 + x + \alpha y + z)^2 + (1 + x + y + \alpha z)^2}{\alpha(\alpha + x + y + z)^2 + (1 + \alpha x + y + z)^2 + (1 + x + y + z)^2 + (1 + x + \alpha y + z)^2 + (1 + x + y + \alpha z)^2}$$

The function f has the following properties:
(1) $f(x, y, z, \alpha) = f(x, z, y, \alpha)$, for any $x, y, z > 0, \alpha > 1$.
(2) $f(1, y, z, \alpha) = 1$, for any $y, z > 0, \alpha > 1$.

From these properties we get the following

Lemma 1   *1. If* $\mathbf{v} = (x, y, z)$ *is a solution to (10) then any vector obtained by permutations of coordinates of* $\mathbf{v}$ *is a solution to (10).*
2. $x = y = z = 1$ *is a solution to (10).*

We are going to find special solutions of (10):
Case: $x \neq 1$, $y = z = 1$. Then from the first equation of (10) we obtain

$$x = f(x, 1, 1, \alpha) = \frac{3 + \alpha x}{\alpha + x + 2} \cdot \frac{\alpha(3 + \alpha x)^2 + (3 + x)^2 + 3(2 + x + \alpha)^2}{(3 + \alpha x)^2 + (3 + x)^2 + (2 + \alpha)(2 + x + \alpha)^2}.$$

This equation can be written as $P_4(x, \alpha) = 0$ with a polynomial P_4 of degree 4. It has root $x = 1$, therefore we reduce our equation to

$$\frac{P_4(x, \alpha)}{x - 1} = (\alpha^2 + \alpha + 3)x^3 - (\alpha^4 - \alpha^3 - 6\alpha^2 - 16\alpha - 23)x^2 -$$

$$(\alpha^4 + 5\alpha^3 - 24\alpha^2 - 52\alpha - 65)x + 9(\alpha^2 + 7\alpha + 7) = 0. \tag{11}$$

Using solutions of cubic equations we obtain the following

Lemma 2   *There exists* $\alpha^* \approx 4.538695$ *such that*

i. *If* $\alpha \in [1, \alpha^*)$ *then Eq. (11) has not positive solutions.*
ii. *If* $\alpha = \alpha^*$ *then Eq. (11) has unique positive solution,* x_1^*.
iii. *If* $\alpha \in (\alpha^*, +\infty)$ *then Eq. (11) has two positive solutions,* x_1^*, x_2^*, *with* $x_1^* < x_2^*$.

Case: $x = y \neq 1$, $z = 1$. Then from the first equation of (10) we obtain

$$x = f(x, x, 1, \alpha) = \frac{2 + (1 + \alpha)x}{\alpha + 2x + 1} \cdot \frac{2(1 + \alpha + 2x)^2 + (1 + \alpha)(2 + (1 + \alpha)x)^2 + 4(1 + x)^2}{(1 + \alpha)(1 + \alpha + 2x)^2 + 2(2 + (1 + \alpha)x)^2 + 4(1 + x)^2}.$$

Since $x \neq 1$ the last equation can be reduced to the following cubic equation

$$4(\alpha^2 + 4\alpha + 5)(x^3 + 1) - (\alpha^4 + 2\alpha^3 - 16\alpha^2 - 50\alpha - 57)x(x + 1) = 0. \tag{12}$$

Solving this equation we get two *positive* solutions:

$$x_3^* = \frac{\alpha^4 - 12\alpha^2 + 2\alpha^3 - 34\alpha - 37 - (\alpha - 1)\sqrt{(\alpha + 3)(1 + \alpha)(\alpha^4 + 2\alpha^3 - 20\alpha^2 - 66\alpha - 77)}}{8(5 + 4\alpha + \alpha^2)},$$

$$x_4^* = \frac{\alpha^4 - 12\alpha^2 + 2\alpha^3 - 34\alpha - 37 + (\alpha - 1)\sqrt{(\alpha + 3)(1 + \alpha)(\alpha^4 + 2\alpha^3 - 20\alpha^2 - 66\alpha - 77)}}{8(5 + 4\alpha + \alpha^2)}.$$

Note that solutions x_3^* and x_4^* exist iff $\alpha^4 + 2\alpha^3 - 20\alpha^2 - 66\alpha - 77 \geq 0$. It is easy to check that there is $\alpha^{**} \approx 5.0806$ such that the last inequality is true for any $\alpha \geq \alpha^{**}$.
Thus, we proved the following

Lemma 3 *The following hold*

- *If $\alpha < \alpha^{**}$ then Eq. (12) has not positive solution.*
- *If $\alpha = \alpha^{**}$ then (12) has one solution, x_3^*.*
- *If $\alpha > \alpha^{**}$ then (12) has two solutions, $x_3^* < x_4^*$.*

Case: $x = y = z \neq 1$. From the first equation of (10) we obtain

$$x = f(x, x, x, \alpha) = \frac{1 + (2 + \alpha)x}{\alpha + 3x} \cdot \frac{(\alpha + 3x)^2 + (2 + \alpha)(1 + (2 + \alpha)x)^2 + (1 + 3x)^2}{\alpha(\alpha + 3x)^2 + 3(1 + (2 + \alpha)x)^2 + (1 + 3x)^2}.$$

This equation can be reduced to the following

$$9(\alpha^2 + 7\alpha + 7)x^3 - (\alpha^4 + 5\alpha^3 - 24\alpha^2 - 52\alpha - 65)x^2$$

$$- (\alpha^4 - \alpha^3 - 6\alpha^2 - 16\alpha - 23)x + \alpha^2 + \alpha + 3 = 0. \tag{13}$$

Comparing (11) and (13) one can see that if x is a solution for one of these equations then $1/x$ will be solution to the second one. Therefore, by Corollary 2 of [6] we conclude that Gibbs measures corresponding to solutions of equation (11) coincide with Gibbs measures corresponding to solutions of (13).

By above results we get the following list of solutions (v_{-1}, v_1, v_2) to (10):

$$\mathbf{v}_{10} = (1, 1, 1), \quad \mathbf{v}_{11} = (x_1^*, 1, 1), \quad \mathbf{v}_{12} = (1, x_1^*, 1), \quad \mathbf{v}_{13} = (1, 1, x_1^*);$$

$$\mathbf{v}_{21} = (x_2^*, 1, 1), \quad \mathbf{v}_{22} = (1, x_2^*, 1), \quad \mathbf{v}_{23} = (1, 1, x_2^*);$$

$$\mathbf{v}_{31} = (1, x_3^*, x_3^*), \quad \mathbf{v}_{32} = (x_3^*, 1, x_3^*), \quad \mathbf{v}_{33} = (x_3^*, x_3^*, 1);$$

$$\mathbf{v}_{41} = (1, x_4^*, x_4^*), \quad \mathbf{v}_{42} = (x_4^*, 1, x_4^*), \quad \mathbf{v}_{43} = (x_4^*, x_4^*, 1).$$

Denote by μ_{ij} the Gibbs measure which, by Theorem 1, corresponds to the solution $\mathbf{v}_{ij}$.

Define critical temperatures

$$T_{1,c} = \frac{1}{\ln \alpha^*}, \quad T_{2,c} = \frac{1}{\ln \alpha^{**}}.$$

Note that $T_{1,c} > T_{2,c}$.

Summarizing above results we obtain the following

Theorem 2 *For the model (3) of DNAs on the Cayley tree of order $k = 2$ the following statements are true*

(1) If the temperature $T > T_{1,c}$ then there is at last one translation-invariant Gibbs measure (TIGM), μ_{10}.

(2) If $T = T_{1,c}$ then there are at least 4 TIGMs μ_{1j}, $j = 0, 1, 2, 3$.

(3) If $T_{2,c} < T < T_{1,c}$ then there are at least 7 TIGMs μ_{1j}, $j = 0, 1, 2, 3$, μ_{2j}, $j = 1, 2, 3$.

(4) If $T = T_{2,c}$ then there are at least 10 TIGMs μ_{1j}, $j = 0, 1, 2, 3$, μ_{ij}, $i = 2, 3$, $j = 1, 2, 3$.

(5) If $T < T_{2,c}$ then there are at least 13 TIGMs μ_{1j}, $j = 0, 1, 2, 3$, μ_{ij}, $i = 2, 3, 4$, $j = 1, 2, 3$.

Analysis of system (9) for the case $k \geq 3$ seems difficult.

4 Markov Chains of TIGMs and Holliday Junction of DNA

We note that the quantities $u_{i,l}$ and $v_{i,l}$ of (6) define a boundary law in the sense of Definition 12.10 of [4]. In our case these quantities mean a boundary law of our biological system of DNAs.

Remark 2 Note that a Gibbs measure can be obtained as an infinite-volume limit of the finite-dimensional Gibbs measures with a boundary condition (configuration) [4]. However, in general, it is not clear what kind of boundary condition is needed to get TIGM (mentioned in Theorem 2, i.e. corresponding to a boundary law) as an infinite volume limit. To answer this question some additional work is needed (see for example [3], where this problem, particularly solved for the usual Potts model on the Cayley tree). For our Potts model of DNA this question will be considered in a separate paper.

For marginals on the two-edge sets which consist of two neighbor edges l, t, considering a boundary law $\{(u_{i,l}, i = -1, 0, 1, 2; v_{j,t}, j = -1, 1, 2), l \notin \mathbf{Z} - \text{path}, t \in \mathbf{Z} - \text{path}\}$, i.e., the solutions of system (6) (this boundary law is normalized at -2, i.e., $h_{-2,l} = 0$), we have in the case of the Potts model that

$$\mu(\sigma(l) = a, \sigma(t) = b) = \frac{1}{Z} z_{a,l} \exp(\beta \delta_{ab}) z_{b,t}, \quad a, b = -2, -1, 0, 1, 2$$

where Z is normalizing factor and $z_{a,l}$ takes values $u_{a,l}$ or $v_{a,l}$ depending on the relation of l to $\mathbf{Z} - \text{path}$.

From this, the relation between the boundary law and the transition matrix for the associated tree-indexed Markov chain (Gibbs measure) is obtained from the formula of the conditional probability. By Theorem 2 we have up to 13 TIGMs, and to each of them we can associate a Markov chain. We are interested to study thermodynamics of Holliday junctions of DNAs, for this reason it will be sufficient to study Markov chains on the subtree consisting edges which are not on a $\mathbf{Z}$-path and Markov chains on the $\mathbf{Z}$-paths. Here we define these two Markov chains:

- Define tree-edge-indexed Markov chain with states $\{-2, -1, 0, 1, 2\}$ with transition matrix $\mathbf{P}^{[l,t]} = \left(P_{ij}^{[l,t]} \right)$, where $P_{ij}^{[l,t]}$ is the probability to go from a state i at edge l to a state j at the neighbor edge t. Using solutions (u_i, v_j) to (9) we write the matrices $\mathbf{P}^{[l,t]} = \left(P_{ij}^{[l,t]} \right)$:

$$\mathbf{P}^{[l,t]} = \mathbf{P}_{(5 \to 5)}^{[l,t]} = \begin{pmatrix} \frac{\alpha}{Z_1} & \frac{u_{-1}}{Z_1} & \frac{u_0}{Z_1} & \frac{u_1}{Z_1} & \frac{u_2}{Z_1} \\ \frac{1}{Z_2} & \frac{\alpha u_{-1}}{Z_2} & \frac{u_0}{Z_2} & \frac{u_1}{Z_2} & \frac{u_2}{Z_2} \\ \frac{1}{Z_3} & \frac{u_{-1}}{Z_3} & \frac{\alpha u_0}{Z_3} & \frac{u_1}{Z_3} & \frac{u_2}{Z_3} \\ \frac{1}{Z_4} & \frac{u_{-1}}{Z_4} & \frac{u_0}{Z_4} & \frac{\alpha u_1}{Z_4} & \frac{u_2}{Z_4} \\ \frac{1}{Z_5} & \frac{u_{-1}}{Z_5} & \frac{u_0}{Z_5} & \frac{u_1}{Z_5} & \frac{\alpha u_2}{Z_1} \end{pmatrix}, \quad \text{if } l, t \notin \mathbf{Z} - \text{path}.$$

Here $Z_1 = \alpha + u_{-1} + u_0 + u_1 + u_2$, $Z_2 = 1 + \alpha u_{-1} + u_0 + u_1 + u_2$, $Z_3 = 1 + u_{-1} + \alpha u_0 + u_1 + u_2$, $Z_4 = 1 + u_{-1} + u_0 + \alpha u_1 + u_2$, $Z_5 = 1 + u_{-1} + u_0 + u_1 + \alpha u_2$.

- Define tree-edge-indexed Markov chain with states $\{-2, -1, 1, 2\}$

$$\mathbf{P}^{[l,t]} = \mathbf{P}_{(4 \to 4)}^{[l,t]} = \begin{pmatrix} \frac{\alpha}{Y_1} & \frac{v_{-1}}{Y_1} & \frac{v_1}{Y_1} & \frac{v_2}{Y_1} \\ \frac{1}{Y_2} & \frac{\alpha v_{-1}}{Y_2} & \frac{v_1}{Y_2} & \frac{v_2}{Y_2} \\ \frac{1}{Y_3} & \frac{v_{-1}}{Y_3} & \frac{\alpha v_1}{Y_3} & \frac{v_2}{Y_3} \\ \frac{1}{Y_4} & \frac{v_{-1}}{Y_4} & \frac{v_1}{Y_4} & \frac{\alpha v_2}{Y_4} \end{pmatrix}, \quad \text{if } l, t \notin \mathbf{Z} - \text{path}.$$

Here $Y_1 = \alpha + v_{-1} + v_1 + v_2$, $Y_2 = 1 + \alpha v_{-1} + v_1 + v_2$, $Y_3 = 1 + v_{-1} + \alpha v_1 + v_2$, $Y_4 = 1 + v_{-1} + v_1 + \alpha v_2$.

We note that each matrix $\mathbf{P}_{(n \to n)}^{[l,t]}$, $n = 4, 5$ is homogenous on the corresponding set of neighbor edges $\langle l, t \rangle$ where it is given, i.e., $\mathbf{P}_{(n \to m)}^{[l,t]}$ does not depend on $\langle l, t \rangle$ itself but only depends on its relation with $\mathbf{Z}-$path.

It is easy to find the following stationary distributions

$$\pi_{(5 \to 5)} = (\pi_{(5 \to 5), -2}, \pi_{(5 \to 5), -1}, \pi_{(5 \to 5), 0}, \pi_{(5 \to 5), 1}, \pi_{(5 \to 5), 2})$$

of the matrix $\mathbf{P}_{(5 \to 5)}^{[l,t]}$:

$$\pi_{(5 \to 5)} = \frac{1}{N} \begin{pmatrix} \alpha + u_{-1} + u_0 + u_1 + u_2 \\ (1 + \alpha u_{-1} + u_0 + u_1 + u_2)u_{-1} \\ (1 + u_{-1} + \alpha u_0 + u_1 + u_2)u_0 \\ (1 + u_{-1} + u_0 + \alpha u_1 + u_2)u_1 \\ (1 + u_{-1} + u_0 + u_1 + \alpha u_2)u_2 \end{pmatrix}^t,$$

where N the normalizing factor.

$$\pi_{(4\to4)} = \frac{1}{M} \begin{pmatrix} \alpha + v_{-1} + v_1 + v_2 \\ (1 + \alpha v_{-1} + v_1 + v_2)v_{-1} \\ (1 + v_{-1} + \alpha v_1 + v_2)v_1 \\ (1 + v_{-1} + v_1 + \alpha v_2)v_2 \end{pmatrix}^t,$$

where M the normalizing factor.

The following is known (see p. 55 of [4]) as ergodic theorem for positive stochastic matrices.

Theorem 3 *Let* $\mathbf{P}$ *be a positive stochastic matrix and* π *the unique probability vector with* $\pi\mathbf{P} = \pi$ *(i.e.* π *is stationary distribution). Then*

$$\lim_{n\to\infty} x\mathbf{P}^n = \pi$$

for all initial vector x.

In the case of non-uniqueness of Gibbs measure (and corresponding Markov chains) we have different stationary states for different measures. These depend on the temperature and on the fixed measure.

As a corollary of Theorem 3 and above formulas of matrices and stationary distributions we obtain the following

Theorem 4 *In a stationary state of the set of DNAs, independently on* $l \notin \mathbf{Z}$-path, *a Holliday junction through* l *does not occur with the following probability (with respect to a measure* μ_{ij} *corresponding to a solution* (u_{-1}, u_0, u_1, u_2) *of system (9))*

$$\pi_{(5\to5),0} = \frac{1}{N}(1 + u_{-1} + \alpha u_0 + u_1 + u_2)u_0.$$

(Consequently, a Holliday junction occurs with probability $1 - \pi_{(5\to5),0}$).

One can see that $\pi_{(5\to5),0}$ is a function of solution and temperature only.

Now we are interested to calculate the limit of stationary distribution vectors $\pi_{(5\to5)}$, $\pi_{(4\to4)}$ in case when temperature $T \to 0$ (i.e. $\beta \to \infty$ and $\alpha \to \infty$).

We calculate the limits for measures μ_{3j} and μ_{4j} mentioned in Theorem 2, because for these measures we have explicit formulas of solutions (see x_3^* and x_4^*), i.e., for these measures we have explicit formulas of solution u_i, $i = -1, 0, 1, 2$; v_j, $j = -1, 1, 2$ which vary with $T = 1/\beta$. By symmetry of measures μ_{3j} and μ_{4j} it will be sufficient to consider only μ_{31} and μ_{41}.

For $k = 2$, by above-mentioned solutions $\mathbf{v}_{31}, \mathbf{v}_{41}$ of (10) from (9) we get

$$u_{-1} = 1, \quad u_0 = 4\left(\frac{1+x_i^*}{1+\alpha+2x_i^*}\right)^2, \quad u_1 = u_2 = \left(\frac{2+(1+\alpha)x_i^*}{1+\alpha+2x_i^*}\right)^2, \quad i = 3, 4.$$
$$(14)$$

Recall that measures μ_{31} and μ_{41} do not exist for $T > T_{2,c}$.

Lemma 4 *The following equalities hold*

$$\lim_{T \to 0} \pi_{(5 \to 5)}^{(10)} = \left(\frac{1}{5}, \frac{1}{5}, \frac{1}{5}, \frac{1}{5}, \frac{1}{5}\right).$$

$$\lim_{T \to 0} \pi_{(5 \to 5)}^{(31)} = \left(\frac{1}{2}, \frac{1}{2}, 0, 0, 0\right), \quad \lim_{T \to 0} \pi_{(5 \to 5)}^{(41)} = \left(0, 0, 0, \frac{1}{2}, \frac{1}{2}\right).$$

$$\lim_{T \to T_{2,c}} \pi_{(5 \to 5)}^{(31)} = (0.2433190909, 0.2433190909, 0.03715503259, 0.2381033927, 0.2381033927).$$

$$\lim_{T \to 0} \pi_{(4 \to 4)}^{(31)} = \left(\frac{1}{2}, \frac{1}{2}, 0, 0\right), \quad \lim_{T \to 0} \pi_{(4 \to 4)}^{(41)} = \left(0, 0, \frac{1}{2}, \frac{1}{2}\right).$$

Proof For measure μ_{10} we have $u_{-2} = u_{-1} = u_0 = u_1 = u_2 = 1$ this by formula $\pi_{(5 \to 5)}$ gives the first limit. Since we know explicit formulas for u_i, v_i (see (14) and x_3^* and x_4^*) the remaining limits follow by simple calculations.

Structure of DNAs in low temperatures.
See (1) for the relation of our spin values and the base pairs of a DNA.
By Lemma 4, in case $T \to 0$, the set of DNAs have the following stationary states (configurations):

Case μ_{10}: All neighboring DNAs connected to each other (Holliday junctions) with probability $\frac{4}{5}$ with state one of $\sigma(l) \in \{-2, -1, 1, 2\}$ for any $l \notin \mathbf{Z}$-path. The sequence of $\pm 1, \pm 2$s, in a DNA on the $\mathbf{Z}$-path, is free, i.e. can be any sequence, with iid and equiprobable ($= 1/4$).

Case μ_{31}: All neighboring DNAs connected to each other (Holliday junctions) with state $\sigma(l) = -2$ or $\sigma(l) = -1$ for any $l \notin \mathbf{Z}$-path. A DNA on a $\mathbf{Z}$-path is constructed with sequence of -2 and -1s and iid and equiprobable ($= 1/2$). Thus the system contains only multiple (countable) branched DNAs (which has tree structure).

Case μ_{41}: All neighboring DNAs connected to each other (Holliday junctions) with state $\sigma(l) = 1$ or $\sigma(l) = 2$ for any $l \notin \mathbf{Z}$-path. A DNA on a $\mathbf{Z}$-path is constructed with sequence of 1 and 2s and iid and equiprobable ($= 1/2$).

In case $T = T_{2,c}$ the set of DNAs have the following stationary states with respect to measure $\mu_{31} = \mu_{41}$: each neighboring DNAs have Holliday junction with probability 0.9628449675 (more precisely, a junction through state -2 (and -1) with probability 0.2433190909 and a junction through state $+1$ (and $+2$) with probability 0.2381033927) and no junction with probability 0.03715503259.

Acknowledgements The author thanks Prof. M.V.Velasco and the Department of Mathematical Analysis, University of Granada, Spain for financial support and kind hospitality during his visit to the university. He thanks referees for helpful comments.

References

1. Alberts, B., Johnson, A., Lewis, J., Raff, M., Roberts, K., Walter, P.: Molecular Biology of the Cell, 4th edn. Garland Science, New York (2002)
2. Friedli, S., Velenik, Y.: Statistical Mechanics of Lattice Systems. A Concrete Mathematical Introduction. Cambridge University Press, Cambridge (2018)
3. Gandolfo, D., Rahmatullaev, M.M., Rozikov, U.A.: Boundary conditions for translation-invariant Gibbs measures of the Potts model on Cayley trees. J. Stat. Phys. **167**(5), 1164–1179 (2017)
4. Georgii, H.O.: Gibbs Measures and Phase Transitions. 2nd edn. de Gruyter Studies in Mathematics, 9. Walter de Gruyter, Berlin (2011)
5. Holliday, R.: A mechanism for gene conversion in fungi. Genet. Res. **5**, 282–304 (1964)
6. Külske, C., Rozikov, U.A., Khakimov, R.M.: Description of the translation-invariant splitting Gibbs measures for the Potts model on a Cayley tree. J. Stat. Phys. **156**(1), 189–200 (2014)
7. Külske, C., Rozikov, U.A.: Fuzzy transformations and extremality of Gibbs measures for the Potts model on a Cayley tree. Random Struct. Algorithms **50**(4), 636–678 (2017)
8. Rozikov, U.A.: Gibbs Measures on Cayley Trees. World Scientific Publishing, Singapore (2013)
9. Rozikov, U.A.: Tree-hierarchy of DNA and distribution of Holliday junctions. J. Math. Biol. **75**(6–7), 1715–1733 (2017)
10. Rozikov, U.A., Ishankulov, F.T.: Description of periodic p-harmonic functions on Cayley trees. Nonlinear Diff. Equ. Appl. **17**(2), 153–160 (2010)
11. Rozikov, U.A., Rahmatullaev, M.M.: On free energies of the Potts model on the Cayley tree. Theor. Math. Phys. **190**(1), 98–108 (2017)
12. Swigon, D.: The Mathematics of DNA Structure, Mechanics, and Dynamics, IMA. Volumes in Mathematics and Its Applications, vol. 150, pp. 293–320 (2009)
13. Thompson, C.: Mathematical Statistical Mechanics. Princeton University Press, Princeton (1972)

Further Developments of the Pluripotential Theory (Survey)

Azimbay Sadullaev

Abstract It is well known that pluripotential theory, constructed in the 1980s, is based on plurisubharmonic (psh) functions and on the Monge-Ampère operator $(dd^c u)^n$. In the 1990s there were many attempts to develop and expand pluripotential theory to broader classes such as the class of m-subharmonic ($m - sh$) functions ($1 \leq m \leq n$). In this paper we will discuss some of the most important results of the theory of $m - sh$ function as well as the difficulties and problems of constructing a potential theory in the class of $m - sh$ functions.

Keywords Pluripotential theory · Plurisubharmonic functions · Operator Monge-Ampere · Pluripolar sets · M-subharmonic functions · Maximal M-subharmonic functions

1 Introduction

It is well known that the classical potential theory is based on the class of subharmonic (sh) functions and on the Laplace operator Δ. The pluripotential theory, constructed in the 80s of the last century, is based on plurisubharmonic (psh) functions and on the Monge-Ampère operator

$$\left(dd^c u \right)^n = const \cdot \| \frac{\partial^2 u}{\partial z_j \partial \bar{z}_k} \| (dd^c |z|^2)^n.$$

Here as usual, $d = \partial + \bar{\partial}$, $d^c = \frac{\partial - \bar{\partial}}{4i}$. During 1976–90, intensive research was carried out in building the pluripotential theory: basic objects of the theory, such as extremal Green function $V^*(z, K)$, $\mathscr{P}$-measure $\omega^*(z, K, D)$, pluripolar sets, capacity values $\mathscr{P}(K, D)$, $C(K, D)$, etc. have been introduced and studied, and the foundation of pluripotential theory was practically built. All the basic fundamental theorems of the theory have been identified and the method of their application has been

A. Sadullaev (✉)
National University of Uzbekistan, Tashkent, Uzbekistan
e-mail: sadullaev@mail.ru

© Springer Nature Switzerland AG 2018
Z. Ibragimov et al. (eds.), *Algebra, Complex Analysis, and Pluripotential Theory*,
Springer Proceedings in Mathematics & Statistics 264,
https://doi.org/10.1007/978-3-030-01144-4_14

developed. Most importantly, these studies have been used successfully in solving various problems that have accumulated in multidimensional complex analysis and in the theory of plurisubharmonic functions. Nowadays, this theory is one of the main directions in complex analysis, being the basic technique of investigating the space of analytic functions of several variables.Here, we will briefly discuss the work closely related to our studies (see, bibliography). For a more complete overview of the works, the reader can find, for example, in [7, 31, 32].

Let us formulate some of the most common applications, to show how the pluripotential theory is important in multidimensional complex analysis:

(a) First, we mention Siciak's theorem [50] on the analogue of the Bernstein-Walsh theorem in space $\mathbf{C}^n$, namely if if $K \subset \mathbf{C}^n$ *is a pluriregular compact and* $f \in C(K)$ *is a continuous function, then the speed of polynomial approximation of* f *is completely described in terms of the Green function* $V^*(z, K)$

$$\varlimsup_{m \to \infty} e_m^{1/m}(f, K) \leq \frac{1}{R} \Leftrightarrow f \in \mathscr{O}\left\{z \in \mathbf{C}^n : V^*(z, K) < \ln R\right\}, \quad R > 1$$

(b) (V. Zakharyuta, [57]). *Suppose that a function* $f(z, w)$ *is separately analytic in*

$$X = [E \times G] \cup [D \times F], \ E \subset D \subset \mathbf{C}^n, \ F \subset G \subset \mathbf{C}^m,$$

i.e., if $f(z, w)$ *holomorphic as a function of w in G for each fixed $z \in E$ and holomorphic as a function of z in D for each fixed $w \in F$, then f extends holomorphically to an open set*

$$\hat{X} = \left\{(z, w) \in D \times G \subset \mathbf{C}^{n+m} : \omega^*(z, K, D) + \omega^*(z, F, G) + 1 < 0\right\},$$

where $\omega^*(z, \cdot, \cdot)$ *is the $\mathscr{P}$-measure.*

(c) **Theorem** (Sadullaev A., Chirka E.M. [49]). *Suppose that a function* $f('z, z_n)$ *is holomorphic in the polydisk $U =' U \times U_n$ in $\mathbf{C}^n$ and for each fixed $'a$ in some nonpluripolar set $E \subset' U$, the function $f('a, z_n)$ of z_n can be continued to a function, holomorphic on the whole plane with the exception of some polar set singularities. Then f can be continued holomorphically to $('U \times)\mathbf{C} \setminus S$, where S is closed pluripolar subset of $'U \times \mathbf{C}$.*

A reader can find proofs of (a)–(c) in the monographs [46–48].

(d) **Local criterion of algebraicity.** Let $A \subset \mathbf{C}^n$ there be a piece of the analytic set.

Problem: *give a criterion for the algebraicity of this piece.* If A is given globally, not locally in $\mathbf{C}^n$, then there are many criteria, Bishop [13] , Rudin [38], (see also [17, 39, 58]), etc.

This type of local problem is not new in complex analysis. There is, for example, Kronecker's criterion, when a germ of function $f(\xi) = a_0 + a_1\xi + a_2\xi^2 + \cdots$ at the point $0 \in \mathbf{C}$ is a rational function.

Theorem [45]. *A piece $A \subset \mathbf{C}^n$, $0 \in A$, is a piece of an algebraic set if and only if*

$$V(z, \overline{B}) \in L^{\infty}(A), \quad B = B(0, r) \cap A \subset\subset A, \quad r > 0.$$

2 $(B)m$-Subharmonic Functions

In the 1990s there were many attempts to develop and expand the pluripotential theory to broader classes of functions. One such class is the m-subharmonic $(m - sh)$ functions $(1 \le m \le n)$. An upper semicontinuous in a domain $D \subset \mathbf{C}^n$ is said to be m-subharmonic in D, $u \in m - sh(D)$, if $dd^c u \wedge \beta^{m-1} \ge 0$, in the generalized sense, as current, i.e.

$$dd^c u \wedge \beta^{m-1}(\omega) = \int u \beta^{m-1} \wedge dd^c \omega \ge 0, \quad \forall \omega \in F^{n-m,n-m}, \quad \omega \ge 0.$$

Here $\beta = dd^c |z|^2$ is the standard volume form of the space $\mathbf{C}^n$ and $F^{n-m,n-m}$ is the space of compactly supported in D, smooth differential forms, bi-degree $(n - m, n - m)$. Note that

$$psh(D) = 1 - sh(D) \subset m - sh(D) \subset n - sh(D) = sh(D).$$

Such functions have an excellent geometric characterizations.

Theorem 1 (Z. Khusanov, B. Abdullaev, 1990, [29, 30]) *An upper semicontinuous function u, defined in a domain $D \subset \mathbf{C}^n$ is $m - sh$ if and only if for any complex plane $\Pi \subset \mathbf{C}^n$, dim $\Pi = m$, the restriction $u|_\Pi \in sh (\Pi \cap D)$.*

$m - sh$ functions and functions closely related to them are considered and applied in various problems in the function theory (see [1, 2, 19, 25, 28, 54]). In a series of papers F.R.Harvey and H.B.Jr. Lawson [20–24] functions closely related to the $m - sh$ functions were applied in the problems of convex geometry, convex hull and minimal surfaces in calibrated geometry.

We note that contrary to the expectations, the operator $(dd^c u)^{n-m+1} \wedge \beta^{m-1}$ is not suitable for the construction of the potential theory in the class of $m - sh$ functions. In particular, the equation $(dd^c u)^{n-m+1} \wedge \beta^{m-1} = 0$ does not determine the maximal $m - sh$ functions. Moreover, for $1 < m < n$ the class $m - sh$ does not correspond to this operator, since the next necessary condition does not hold: $m - sh \not\subset \{(dd^c u)^{n-m+1} \wedge \beta^{m-1} \ge 0\}$. For example, $(dd^c u)^2 \wedge \beta = -\frac{\beta^3}{3} < 0$ for $u(z) = -|z_1|^2 + |z_2|^2 + |z_3|^2 \in 2 - sh (\mathbf{C}^3)$. So that the class $m - sh$ is not right for the operator $(dd^c u)^{n-m+1} \wedge \beta^{m-1}$. In this way B. Abdullaev suggested to use a subclass of the class $m - sh$:

$$(A)m - sh = \left\{ u \in m - sh : (dd^c u)^{n-m+1} \wedge \beta^{m-1} \ge 0 \right\} \subset m - sh$$

and Z.Błocki ([14], 2005, see also [18]) proposed using a class of functions

$$(B)m - sh =$$

$$= \left\{ u \in m - sh : (dd^c u) \wedge \beta^{n-1} \geq 0, \ (dd^c u)^2 \wedge \beta^{n-2} \geq 0, \dots, (dd^c u)^{n-m+1} \wedge \beta^{m-1} \geq 0 \right\}.$$

Note that the definitions of $(A)m - sh$, $(B)m - sh$ are applicable for functions, $u \in C^2(D)$. In the general case, for upper-semicontinuous functions u these classes can be defined in the generalized sense. Z.Błocki proved a series of necessary properties of the class $(B)m - sh$. In particular, operators $(dd^c u)^{n-k+1} \wedge \beta^{k-1}$ defined initially in the class $C^2(D)$, may be expanded into a class $C(D)$, which allowed to build potential theory in the class $(B)m - sh$, based on the operator $(dd^c u)^{n-m+1} \wedge \beta^{m-1}$.

The potential theory in the class of $(B)m - sh$ functions was constructed in the works of Abdullaev-Sadullaev [3, 4], where all the main potential properties of this class are proved. Using solely the class $(B)m - sh(D) \cap C(D)$, we define the m-capacity $C_m(E, D)$ as follows. let K be a compact set in a domain $D \subset \mathbf{C}^n$. Then

$$C_m(K) = C_m(K, D) =$$

$$\inf \left\{ \int_D (dd^c u)^m \wedge \beta^{n-m} : \ u \in m - sh(D) \cap C(D), \ u|_K \leq -1, \ \underline{\lim}_{z \to \partial D} u(z) \geq 0 \right\}$$

is called the m-capacity of the condenser (K, D). The value $C_m(K)$ has all the basic properties of capacities. Moreover, $C_m(K) = 0 \Leftrightarrow K$ is m-polar. The fundamental theorems of potential theory in the class of $(B)m - sh(D)$ are listed below:

Theorem 2 *The set I_K of irregular points of arbitrary compact set K has zero capacity: $C_m(K) = 0$ i.e., I_K is an m-polar set.*

Theorem 3 *Let $\{u_j\}$ be an increasing sequence of m-functions such that the function $u(z) = \lim_{m \to \infty} u_j(z)$ is locally bounded from above. Then the set $\{u(z) < u^*(z)\}$, where u^* is the regularization of u, is m-polar in D.*

Theorem 4 *A locally m-polar set is globally m-polar in $\mathbf{C}^n$, i.e., if a set $E \subset \mathbf{C}^n$ is locally m-polar, then there exists a function $u(z) \in m - sh(\mathbf{C}^n)$, such that $u \not\equiv -\infty$, $u|_E = -\infty$.*

Theorem 5 (quasi-continuity, analogue of Luzin's theorem) *An arbitrary m-subharmonic function is continuous almost everywhere with respect to the capacity, i.e. if $u \in m - sh(D)$, then for arbitrary $\varepsilon > 0$ there exists an open set $U \subset D$ such that its capacity $C_m(U) < \varepsilon$ and the function u is continuous in $D \setminus U$.*

Theorem 6 $(B)m - sh(D) \subset (A)m - sh(D) \underset{\neq}{\subset} m - sh(D)$.

Remark 1 It seems, in fact, the classes $(B)m - sh(D)$, $(A)m - sh(D)$ coincide, $(B)m - sh(D) = (A)m - sh(D)$. At least for that reason, it can be checked for $m = 2$ and $u(z) \in C^2 \cap 2 - sh$.

Proposition 1 *If $u(z) \in C^2 \cap 2 - sh$ and $(dd^c u)^{n-1} \wedge \beta \geq 0$, then*

$$(dd^c u) \wedge \beta^{n-1} \geq 0, \ (dd^c u)^2 \wedge \beta^{n-2} \geq 0, \ldots, (dd^c u)^{n-2} \wedge \beta^2 \geq 0.$$

O u t l i n e   o f   t h e   p r o o f. Let $u(z) \in C^2 \cap 2 - sh$ and $\lambda_1 \leq \lambda_2 \leq \cdots \leq \lambda_n$ denotes the eigenvalues of the Hermitian matrix $\left(\frac{\partial^2 u}{\partial z_j \partial \bar{z}_k} \right)$ in the fixed point $o \in D$. If $\lambda_1 \geq 0$, the Proposition is trivial. We can assume $\lambda_1 = -1$. Then $\lambda_2 \geq 1$, since $m = 2$ and $dd^c u \wedge \beta \geq 0$, i.e. $\lambda_i + \lambda_j \geq 0$, $1 \leq i < j \leq n$. Note that for arbitrary fixed $1 \leq k \leq n$

$$(dd^c u)^k \wedge \beta^{n-k} = \left(\frac{i}{2} \right)^k (\lambda_1 dz_1 \wedge d\bar{z}_1 + \lambda_2 dz_2 \wedge d\bar{z}_2 + \cdots + \lambda_n dz_n \wedge d\bar{z}_n)^k \wedge \beta^{n-k} =$$

$$= k! \left(\frac{i}{2} \right)^k \sum_{1 \leq j_1 < \cdots < j_k \leq n} \lambda_{j_1} \ldots \lambda_{j_k} dz_{j_1} \wedge d\bar{z}_{j_1} \wedge \cdots \wedge dz_{j_k} \wedge d\bar{z}_{j_k} \wedge$$

$$\wedge (n - k)! \left(\frac{i}{2} \right)^{n-k} \sum_{1 \leq s_1 < \cdots < s_{n-k} \leq n} dz_{s_1} \wedge d\bar{z}_{s_1} \wedge \cdots \wedge dz_{s_{n-k}} \wedge d\bar{z}_{s_{n-k}} =$$

$$= k!(n - k)! \left(\frac{i}{2} \right)^n \sum_{1 \leq j_1 < \cdots < j_k \leq n} \lambda_{j_1} \ldots \lambda_{j_k} dz_1 \wedge d\bar{z}_1 \wedge \cdots \wedge dz_n \wedge d\bar{z}_n =$$

$$= \frac{k!(n - k)!}{n!} \left[\sum_{2 \leq j_1 < \cdots < j_k \leq n} \lambda_{j_1} \ldots \lambda_{j_k} - \sum_{2 \leq j_2 < \cdots < j_k \leq n} \lambda_{j_2} \ldots \lambda_{j_k} \right] \beta^n.$$

In particular

$$(dd^c u)^{n-1} \wedge \beta = \frac{1}{n} \left[\sum_{2 \leq j_1 < \cdots < j_{n-1} \leq n} \lambda_{j_1} \ldots \lambda_{j_{n-1}} - \sum_{2 \leq j_2 < \cdots < j_{n-1} \leq n} \lambda_{j_2} \ldots \lambda_{j_{n-1}} \right] \beta^n =$$

$$= \frac{1}{n} \left[\lambda_2 \ldots \lambda_n - \sum_{2 \leq j_2 < \cdots < j_{n-1} \leq n} \lambda_{j_2} \ldots \lambda_{j_{n-1}} \right] \beta^n.$$

Now the proof of the Proposition follows from the next Lemma

Lemma 1 *For arbitrary positive numbers $\mu_1, \ldots, \mu_n \geq 0$ if*

$$\mu_1 \ldots \mu_n - \sum_{1 \leq j_1 < \cdots < j_{n-1} \leq n} \mu_{j_1} \ldots \mu_{j_{n-1}} \geq 0$$

then

$$\sum_{1 \leq j_1 < \cdots < j_k \leq n} \mu_{j_1} \dots \mu_{j_k} - \sum_{1 \leq j_1 < \cdots < j_{k-1} \leq n} \mu_{j_1} \dots \mu_{j_{k-1}} \geq 0$$

I n   f a c t, we use the following Newton's inequality (see, for example [56]). Suppose $\mu_1, \dots, \mu_n \in \mathbf{R}$ are real numbers and let

$$\sigma_k(\mu) = \sum_{1 \leq j_1 < \cdots < j_k \leq n} \mu_{j_1} \cdots \mu_{j_k}$$

denote the kth elementary symmetric function in $\mu_1, \dots, \mu_n$. Then the elementary symmetric means, given by

$$S_k(\mu) = \frac{\sigma_k(\mu)}{\binom{n}{k}}$$

satisfy the inequality

$$S_{k-1}(\mu) S_{k+1}(\mu) \leq S_k^2(\mu).$$

If all the numbers μ_j are nonzero, then equality holds if and only if all the numbers μ_j are equal. If all $\mu_j \geq 0$ then

$$\sigma_{k-1}(\mu) \sigma_{k+1}(\mu) \leq \sigma_k^2(\mu).$$

Assumption of Lemma 1 is $\sigma_{n-1}(\mu) \leq \sigma_n(\mu)$. Plugging this into the inequality above for $k = n - 1$ and using that all $\mu_j \geq 0$, we get $\sigma_{n-2}(\mu)\sigma_n(\mu) \leq \sigma_{n-1}^2(\mu) \leq \sigma_{n-1}(\mu)\sigma_n(\mu)$, i.e. $\sigma_{n-2}(\mu) \leq \sigma_{n-1}(\mu)$. The lemma follows then from iteration $k = n - 2, n - 3, \dots$.

3   Potential Theory in the Class $m - sh$ Functions

Our next goal is to study the potential properties of the class $m - sh(D)$. The main objects of the potential theory are polar sets and the $\mathscr{P}$-measure in the class of $m - sh$ functions. For a set $E \subset D \subset \mathbf{C}^n$ we consider the class of functions

$$\mathscr{U}(E, D) = \{ u \in m - sh(D) : u|_D \leq 0, \ u|_E \leq -1 \}$$

and function

$$\omega(z, E, D) = \sup \{ u(w) : u \in \mathscr{U}(E, D) \}.$$

A regularization $\omega^*(z, E, D) = \varlimsup_{w \to z} \omega(w, E, D)$ is called $\mathscr{P}_m$-measure of the set E with respect to a domain D (we naturally assume, that D is m-regular, i.e. there is a $\rho(z) \in m - sh(D) : \rho(z) < 0, \ \lim_{z \to \partial D} \rho(z) = 0$).

$\mathscr{P}_m$-measure possesses many properties of the well-known harmonic measures or $\mathscr{P}$-measure of the pluripotential theory. In particular, $\mathscr{P}_m$-measure $\omega^*(z, E, D) \in m - sh(D)$, either vanishes nowhere or is identically zero. Moreover, $\omega^*(z, E, D) \equiv 0$ if and only if E is m-polar in D, i.e. $\exists u(z) \in m - sh(D) : u \not\equiv -\infty, u|_E \equiv -\infty$. Note, that for a sufficiently rich class of compacts, so-called m-regular compacts, the $\mathscr{P}$-measure is continuous in D.

Next we introduce the notions of $\mathscr{P}_m$-capacity of subset $E \subset D$ [40–44] (see also [35]):

$$\mathscr{P}_m(E, D) = - \int_D \omega^*(z, E, D) dV.$$

$\mathscr{P}_m(E, D) \geq 0$ is positive and $\mathscr{P}_m(E, D) = 0$ if and only if E is m-polar in D. $\mathscr{P}_m$-capacity is a monotonic and countable subadditive set function. Moreover, for any set $E \subset D$ and for any $\varepsilon > 0$ there exists open set $U \supset E$ such that

$$\mathscr{P}_m(U, D) - \mathscr{P}_m(E, D) < \varepsilon.$$

The capacity $\mathscr{P}_m(E, D)$ is continuous as a set function i.e., for any decreasing sequence of compact sets $K_1 \supset K_2 \supset \ldots$ the equality

$$\mathscr{P}_m \left(\bigcap_{j=1}^{\infty} K_j, D \right) = \lim_{j \to \infty} \mathscr{P}_m \left(K_j, D \right)$$

holds and for any increasing sequence of open sets $G_1 \subset G_2 \subset \ldots$ the equality

$$\mathscr{P}_m \left(\bigcup_{j=1}^{\infty} G_j, D \right) = \lim_{j \to \infty} \mathscr{P}_m \left(G_j, D \right)$$

holds.

Consequently, $\mathscr{P}_m(E, D)$ has all properties of Choquet measurability so that any Borel set is measurable by $\mathscr{P}_m$-capacity. Thus, if $E \subset D$ is a Borel set then its outer and inner capacities coincide:

$$\mathscr{P}_m^*(E, D) = \mathscr{P}_m * (E, D),$$

where

$$\mathscr{P}_m^*(E, D) = \inf \{ \mathscr{P}_m(U, D) : U \supset E - \text{open} \},$$

$$\mathscr{P}_m * (E, D) = \sup \{ \mathscr{P}_m(K, D) : K \subset E - \text{compact} \}.$$

One of the main objects of the class $m - sh$ functions are maximal functions, which are analogues of harmonic functions.

Definition 1 A function $u(z) \in m - sh(D)$ is said to be maximal in a domain $D \subset \mathbb{C}^n$ if the maximum principle holds, that is, if $v \in m - sh(D)$: $\lim\limits_{z \to \partial D} (u(z) - v(z)) \geq 0$, then $u(z) \geq v(z)$, $\forall z \in D$.

Note that the function $u(z) \in m - sh(D)$ is maximal if and only if for any domain $G \subset\subset D$ the inequality $u(z) \geq v(z)$ in G is satisfied for all functions $v \in m - sh(D)$: $u|_{\partial G} \geq v|_{\partial G}$.

Regarding the class $(B)m - sh(D) \cap L^\infty_{loc}(D)$, the maximal functions are characterized by the operator $(dd^c u)^{n-m+1} \wedge \beta^{m-1}$, i.e. $u \in (B)m - sh(D) \cap L^\infty_{loc}(D)$ is maximal in the class $(B)m - sh(D) \cap L^\infty_{loc}(D)$ if and only if $(dd^c u)^{n-m+1} \wedge \beta^{m-1} = 0$.

To find out the geometric nature of the maximal $m - sh$ functions, we calculate $dd^c u \wedge \beta^{m-1}$ in terms of eigenvalues (at fixed point $z \in D$) of the complex Hessian $\left(\frac{\partial^2 u}{\partial z_j \partial \bar{z}_k}\right)$ of u, which is hermitian matrix. After a suitable unitary coordinate transformation, which does not change $\beta = dd^c |z|^2$, the operator $dd^c u$ can be written in the diagonal form:
$dd^c u = \frac{i}{2}[\lambda_1 dz_1 \wedge d\bar{z}_1 + \cdots + \lambda_n dz_n \wedge d\bar{z}_n]$, where the $\lambda_j = \lambda_j(z) \in \mathbb{R}^n$ are eigenvalues. We have

$$dd^c u \wedge \beta^{m-1} =$$

$$= \left(\frac{i}{2}\right)^m \sum_{1 \leq j_1 < \cdots < j_m \leq n} \left(\lambda_{j_1} + \cdots + \lambda_{j_m}\right) dz_{j_1} \wedge d\bar{z}_{j_1} \wedge \cdots \wedge dz_{j_m} \wedge d\bar{z}_{j_m}. \quad (1)$$

Positivity of the form $dd^c u \wedge \beta^{m-1}$ means that all the coefficients (1) are positive,

$$\lambda_{j_1} + \cdots + \lambda_{j_m} \geq 0, \ 1 \leq j_1 < j_2 < \cdots < j_m \leq n.$$

We set

$$\mathcal{M}_u(z) = \left[\prod_{1 \leq j_1 < \cdots < j_m \leq n} \left(\lambda_{j_1}(z) + \cdots + \lambda_{j_m}(z)\right) \right]^{\alpha_m}, \quad (2)$$

where $\alpha_m = \frac{m!(n-m+1)!}{n!}$ is a degree. We choose this degree to make the operator $\mathcal{M}_u(z)$ (a priori) locally bounded on average, i.e.

$$\forall K \subset\subset D \ \exists C(K) : \int_K \mathcal{M}_u(z) dV \leq C(K) \ \forall u \in C^2 \cap m - sh(D), \ |u| \leq 1.$$

It is clear that $\mathcal{M}_u = \lambda_1 + \cdots + \lambda_n = \Delta u$ for $m = n$ and $\mathcal{M}_u = \lambda_1 \ldots \lambda_n$ for $m = 1$. The operator $\mathcal{M}_u$ is an operator in eigenvalues, symmetric, positive in the class $m - sh(D) \cap C^2(D)$. Unlike of the previous cases, it is not a simple differential operator.

What does $\mathcal{M}_u(z) = 0$ **mean at the point** $z^0 \in D$? It means, at least one of the factors in (2) at the point z^0 is zero. For example, $\lambda_1(z^0) + \cdots + \lambda_m(z^0) = 0$. We note, that by using the the the unitary transformation T, the operator $dd^c u$ is reduced to the form $dd^c u = \frac{i}{2}[\lambda_1(z^0)dz_1 \wedge d\bar{z}_1 + \cdots + \lambda_n(z^0)dz_n \wedge d\bar{z}_n]$.

Therefore, the function $u(z)$ in the direction of the m-dimensional plane

$$\Pi = \{z^0\} + T^{-1}\{z_{m+1} = \cdots = z_n = 0\}, \quad z^0 \in \Pi, \quad \dim \Pi = m,$$

is "harmonic", $\Delta u|_\Pi = 0$ in the fixed point $z = z^0$.

It seems to us that in the study of potential properties of the class $m - sh(D)$ the operator $\mathcal{M}_u(z)$ plays the same role, as the Monge-Ampère operator $(dd^c u)^n$ for the class of psh functions. In this direction, there are only a few positive results.

Problem 1 *If $\mathcal{M}_u(z) \equiv 0$ in D, then in D there is field $\mathcal{I}$ of the directions of the m-dimensional planes $\Pi \ni z^0 : u|_\Pi$ is harmonic in the point z^0. Will the domain D be fibered by the integral surfaces of this field? Namely, for any point $z^0 \in D$ there is a surface $S \ni z^0$, $\dim S = m$ such that, the restriction $u|_S$ is harmonic on S with respect to the induced metric.*

A similar result in the class of psh functions for the Monge-Ampère operator $(dd^c u)^n$ was proved in the paper of E. Bedford and M. Kalka [9].

Theorem 7 (see [34]) *The function $u(z) \in C^2 \cap m - sh(D)$ is maximal in the domain $D \subset \mathbf{C}^n$ if and only if $\mathcal{M}_u(z) \equiv 0$.*

In fact, a more general theorem is true (see below, Theorem 9).

4 The Dirichlet Problem

In this section, we will consider the Dirichlet problem for the equation

$$\mathcal{M}_u(z) = \psi(z), \quad u(z) \in m - sh(D), \quad u|_{\partial D} = \varphi(\xi),$$

$$\psi(z) \in C(\overline{D}), \quad \varphi(\xi) \in C(\partial D), \quad \psi(z) \geq 0 \tag{3}$$

We assume that the domain D is bounded, strictly m-pseudoconvex, i.e., in a neighborhood of the closure $\overline{D}$ there exists a strictly $m - sh$ function $\rho(z)$ such that $d\rho|_{\partial D} \neq 0$, $D = \{\rho(z) < 0\}$. To find a solution of (3) we will be guided by the general theory of the existence of solutions of the Dirichlet problem for the equation in the Hessians, as outlined in the works of the L. Caffarelli, L. Nirenberg, J.Spruck, S.Y. Li, X.J. Wang, etc. [16, 36, 37, 55]. For this we denote

$$f(\lambda) = \left[\prod_{1 \leq j_1 < \cdots < j_m \leq n} \left(\lambda_{j_1} + \cdots + \lambda_{j_m}\right) \right]^{\alpha_m}, \quad \lambda = (\lambda_1 + \cdots + \lambda_n) \in \mathbf{R}^n.$$

Then, $\mathcal{M}_u(z) = f(\lambda_1(z), \lambda_2(z) \ldots, \lambda_n(z))$, where $\lambda_1(z), \lambda_2(z) \ldots, \lambda_n(z)$ are the eigenvalues of the complex Hessian matrix $\left(\frac{\partial^2 u}{\partial z_j \partial \bar{z}_k}\right)$. One can check, the function $f(\lambda)$ is positive, concave and strictly increasing on the convex cone

$$\Gamma = \{\lambda = (\lambda_1, \lambda_2, \ldots, \lambda_n) \in \mathbf{R}^n : \lambda_{j_1} + \lambda_{j_2} + \cdots + \lambda_{j_m} \geq 0, \ \forall 1 \leq j_1 < j_2 < \cdots < j_m \leq n\}.$$

Note, that Γ is symmetric in $\lambda = (\lambda_1, \lambda_2, \ldots, \lambda_n)$ with vertex in the point 0. Moreover, the function $f(\lambda)$ satisfies the following properties:

(a) $f(\lambda) = 0 \ \forall \lambda \in \partial\Gamma$;

(b) *For any compact set $K \subset\subset \Gamma$ we have*

$$\lim_{R \to \infty} f(\lambda_1, \lambda_2, \ldots, \lambda_{n-1}, \lambda_n + R) = \infty, \quad \lim_{R \to \infty} f(\lambda R) = \infty$$

uniformly in $\lambda \in K$.

The following theorems follow from these properties of $f(\lambda)$.

Theorem 8 (see [36]) *If D is strictly m-pseudoconvex and $\psi(z) \in C^\infty(\overline{D})$, $\psi(z) > 0$, $\varphi(\xi) \in C^\infty(\partial D)$, then (3) has a unique solution $u(z) \in m - sh(D) \cap C^\infty(\overline{D})$.*

Theorem 9 (Maximum principle, see [36]) *Let $u, v \in m - sh(D) \cap C^2(\overline{D})$: $\mathbf{M}_u(z) \leq \mathbf{M}_v(z) \ \forall z \in D$. Then $u|_{\partial D} \geq v|_{\partial D} \Rightarrow u|_D \geq v|_D$*

The Dirichlet problem for the degenerate equation

$$\mathcal{M}_u(z) = 0, \ u(z) \in m - sh(D), \ u|_{\partial D} = \varphi(\xi), \ \varphi(\xi) \in C(\partial D) \tag{4}$$

is more complicated. However, for the class of *psh* functions (case $m = 1$) the solution of (4), namely $(dd^c u)^n = 0$, $u(z) \in psh(D)$, $u|_{\partial D} = \varphi(\xi)$, $\varphi(\xi) \in C(\partial D)$ is well studied. When $D = \{|z| < 1\}$-ball and $\varphi(\xi) \in C^2(\partial D)$ E. Bedford and B.A. Taylor [10, 12] proved that the solution of (4) belongs to the class $C^{1,1}$. A similar result will also occur in more general settings (see, for example, [15]). At the same time, the example of E. Bedford and J.E. Fornaess [8] shows that, generally speaking, the solutions of (4) does not have a higher smoothness, say C^2.

To construct a solution of the degenerate Eq. (4) we consider the sequence

$$\varphi_k(\xi) \in C^\infty \partial D, \ \varphi_k(\xi) \geq \varphi_{k+1}(\xi), \ \| \varphi_k - \varphi \| \leq \frac{1}{k}, \ k = 1, 2, \ldots,$$

and by the Theorem 8, we find the solutions

$$u_k \in C^\infty(\overline{D}), \ \mathcal{M}_{u_k}(z) = \frac{1}{k}, \ u_k|_{\partial D} = \varphi_k.$$

Theorem 10 *The sequence $u_k(z)$ converges uniformly in $\overline{D}$. Its limit $u(z) = \lim\limits_{k \to \infty} u_k(z)$ is maximal function in D, $u(z) \in C(\overline{D})$, $u|_{\partial D} = \varphi$.*

This limit is naturally called the solution of the Dirichlet problem (4), although the operator $\mathscr{M}_u$ is not yet defined for continuous functions.

Proof Since $\varphi_{k+1}(\xi) \leq \varphi_k(\xi) \leq \varphi_{k+1}(\xi) + \frac{1}{k}$, we have $u_{k+1}|_{\partial D} \leq u_k|_{\partial D} \leq u_{k+1}|_{\partial D} + \frac{1}{k}$. It follows from

$$\mathscr{M}_{u_k}(z) = \frac{1}{k} > \frac{1}{k+1} = \mathscr{M}_{u_{k+1}}(z)$$

and Theorem 9 , that $u_k|_D \leq u_{k+1}|_D + \frac{1}{k}$.

On the other hand, we fix the index k and consider an auxiliary function $v_{k+1} = u_{k+1} + \delta_k \left(|z|^2 - R^2 \right)$, where $R > 0$ are a fixed numbers such, that $D \subset \{ |z| < R \}$ and

$$\delta_k = \frac{\left[\left(1 + \frac{1}{k} \right)^{\frac{1}{\alpha_m}} - 1 \right]^{\alpha_m}}{m(k+1)} \simeq \frac{1}{mk^{\alpha_m}(k+1)}.$$

If $\lambda_1(z), \ldots, \lambda_n(z)$ are the eigenvalues of the complex Hessian matrix $\left(\frac{\partial^2 u_{k+1}}{\partial z_j \partial \bar{z}_k} \right)$, then $\lambda_1(z) + \delta_k, \ldots, \lambda_n(z) + \delta_k$ are the eigenvalues of the complex Hessian matrix $\left(\frac{\partial^2 v_{k+1}}{\partial z_j \partial \bar{z}_k} \right)$. Consequently,

$$\mathscr{M}_{v_{k+1}}(z) = \left[\prod_{1 \leq j_1 < \cdots < j_m \leq n} \left(\lambda_{j_1}(z) + \cdots + \lambda_{j_m}(z) + m\delta_k \right) \right]^{\alpha_m} \geq$$

$$\geq \left[\prod_{1 \leq j_1 < \cdots < j_m \leq n} \left(\lambda_{j_1}(z) + \cdots + \lambda_{j_m}(z) \right) + (m\delta_k)^{\frac{1}{\alpha_m}} \right]^{\alpha_m} =$$

$$= \mathscr{M}_{u_{k+1}}(z) \left[1 + \left(\frac{m\delta_k}{\mathscr{M}_{u_{k+1}}(z)} \right)^{\frac{1}{\alpha_m}} \right]^{\alpha_m} = \frac{1}{k+1} \left[1 + (m(k+1)\delta_k)^{\frac{1}{\alpha_m}} \right]^{\alpha_m} = \frac{1}{k}.$$

Since $v_{k+1}|_{\partial D} \leq u_k|_{\partial D}$, it follows from Theorem 9, that $v_{k+1}|_D \leq u_k|_D$, i.e. $u_{k+1}|_D - \delta_k R^2 \leq u_k|_D$. Together with $u_k|_D \leq u_{k+1}|_D + \frac{1}{k}$, it gives us $u_k|_D - \frac{1}{k} \leq u_{k+1}|_D \leq u_k|_D + \delta_k R^2$, which implies the uniform convergence of $u_k(z)$ in $\overline{D}$.

It remains to prove the maximality of the function $u(z)$. Suppose that there exist a ball $B = B(0, r) \subset\subset D$ and a function $v(z) \in m - sh(D)$: $v|_{\partial B} < u|_{\partial B}$, but $\exists z^0 \in B$: $v(z^0) > u(z^0)$. Using the approximation $v_j(z) \downarrow v(z)$, $v_j \in C^\infty$, we can assume that, $v \in C^\infty$. Then $\exists \delta > 0$: $v(z^0) + \delta \left(|z^0|^2 - R^2 \right) > u(z^0)$, where $D \subset \{ |z| < R \}$. Put $w(z) = v(z) + \delta \left(|z|^2 - R^2 \right)$. It is clear, that $w(z)|_{\partial D} < u(z)|_{\partial D}$, $w(z^0) > u(z^0)$, $\mathscr{M}_w(z) \geq \delta m$. Since the sequence $u_k(z)$ uniformly converges in $\overline{D}$, $u(z) = \lim_{k \to \infty} u_k(z)$, then

$$\exists k_0 > \frac{1}{\delta m} : \ w(z)|_{\partial B} < u_k(z)|_{\partial B}, \ w(z^0) > u(z^0), \ \mathscr{M}_{u_k}(z) = \frac{1}{k}, \ k \geq k_0.$$

This is a contraduction since by the Theorem 9 we have $w(z)|_B < u_k(z)|_B$. *The proof is completed.*

Remark 2 We could build a solution of Eq. (4) using Bremermann method, that

$$\tilde{u}(z) = \sup \left\{ v(z) \in m - sh(D) \cap C(\overline{D}) : \ v|_{\partial D} \leq \varphi(\xi), \ \varphi \in C(\partial D) \right\}.$$

It is not difficult to see, that it coincides with the function $u(z)$, constructed in Theorem 10.

5 The Problem of Defining the Operator $\mathscr{M}_u(z)$ in the Class $C \cap m - sh(D)$

Expansion of the operator $\mathscr{M}_u(z)$ from the class C^2 to the class $C \cap m - sh(D)$ is one of fundamental problems in the construction of potential theory in the class of m-subharmonic functions. Indeed, for a rich class of (regular) compact sets $K \subset D$, the $\mathscr{P}_m$ -measure $\omega^*(z, K, D)$ is a continuous function in D.

This problem is not new. For example, in the real analysis, in the theory of convex functions in $\mathbf{R}^n$ A.D. Aleksandrov [5], I.J. Bakelman [6] showed that the Monge-Ampère operator

$$\mathscr{M}\mathscr{A}_u(x) = \prod_j \lambda_j(x)$$

defined in the class C^2, continuously extends into class of continuous convex functions, $C \cap \operatorname{convex}(D)$ as a Borel measure. Here $\lambda_j(x)$, $j = 1, 2, \ldots, n$, are the eigenvalues of the matrix $D^2 u(x)$ at a fixed point x. Furthermore, Neil S. Trudinger, Xu-Jia Wang [51–53] (see also [16]) prove an analogue of this result for the operator in the Hessian

$$S_m(\lambda) = \sum_{1 \leq j_1 < \cdots < j_m \leq n} \lambda_{j_1} \ldots \lambda_{j_m}.$$

We return to our operator

$$\mathscr{M}(z) = \left[\prod_{1 \leq j_1 < \cdots < j_m \leq n} \left(\lambda_{j_1}(z) + \cdots + \lambda_{j_m}(z) \right) \right]^{\alpha_m}.$$

The problem of extending the operator $\mathscr{M}_u(z)$ to the class $C \cap m - sh(D)$ is related to the following problem

Problem 2 (comparison principle) *Let* $u, v \in C^2(\overline{D}) \cap m - sh(D) : u|_D \geq v|_D,$ $u|_{\partial D} = v|_{\partial D},$ *where* ∂D *is a smooth surface. Prove that the following inequality holds*

$$\int_D \mathcal{M}_u(z) dV \leq \int_D \mathcal{M}_v(z) dV \tag{5}$$

For plurisubharmonic functions and for the Monge-Ampère operator, the comparison principle was proved by Bedford-Taylor [10, 11], and for m-convex functions in $\mathbf{R}^n$ inequalities type (5) were considered by many authors (Krylov, Ivochkina, Trudinger, Wang, and others (see [26, 27, 33, 52, 53]).

One estimation. Let $u \in C^2(D) \cap m - sh(D)$ and $\lambda = (\lambda_1, \lambda_2, \ldots, \lambda_n)$ its eigenvalues. Put

$$N = \max \left\{\lambda_{j_1} + \lambda_{j_2} + \cdots + \lambda_{j_m}\right\} : \ 1 \leq j_1 < j_2 < \cdots < j_m \leq n.$$

Then

$$\sum_{1 \leq j_1 < \cdots < j_m \leq n} (\lambda_{j_1} + \cdots + \lambda_{j_m}) = \frac{(m-1)!(n-m+1)!}{n!}(\lambda_1 + \lambda_2 + \cdots + \lambda_n) = \frac{(m-1)!(n-m+1)!}{n!}\Delta u.$$

On the other hand,

$$\sum_{1 \leq j_1 < \cdots < j_m \leq n} \left(\lambda_{j_1} + \cdots + \lambda_{j_m}\right) \leq N \cdot \frac{m!(n-m)!}{n!}$$

and

$$\Delta u \leq \frac{\frac{(m-1)!(n-m+1)!}{n!}}{\frac{m!(n-m)!}{n!}} \max \left\{\lambda_{j_1} + \lambda_{j_2} + \cdots + \lambda_{j_m}\right\} : \ 1 \leq j_1 < j_2 < \cdots < j_m \leq n.$$

Hence,

$$\lambda_{j_1} + \lambda_{j_2} + \cdots + \lambda_{j_m} \leq N \leq \sum_{1 \leq j_1 < \cdots < j_m \leq n} \left(\lambda_{j_1} + \cdots + \lambda_{j_m}\right) =$$

$$= \frac{(m-1)!(n-m+1)!}{n!}\Delta u, \ 1 \leq j_1 < j_2 < \cdots < j_m \leq n.$$

Therefore,

$$\mathcal{M}_u(z) = \left[\prod_{1 \leq j_1 < \cdots < j_m \leq n} \left(\lambda_{j_1}(z) + \cdots + \lambda_{j_m}(z)\right)\right]^{\alpha_m} \leq N^{n-m+1} \leq const(\Delta u(z))^{n-m+1}.$$

Acknowledgements I would like to express my warm thanks to the referee of this paper for numerous corrections and for witty simplification of the proof of Lemma 1, using Newton's inequality.

References

1. Abdullayev, B.I.: Subharmonic functions on complex Hyperplanes of $\mathbb{C}^n$. J. Sib. Fed. Univ. Math. Phys. Krasnoyarsk **6**(4), 409–416 (2013)
2. Abdullayev, B.I.: $\mathscr{P}$-measure in the class of $m - wsh$ functions. J. Sib. Fed. Univ. Math. Phys. Krasnoyarsk **7**(1), 3–9 (2014)
3. Abdullaev, B., Sadullaev, A.: Potential theory in the class of m-subharmonic functions. Proc. Steklov Inst. Math. RAN **279**, 155–180 (2012)
4. Abdullaev, B., Sadullaev, A.: Capacities and hessians in the class of m-subharmonic functions. Dokl. Math. RAN **87**(1), 88–90 (2013)
5. Aleksandrov, A.D.: Dirichlet problem for the equation det $(z_{i,j}) = \varphi$. Vestnik Leningrad Univ. **13**, 5–24 (1958)
6. Bakelman, I.J.: Convex Analysis and Nonlinear Geometric Elliptic Equations. Springer, Berlin (1994)
7. Bedford, E.: Survey of pluripotential theory. Several Complex Variables. Mathematical Notes, vol. 38, pp. 48–95 (1993)
8. Bedford, E., Fornaess, J.E.: Counterexamples to regularity for the complex Monge-Ampère equation. Invent. Math. **50**, 129–134 (1979)
9. Bedford, E., Kalka, M.: Foliations and complex Monge-Ampère equation. Comm. Pure Appl. Math. **XXX**, 543–571 (1977)
10. Bedford, E., Taylor, B.A.: The Dirichlet problem for a complex Monge- Ampère equations. Invent. Math. **37**(1), 1–44 (1976)
11. Bedford, E., Taylor, B.A.: Variational properties of the complex Monge- Ampère equation. I. Dirichlet principle. Duke Math. J. **45**(2), 375–403 (1978); II. Intrisic norms. Am. J. Math. **101**, 1131–1166 (1979)
12. Bedford, E., Taylor, B.A.: A new capacity for plurisubharmonic functions. Acta Math. **149**(1–2), 1–40 (1982)
13. Bishop, E.: Condition for the analyticity of certain sets. Michigan Math. J. **11**, 289–304 (1964)
14. Błocki, Z.: Weak solutions to the complex Hessian equation. Ann. Inst. Fourier, Grenoble, 55, **5**, 1735–1756 (2005)
15. Caffarelli, L., Kohn, J.J., Nirenberg, L., Spruck, J.: The Dirichlet problem for non linear second order elliptic equations, II. Complex Monge-Ampère and Uniformly Elliptic equations. Commun. Pure Appl. Math. **38**, 209–252 (1985)
16. Caffarelli, L., Nirenberg, L., Spruck, J.: The Dirichlet problem for non linear second order elliptic equations, III. Functions of eigenvalues of the Hessian. Acta Math. **155**, 261–301 (1985)
17. Demailly, J.P.: Measures de Monge - Ampère et caracterisation des varietes algebriques affines. Memoires de la societe mathematique de France **19**, 1–125 (1985)
18. Dinew, S., Kołodziej, S.: A priori estimates for the complex Hessian equation Anal. PDE, **17**, 227–244 (2014)
19. Drnovšek, B.D., Forstnerič, F.: Minimal hulls of compact sets in $\mathbb{R}^3$. Trans. Am. Math. Soc. **368**(10), 7477–7506, October 2016. http://dx.doi.org/10.1090/tran/6777. Article electronically published on December 14, 2015
20. Harvey Jr., F.R., Lawson, H.B.: Calibrated geometries. Acta Mathematica **148**, 47–157 (1982)
21. Harvey, F.R., Jr, L.H.B.: An introduction to potential theory in calibrated geometry. Amer. J. Math. **131**(4), 893–944 (2009)
22. Harvey, F.R., Jr, L.H.B.: Duality of positive currents and plurisubharmonic functions in calibrated geometry. Amer. J. Math. **131**(5), 1211–1240 (2009)

23. Harvey, F.R., Lawson, H.B.: Plurisubharmonicity in a general geometric context. Geom. Anal. I, 363–401 (2010)
24. Harvey Jr., F.R., Lawson, H.B.: Geometric plurisubharmonicity and convexity - an introduction. Adv. Math. **230**, 2428–2456 (2012)
25. Ho, L.H.: ∂-problem on weakly q-convex domains. Math. Ann. **290**, 3–18 (1991)
26. Ivochkina, N.: Solution of the Dirichlet problem for a Monge-Ampère type equations. Math. USSR Sbornic **128**(3), 403–415 (1985)
27. Ivochkina, N., Trudinger, N.S., Wang, X.-J.: The Dirichlet problem for degenerate Hessian equations. Comm. Partial Diff. Eq. **29**, 219–235 (2004)
28. Joyce, D.: Riemannian holonomy groups and calibrated geometry. Oxford Graduate Texts in Mathematics 12, OUP (2007)
29. Khusanov, Z.: Capacity properties of $q-$ subharmonic functions, I, Izv. Akad. Nauk. UzSSR, Ser. Fiz.-Mat. **1**, 41–45 (1990)
30. Khusanov, Z.: Capacity properties of $q-$ subharmonic fonctions, II, Izv. Akad. Nauk. UzSSR, Ser. Fiz.-Mat. **5**, 28–33 (1990)
31. Klimek, M.: Pluripotential Theory. Clarendon Press, Oxford (1991)
32. Kołodziej, S.: The complex Monge-Ampère equation and pluripotential theory. Mem. Amer. Math. Soc. **178**, 64p (2005)
33. Krylov, N.V.: A bounded non-homogeneous elliptic and parabolic equations in a domain. Izv. Akad. Nauk SSSR, ser. Mat. **47**(1), 75–108 (1983)
34. Le Mau, H., Xuan Hong, N.: Maximal $q-$ subharmonicity in $\mathbf{C}^n$. Vietnam J. Math. **41**, 1–10 (2013)
35. Levenberg, N., Taylor, B.A.: Comparison of capacities in $\mathbf{C}^n$. Lect. Notes Math. **1094**, 162–172 (1984)
36. Li, S.Y.: On the Dirichlet problems for symmetric function equations of the eigenvalues of the complex Hessian. Asian J. Math. **8**, 87–106 (2004)
37. Lu, HCh.: Solutions to degenerate Hessian equations. Jurnal de Mathematique Pures et Appliques **100**(6), 785–805 (2013)
38. Rudin, W.: A geometric criterion for algebraic varieties. J. Math. Mech. **17**, 671–683 (1968)
39. Sadullaev, A.: A criterion for algebraic varieties of analytic sets. Funct. Anal. Appl. **6**(1), 85–86 (1972)
40. Sadullaev, A., Defect divisors in the sense of Valiron. Russian Math. Sb. **108** (150):4, 567–580 (1979)
41. Sadullaev, A.: Locally and globally $\mathscr{P}-$ regular compacta in $\mathbf{C}^n$. Dokl. Akad. Nauk SSSR **250**(6), 1324–1327 (1980)
42. Sadullaev, A.: Operator $(dd^c u)^n$ and condenser capacity. Dokl. Akad. Nauk SSSR **251**(1), 44–57 (1980)
43. Sadullaev, A.: $\mathscr{P}-$ regularity of sets in $\mathbf{C}^n$. Lect. Notes. Math. **798**, 402–407 (1980)
44. Sadullaev, A.: Plurisubharmonic measures and capacities on complex manifolds. Russian Math. Surveys **36**(4), 61–119 (1981)
45. Sadullaev, A.: Estimates of polynomials on analytic sets. Izv. Akad. Nauk SSSR, ser. Math. **46**(3), 524–534 (1982)
46. Sadullaev, A.: Rational approximation and pluripolar sets. Math. USSR Sbornic **47**(1), 91–113 (1984)
47. Sadullaev, A.: Plurisubharmonic functions, Several Complex variables II. Springer, Berlin. Encyclopedia of Math. Sc. pp. 59–106 (1994)
48. Sadullaev, A.: Pluripotential Theory. Applications. Palmarium Akademic Publishing, Germany (2012)
49. Sadullaev, A., Chirka, E.M.: On continuation of functions with polar singularities. Math. USSR Sbornic **60**(2), 377–384 (1988)
50. Siciak, J.: On some extremal functions and their applications in the theory of analytic functions of several complex variables. Trans. Amer. Math. Soc. **105**(2), 322–357 (1962)
51. Trudinger, N.S.: On the Dirichlet problem for Hessian equation. Acta Math. **175**, 151–164 (1995)

52. Trudinger, N.S., Wang, X.J.: Hessian measures, topological methods in nonlinear analysis. J. Juliusz Schauder Center **10**, 225–239 (1997)
53. Trudinger, N.S., Wang, X.J.: Hessian measures II. Ann. Math. **150**, 579–604 (1999)
54. Verbitsky, M.: Plurisubharmonic functions in calibrated geometry and convexity. Mathematische Zeitschrift **264**(4), 939–957 (2010)
55. Wang, X.J.: The k-Hessian equation. Lecture Notes in Math. Springer, Berlin, vol. 1977, 177–252 (2009)
56. Whiteley, J.N.: On Newton's inequality for real polynomials. Am. Math. Mon. **76**(8), 905–909 (1969)
57. Zakharyuta, V.P.: Extremal plurisubharmonic functions, Hilbert scales and isomorphisms of spaces of analytic functions, I,II. Theory of Functions, Functional analysis and Applications, (1974), 19, 133–157, (1974), 65–83
58. Zeriahi, A.: A criterion of algebraicity for Lelong classes and analytic sets. Acta Math. **184**, 113–143 (2000)

Properties of Solutions of the Cauchy Problem for Degenerate Nonlinear Cross Systems with Convective Transfer and Absorption

Sh. A. Sadullaeva and M. B. Khojimurodova

Abstract In this paper the Cauchy problem for nonlinear systems is considered. The conditions of existence of the solutions on time for the problem Cauchy are given. Moreover the properties of the finite velocity of a propagation and localization of the disturbance, an asymptotic of self-similar solutions will be defined. The results of numerical solutions will be carried out and on the basis of calculations some necessary statements will be given.

Keywords Nonlinear systems · Finite velocity · Localization · Asymptotic Self-similar

1 Introduction

In the domain $Q = \{(t, x) : t > 0, x \in R^N\}$ we investigate properties of the process of a nonlinear diffusion-reaction with heterogeneous density:

$$
\begin{aligned}
\frac{\partial u}{\partial t} &= div\left(v^{m_1-1}\left|\nabla u^k\right|^{p-2}\nabla u\right) - div(c(t)u) - \gamma_1(t)u, \\
\frac{\partial v}{\partial t} &= div\left(u^{m_2-1}\left|\nabla v^k\right|^{p-2}\nabla v\right) - div(c(t)v) - \gamma_2(t)v,
\end{aligned}
\tag{1}
$$

$$
u(0, x) = u_0(x) \geq 0, v(0, x) = v_0(x) \geq 0, \ x \in R^N
\tag{2}
$$

where $k \geq 1$, p, m_i, $i = 1, 2$ - given positive numbers, $\nabla(.) - \underset{x}{grad}(.)$, functions $u_0(x)$, $v_0(x) \geq 0$, $x \in R^N$, $0 < \gamma_i(t) \in C(0, \infty)$, $i = 1, 2$.

Sh. A. Sadullaeva (✉) · M. B. Khojimurodova
Tashkent University of Information Technologies, TUIT, Tashkent, Uzbekistan
e-mail: orif_sh@list.ru

M. B. Khojimurodova
e-mail: mohim15-85@mail.ru

© Springer Nature Switzerland AG 2018   183
Z. Ibragimov et al. (eds.), *Algebra, Complex Analysis, and Pluripotential Theory*,
Springer Proceedings in Mathematics & Statistics 264,
https://doi.org/10.1007/978-3-030-01144-4_15

The system (1) describes a set of physical processes, for example process of mutual reaction - diffusions, heat conductivity, a polytropical filtration of a liquid and gas in the nonlinear environment whose capacity equal $\gamma_1(t)u$, $\gamma_2(t)v$. Particular cases ($k = 1$, $p = 2$) of the system were considered in works [1, 2, 4, 5]. The conditions of existence of the solutions on time for the problem Cauchy are given.

The system (1) in the domain, where $u = v = 0$ -is degenerated, and in the domain of degeneration it may not have the classical solution. Therefore the weak solutions of the system (1) having physical sense: $0 \le u, v \in C(Q)$ and $v^{m_1-1}\left|\nabla u^k\right|^{p-2}\nabla u$, $u^{m_2-1}\left|\nabla v^k\right|^{p-2}\nabla v \in C(Q)$ - satisfying to some integral identity are studied. The problem (1)–(2) for different values of parameters has been studied by many authors [3, 6–12, 15, 16].

2 Properties of the Finite Velocity of a Propagation and Localization of the Disturbance

Although the way of construction of self-similar equation for system (1) is less offered, method of nonlinear splitting facilitates considerably the research of qualitative properties of solutions of problem (1)–(2).

$$
\begin{aligned}
u_+(t, x) &= (T + t)^{-\alpha_1} f(\xi), \\
v_+(t, x) &= (T + t)^{-\alpha_2} \psi(\xi), \\
\xi &= \frac{|x|}{[\tau(t)]^{1/p}}
\end{aligned}
\tag{3}
$$

where

$$
\tau(t) = \frac{(T + t)^{1-[(m_1-1)\alpha_2+(k(p-2))\alpha_1]}}{1 - [(m_1 - 1)\alpha_2 + (k(p - 2))\alpha_1]}.
$$

If

$$
(m_1 - 1)\alpha_2 + (k(p - 2))\alpha_1 = (m_2 - 1)\alpha_1 + k(p - 2)\alpha_2,
$$

$$
f(\xi) = \left(a - \xi^\gamma\right)_+^{q_1}, \quad \psi(\xi) = \left(a - \xi^\gamma\right)_+^{q_2}, \quad a > 0, \quad \gamma = p/(p - 1), \quad (n)_+ = \max(0, n),
$$

$$
q_1 = \frac{(p - 1)(k(p - 2) - (m_1 - 1))}{q}, \quad q_2 = \frac{(p - 1)(k(p - 2) - (m_2 - 1))}{q},
$$

$$
q = [k(p - 2)]^2 - (m_1 - 1)(m_2 - 1).
$$

The numbers are given by the following system of algebraic equations

$$
k(p - 2)q_1 + (m_1 - 1)q_2 = p - 1,
$$

$$
(m_2 - 1)q_1 + k(p - 2)q_2 = p - 1.
$$

The solution have phenomena of the finite velocity of a propagation and localization:

$$u(t, x), \ v(t, x) \equiv 0$$

when

$$|\xi| \geq l(t) = a^{(p-1)/p}[\tau(t)]^{1/p}, \quad 1 - ((m_1 - 1)\alpha_2 + k(p - 2)\alpha_1) > 0.$$

The behavior of the front of indignation $x_f(t)$ in one-dimensional case is defined by the equality

$$x_f(t) = \int_0^t c(y)dy - a^{(p-1)/p}[\tau(t)]^{1/p}.$$

Therefore the front of indignation when

$$\int_0^t c(y)dy > a^{(p-1)/p}[\tau(t)]^{1/p},$$

grows in a direction of the right axis, filling all right plane, and when

$$\int_0^t c(y)dy < a^{(p-1)/p}[\tau(t)]^{1/p}.$$

The front plane moves on the left filling all the left at increase. In a special case when the condition is satisfied

$$\int_0^t c(y)dy = a^{(p-1)/p}[\tau(t)]^{1/p},$$

the movement of the front of indignation stops. Thus, there happens a full stop of the front of indignations. This new property takes place only when convective transfer with velocity dependent on time is present. Let's notice, if $k = 1$, $m_1 = m_2$, $p = 2$ then the system (1) turns to one equation with the solution type of Zeldovich–Barenblatt

$$u(t, x) = (T + t)^{-\alpha} \bar{f}(\xi), \quad \alpha = \frac{N}{2 + (m - 1)N}, \quad \xi = \frac{|x|}{[\tau(t)]^{1/p}},$$

$$\bar{f}(\xi) = A \left(a - \xi^{p/(p-1)}\right)_+^{1/(m-1)}, \quad [\tau(t)]^{1/p} = (T + t)^{1/(2+(m-1)N)}.$$

The method of construction of the solution which is offered above, allows to establish the new nonlinear effects for other nonlinear degenerating systems, thanks to theorems of comparison of solutions.

3 Localisation of Solutions of System (1)

The system (1) has the solution

$$u(t, x) = (T + \tau(t))^{-\alpha_1} u_1(t) f_3(\xi),$$

$$v(t, x) = (T + \tau(t))^{-\alpha_2} v_1(t) \psi_3(\xi),$$

$$\xi = \frac{|\eta|}{[T + \tau(t)]^{1/p}},$$

$$u_1(t) = (T + \tau(t))^{-\alpha_1}, \quad v_1(t) = (T + \tau(t))^{-\alpha_2},$$

$$\tau(t) = \int_0^t v_1^{m_1-1}(\eta) u_1^{p-2}(\eta) d\eta = \int_0^t u_1^{m_1-1}(\eta) v_1^{p-2}(\eta) d\eta,$$

where $f_3(\xi)$, $\psi_3(\xi)$- are functions defined above. Functions u, v-possess the property

$$u(t, x), \ v(t, x) \equiv 0, \quad \text{if} \ \ |\xi| \geq a^{(p-1)/p} [\tau(t)]^{1/p},$$

or

$$\left| \int_0^t c(y) dy - x \right| \geq l(t) = a^{(p-1)/p} [\tau(t)]^{1/p}.$$

Therefore, conditions of localization of the solutions will be the following conditions

$$\tau(t) < \infty, \quad \int_0^t c(y) dy < \infty, \quad \forall t > 0.$$

On the next stage on the basis of construction of the self-similar solutions the type of Zeldovich–Kompanejts we will show new property of the solution of the system (1), namely property FVPD. For this purpose for the solution of system (1) we search in a the kind of

$$\begin{aligned} u(t, x) &= (T + t)^{-\alpha_1} w(\tau(t), x), \\ v(t, x) &= (T + t)^{-\alpha_2} z(\tau(t), x). \end{aligned} \tag{4}$$

Then, substituting (4) in (1), after simple calculations we choose function $\tau(t)$ as follows

$$\tau(t) = \int (T+t)^{-(m_1-1)\alpha_1-k(p-2)\alpha_2}dt = \int (T+t)^{-(m_1-1)\alpha_2-k(p-2)\alpha_1}dt.$$

At $1-(m_1-1)\alpha_2-k(p-2)\alpha_1 > 0$ calculation of the integral for $\tau(t)$ will give the following result

$$\tau(t) = \frac{(T+t)^{1-(m_1-1)\alpha_2-k(p-2)\alpha_1}}{1-(m_1-1)\alpha_2-k(p-2)\alpha_1}.$$

Then we will have the system

$$
\begin{aligned}
\frac{\partial w}{\partial \tau} &= div\left(z^{m_1-1}\,|\nabla w|^{p-2}\,\nabla w\right) + \alpha_1(T+t)^{p_1}w, \\
\frac{\partial z}{\partial \tau} &= div\left(w^{m_2-1}\,|\nabla z|^{p-2}\,\nabla z\right) + \alpha_2(T+t)^{p_2}z,
\end{aligned}
\tag{5}
$$

$$p_1 = -1 + (m_1-1)\alpha_2 + k(p-2)\alpha_1,$$

$$p_2 = -1 + (m_1-1)\alpha_1 + k(p-2)\alpha_2.$$

The system can be rewritten in a kind of

$$
\begin{aligned}
\frac{\partial w}{\partial \tau} &= div\left(z^{m_1-1}\,|\nabla w|^{p-2}\,\nabla w\right) + \frac{\alpha_1}{[1-(m_1-1)\alpha_2+k(p-2)\alpha_1]\tau(t)}w, \\
\frac{\partial z}{\partial \tau} &= div\left(w^{m_2-1}\,|\nabla z|^{p-2}\,\nabla z\right) + \frac{\alpha_2}{[1-(m_2-1)\alpha_1+k(p-2)\alpha_2]\tau(t)}z,
\end{aligned}
$$

Further, believing

$$w\left(\tau(t),x\right) = f(\xi), \quad v\left(\tau(t),x\right) = \psi(\xi), \quad \xi = \frac{|x|}{[\tau(t)]^{1/p}} \tag{6}$$

And substituting (6) in (5), after simple calculations, at condition performance

$$(m_1-1)\alpha_2 + k(p-2)\alpha_1 = (m_2-1)\alpha_1 + k(p-2)\alpha_2 \tag{7}$$

for the functions $f(\xi)$, $\psi(\xi)$ we will have the following degenerate self-similar equation

$$
\begin{aligned}
\xi^{1-N}\frac{d}{d\xi}\left(\xi^{N-1}\psi^{m_1-1}\left|\frac{df^k}{d\xi}\right|^{p-2}\frac{df}{d\xi}\right) + \frac{\xi}{p}\frac{df}{d\xi} + b_1f = 0 \\
\xi^{1-N}\frac{d}{d\xi}\left(\xi^{N-1}f^{m_2-1}\left|\frac{d\psi^k}{d\xi}\right|^{p-2}\frac{d\psi}{d\xi}\right) + \frac{\xi}{p}\frac{d\psi}{d\xi} + b_2\psi = 0
\end{aligned}
\tag{8}
$$

where

$$b_1 = \alpha_1/[1-(m_1-1)\alpha_2-k(p-2)\alpha_1], \quad b_2 = \alpha_2/[1-(m_2-1)\alpha_1-k(p-2)\alpha_2].$$

Theorem 1 *If were* $b_1 \leq \frac{N}{p}$, $i = 1, 2$, *and* A_1, A_2 *-are the solutions of algebraic equation system*

$$A_1^{k(p-2)} A_2^{m_1-1} = 1/p \, (\gamma k \gamma_1)^{p-1}, \quad A_1^{m_2-1} A_2^{k(p-2)} = 1/p \, (\gamma k \gamma_2)^{p-1}$$

and

$$u_0(x) = A_1 T^{-\alpha_1} f_3(\xi)_{t=0}, \quad v_0(x) = A_2 T^{-\alpha_2} \psi_3(\xi)_{t=0}, \quad x \in R^N.$$

Then the solution of the system (1) possesses the property of FVPD.

Theorem 2 *Let* $b_1 \leq \frac{N}{p}$, $i = 1, 2$, $1 - [(m_1 - 1)\alpha_2 + k(p - 2)\alpha_1] > 0$ *as well as* A_1, A_2- *the solutions of algebraic equation system*

$$A_1^{k(p-2)} A_2^{m_1-1} = 1/p \, (\gamma k \gamma_1)^{p-1}, \quad A_1^{m_2-1} A_2^{k(p-2)} = 1/p \, (\gamma k \gamma_2)^{p-1}$$

and
$$u_0(x) = A_1 T^{-\alpha_1} f_3(\xi)_{t=0}, \quad v_0(x) = A_2 T^{-\alpha_2} \psi_3(\xi)_{t=0}, \quad x \in R^N.$$

Then the solution of the system will be spatially localised, if the following conditions is true at $t > 0$

$$\int_0^t c(y)dy < \infty, \quad \tau(t) < \infty.$$

4 Asymptotic of Self-similar Solutions

Theorem 3 *Let's* $q_1 > 0$, $q_2 > 0$. *Then the solution of the system (8) with the compact carrier at* $\eta \to \infty$ $\left(\eta = -\ln\left(a - \xi^{p/(p-1)}\right)\right)$ *has asymptotical representation*

$$f(\xi) = A_5 f_3(\xi) \, (1 + o(1)), \quad \psi(\xi) = A_6 \psi_3(\xi) \, (1 + o(1)),$$

where $A_i > 0$, $i = 5, 6$ - *are the solutions of system of the algebraic equations*

$$A_1^{k(p-2)} A_2^{m_1-1} = 1/p \, (\gamma k \gamma_1)^{p-1}, \quad A_1^{m_2-1} A_2^{k(p-2)} = 1/p \, (\gamma k \gamma_2)^{p-1}.$$

From this theorem for the position of free boundary it is had $|\xi| \to l\,(t) = a^{(p-1)/p}$
$[\tau(t)]^{1/p}$, *i.e.* $\left| \int_0^t c(y)dy - x \right| \to a^{(p-1)/p}[\tau(t)]^{1/p}$,

where $\tau(t) = \int_0^t [u_1(y)]^{k(p-2)+(m_1-1)}dy$, $u_1(t) = \exp\left(-\int_0^t \gamma_1(y)dy\right)$.

Theorem 4 *If* $q_i < 0,$ $\quad N + kqq_i < 0,$ $\quad i = 1, 2.$ *Then the solution of system of disappearing on infinity (8) at* $\eta \to \infty$ $\;\left(\eta = \ln\left(a + \xi^{p/(p-1)}\right)\right)$ *will have the asymptotical* *representation* $\quad f(\xi) = A_7 (a + \xi^\gamma)^{q_1} (1 + o(1))$ $\;\psi(\xi) = A_8$ $(a + \xi^\gamma)^{q_2} (1 + o(1)),$ *where factors* $A_i > 0,$ $\;i = 7, 8$- *the solutions of system of the algebraic equations*

$$A_7^{k(p-2)} A_8^{m_1-1} = \left[-\frac{1}{\gamma^{p-1} p(N + k\gamma q_1)} + b_1 \right] (|k\gamma q_1|)^{2-p},$$

$$A_7^{m_2-1} A_8^{k(p-2)} = \left[-\frac{1}{\gamma^{p-1} p(N + k\gamma q_2)} + b_2 \right] (|k\gamma q_2|)^{2-p}.$$

5 Results of Numerical Experiments and Visualization

At the numerical solution of a problem, the equation was approximated on a grid under the implicit circuit of variable directions (for a multidimensional case) in a combination to the method of balance. Iterative processes were constructed based on the methods of Picard, Newton and a special method. Results of computational experiments show, that all listed iterative methods are effective for the solution of nonlinear problems and leads to the nonlinear effects if we use as initial approximation the solutions of self similar equations constructed by the method of nonlinear splitting and by the method of standard equation. As it was expected, for achievement of identical accuracy the method of Newton demands smaller quantity of iterations, than methods of Picard and special method due to a successful choice of an initial approximation. We observe that in each considered cases Newton's method has the best convergence due to a good initial approximation. The result of numerical experiments are presented in visual form with animation.

References

1. Aripov, M.: Asymptotic of the solutions of the non-Newton polytrophic filtration equation. ZAMM **80**(3), 767–768 (2000)
2. Aripov, M.: One invariant group method for the quasilinear equations and their system. In: Proceedings of the International Conference on Mathematics and its Applications in the New Millennium. Malaysia, pp. 535–543 (2000)
3. Aripov, M., Muhammadiev, J.: Asymptotic behavior of auto model solutions for one system of quasilinear equations of parabolic type. Buletin Stiintific-Universitatea din Pitesti, Seria Matematica si Informatica. **3**, 19–40 (1999)
4. Aripov, M., Sadullaeva, ShA: An asymptotic analysis of a self-similar solution for the double nonlinear reaction-diffusion system. J. Nanosyst. Phys. Chem. Math. **6**(6), 793–802 (2015)
5. Aripov, M., Sadullaeva, ShA: Qualitative properties of solutions of a doubly nonlinear reaction-diffusion system with a source. J. Appl. Math. Phys. **3**, 1090–1099 (2015)

6. Cho, Chien-Hong: On the computation of the numerical blow-up time. Jpn. J. Ind. Appl. Math. **30**(2), 331–349 (2013)
7. Deng, K., Levine, H.A.: The role of critical exponents in blow-up theorems: The sequel. J. Math. Anal. Appl. **243**, 85–126 (2000)
8. Ferreira, Rael, Perez-Llanos, Mayte: Blow-up for the non-local -Laplacian equation with a reaction term. Nonlinear Anal.: Theory Methods Appl. **75**(14), 5499–5522 (2012)
9. Jiang, Z.X., Zheng, S.N.: Doubly degenerate parabolic equation with nonlinear inner sources or boundary flux. Doctor Thesis, Dalian University of Technology, In China, (2009)
10. Martynenko, A.V., Tedeev, A.F.: The Cauchy problem for a quasilinear parabolic equation with a source and inhomogeneous density. Comput. Math. Math. Phys. **47**(2), 238–248 (2007)
11. Martynenko, A.V., Tedeev, A.F.: On the behavior of solutions to the Cauchy problem for a degenerate parabolic equation with inhomogeneous density and a source. Comput. Math. Math. Phys. **48**(7), 1145–1160 (2008)
12. Mu, C., Zheng, P.: Dengming Liu, Localization of solutions to a doubly degenerate parabolic equation with a strongly nonlinear source. Commun. Contemp. Math. **14**, 1250018 [18 pages]. https://doi.org/10.1142/S0219199712500186
13. Sadullaeva, ShA: Numerical investigation of solutions to a reaction-diffusion system with variable density. Journal Sib. Fed. Univ. Math. Phys., J. Sib. Fed. Univ. Math. Phys. **9**(1), 90–101 (2016)
14. Samarskii, A.A., Galaktionov, V.A., Kurduomov, S.P., Mikhailov, A.P.: Blowe-up in quasilinear parabolic equations, vol. 4, p. 535. Walter de Grueter, Berlin (1995)
15. Tedeyev, A.F.: Conditions for the existence and nonexistence of a compact support in time of solutions of the Cauchy problem for quasilinear degenerate parabolic equations. Sib. Math. Jour. **45**(1), 189–200 (2004)
16. Vazquez, J.L.: The Porous Medium Equation. Mathematical Theory, Oxford Mathematical Monographs, p. 430. The Clarendon Press, Oxford University Press, Oxford (2007)

Class R of Gonchar in $\mathbf{C}^n$

Azimbay Sadullaev and Zafar Ibragimov

Abstract The class R in the complex space $\mathbf{C}^n$ was introduced and investigated by A. A. Gonchar. This class and its properties have found a number of applications in approximation theory, in problems of analytic continuations of functions, in descriptions of the structures of singular sets of analytic functions, and in pluripotential theory. In this paper we study further properties of this class.

Keywords Rational approximation · Pluripolar set · Fine-analytic function · Fine topolgy · Plurisubharmonic function

1 Introduction

In this paper we denote by $K \subset \mathbf{C}^n$ a nonpluripolar compact subset of the n-dimensional complex space $\mathbf{C}^n$. We say that a continuous function $f \in C(K)$ belongs to the class $R = R_K$ if it admits rapid rational approximation in K. That is,

$$\lim_{m \to \infty} \rho_m^{1/m}(f, K) = 0, \qquad (1)$$

where $\rho_m(f, K) = \inf \left\{ \|r_m - f\|_K : \deg r_m \leq m \right\}$ is the deviation of f from the class of rational functions of degree m. We note that on the complex plane $\mathbf{C}$ (case $n = 1$), for any domain $D \subset \mathbf{C}$ there exists a function $f \in O(D) : f \in R_K \; \forall K \subset\subset D$ such that f does not holomorphically continue outside D. One of the properties of the functions from the class R, which was proved by A. A. Gonchar, is their single-valuedness in $\mathbf{C}^n$, i.e., its maximum natural Weierstrass domain of exis-

A. Sadullaev
Department of Mathematics, National University of Uzbekistan,
Tashkent, Uzbekistan 100021
e-mail: sadullaev@mail.ru

Z. Ibragimov (✉)
Department of Mathematics, Urgench State University, Khorezm, Urgench,
Uzbekistan 220100
e-mail: z.ibragim@gmail.com

© Springer Nature Switzerland AG 2018

Z. Ibragimov et al. (eds.), *Algebra, Complex Analysis, and Pluripotential Theory*,
Springer Proceedings in Mathematics & Statistics 264,
https://doi.org/10.1007/978-3-030-01144-4_16

tence is one-sheeted, $W_f \subset \mathbf{C}^n$ [12–14]. The class R and its properties have found a number of applications in multidimensional complex analysis, in approximation theory, in problems of analytic continuation of functions with thin singularities [15, 26, 27], and in pluripotential theory [16, 21–25].

In the paper [28] the authors considered germs of a finite order holomorphic function $f \in R^0$ in a neighborhood of a point $0 \in \mathbf{C}$ and found a connection between such functions and fine-analytic (finely analytic) functions in the whole plane $\mathbf{C}$. The notion of fine-analyticity was introduced and used by B. Fuglede [10, 11] for a class of functions that have the Mergelyan property in fine neighborhoods of a point. In work of A. Edigarian and J. Wiegerinck [4, 5], A. Edigarian, S. El Marzguioui and J. Wiegerinck [6], S. El Marzguioui and J. Wiegerinck [20], J. Wiegerinck [29], T. Edlund [7], T. Edlund and B. Jöricke [8] fine-analytic functions were studied for their applications in pluripotential theory, more precisely, in the description of pluripolar hulls, in the establishment of pluripolarity of graphs $\Gamma_f = \{w = f(z)\}$. On pluripolar hulls of analytic sets and graphs $\Gamma_f = \{w = f(z)\}$ see also the papers of N. Levenberg, G. Martin and E. Poletsky [17], N. Levenberg and E. Poletsky [18]. In the paper [8], T. Edlund and B. Jöricke first noted that fine-analytic continuation is connected to the pluripolar hull of graphs of analytic functions.

The goal of this work is to study properties of functions from the class R_K for arbitrary nonpluripolar compact set $K \subset \mathbf{C}^n$. More precisely, we study the following question: if $f \in R_K$ then, what additional properties does f have outside of K? We note that if in Eq. (1) instead of rational approximations we consider polynomial approximations, i.e.,

$$\lim_{m \to \infty} e_m^{1/m}(f, K) = 0, \tag{2}$$

where $e_m(f, K) = \inf\{\|p_m - f\|_K : p_m - polynomials, \deg p_m \leq m\}$, then it is clear that f continues to the whole space $\mathbf{C}^n$ as an entire function. The main results of this work are the following three theorems.

Theorem 1 *If a function $f \in R_K$ is of finite order s, i.e., there is a sequence of rational functions*

$$r_m(z) = \frac{p_m(z)}{q_m(z)}, \quad \deg r_m \leq m, \quad m = 1, 2, \ldots,$$

such that

$$\|f - r_m\|_K^{1/m} \leq \frac{1}{m^{1/s}}, \quad m \geq m', \ s < \infty,$$

then f fine-analytically (see Definition 2) continues to the whole space $\mathbf{C}^n$. That is, there is a fine-analytic function $\tilde{f}$ on $\mathbf{C}^n$ such that $\tilde{f}|_K \equiv f$.

Theorem 2 *In the assumptions of Theorem 1, the sequence $\{r_m(z)\}$ rapidly converges in capacity to the function $\tilde{f}$ in the whole $\mathbf{C}^n$, i.e.,*

$$\lim_{m \to \infty} C \left\{ z \in B\,(0, R) : \ \left| r_m\,(z) - \tilde{f}\,(z) \right|^{1/m} > \varepsilon \right\} = 0, \ \ \forall \varepsilon > 0, \ \forall R > 0.$$

The class R_K of finite order has also a following property.

Theorem 3 *Let $f \in R_K$ be a function of finite order and let $\tilde{f}$ be its fine-analytic continuation in $\mathbb{C}^n$. Then any sequence of rational functions $r_m^*\,(z)$, $\deg r_m^* \leq m$, which rapidly converges to the function $f\,(z)$ on a compact set K will rapidly converge in capacity to the function $\tilde{f}$, in the whole $\mathbb{C}^n$.*

We note that this theorem does not require finite order rapid convergence of $r_m^*\,(z)$. The techniques of proof of these main theorems is close to the case of $n = 1$, but for arbitrary n we use pluripotential theory. Moreover, in connection with the complicated complex structure of compacta on which rational approximation is possible, we have to use a weaker definition of fine-analytic functions (see Definition 2 below).

2 Definitions

We first recall the definition of fine-analytic functions. In the plane $\mathbb{C}$ they are determined by means of the fine topology (see, for example [1–3]). The (pluri-)fine topology in $\mathbb{C}^n$ is the weakest topology in which all plurisubharmonic (*psh*) functions are continuous. This topology is generated by the sets of the form

$$\{z \in B : \ u\,(z) > 0, \ \ u \in psh\ (B)\},$$

where $B = B\,(a, \varepsilon) : \ \{|z - a| < \varepsilon\}$ is a ball. A fine open neighborhood of a point a is an open set $V \ni a$, in the fine topology for which the complement $W = \mathbb{C}^n \backslash V$ is thin at the point a, i.e., $\exists B = B(a, \varepsilon)$, $\exists u \in psh\,(B) : \ \lim_{z \to a} \sup_{z \in B \cap W} u\,(z) < u\,(a)$. V is a fine neighborhood of a if it contains a fine open neighborhood $U \ni a$.

Definition 1 A function f is *fine-analytic quasi everywhere in a planar domain* $D \subset \mathbb{C}$, (in short *fine-analytic*, because there is no risk of confusion with (fine)-analytic functions on all of D), if

1. it is defined almost everywhere with respect to the capacity in D, i.e., outside of some polar set $E \subset D$ it assumes finite values;
2. for each point $z^0 \in D \backslash E$ there is a closed fine neighborhood F of z^0, such that $f|_F \in R\,(F)$ or, equivalently, the restriction $f|_F$ is uniformly approximated by rational functions on F.

As mentioned in the work of Bedford-Taylor [1], in contrast to the planar case, in the multidimensional space $\mathbb{C}^n$, $n > 1$, the notion of thin set (i.e., a set that is thin in any of its point) does not coincide with the concept of a pluripolar set. Moreover, because of the intricate structure of $R\,(F)$ for $F \subset \mathbb{C}^n$, it is convenient for us to define fine-analytic functions in $\mathbb{C}^n$ as follows.

Definition 2 A function $f(z)$ is called fine-analytic in a domain $D \subset \mathbf{C}^n$, if there is an increasing sequence of closed sets $F_j \subset F_{j+1} \subset D$, $j = 1, 2, \ldots$, such that

1. for an arbitrary ball $B \subset\subset D$ the condenser capacity $C(B \cap F_j) \to 0$ as $j \to \infty$. Note it follows that $E := D \setminus \cup_j F_j$ is pluripolar.
2. $f(z)$ assumes a finite value for all z in $\cup_j F_j$;
3. for each ball $B \subset\subset D$ and for each number j, the restriction $f|_{F_j \cap \bar{B}} \in R(F_j \cap \bar{B})$,

   (i.e., $f|_{F_j \cap \bar{B}}$ can be uniformly approximated by rational functions on $F_j \cap \bar{B}$).

Example 1 We should point out that in $\mathbf{C}^n$, $n > 1$, we cannot give a definition of fine-analytic functions as in Definition 1. The function $r(z_1, z_2) = z_2^2/z_1$ if $(z_1, z_2) \neq (0, 0)$ and $r(0, 0) = 0$ if $(z_1, z_2) = (0, 0)$ is fine-analytic in $\mathbf{C}^2$ in the sense of Definition 2, but it is not fine-analytic in the sense of Definition 1: the pluripolar set $\{z_2 = 0\}$ is not thin at the point $(0, 0) \in \mathbf{C}^2$.

Remark 1 It is possible that for $n = 1$, Definitions 1 and 2 coincide. However for $n = 1$, replacing (1) in Definition 2 by the condition that $E = D \setminus \cup_j F_j$ is a polar set, this modified definition gives a different class of functions than finely holomorphic functions on the fine domain $\cup_j F_j$ as defined in [9], even if the polar set E is empty.

As an example, let Δ be the unit disc and γ an arc of positive measure connecting -1 and 1. Let $f = \int_\gamma 1/(\zeta - z) dm$. Then f is continuous on $\mathbf{C}$ and holomorphic outside γ. The set $\Delta \setminus \gamma$ has two components. Let F_0 be the union of the upper component with γ, and let

$$F_j = \{F_0 \cap \{|z| \le 1 - 1/j\} \cup \{z \in \Delta : d(z, F_0) \ge 1/j, |z| \le 1 - 1/j\}.$$

Then f is even in $P(F_j)$, $j = 1, 2, \ldots$; $\cup_j F_j = \Delta$; and $E = \Delta \setminus \cup_j F_j = \emptyset$. However, f is not fine analytic in the sense of Definition 1.

Definition 3 We say that a function $f \in R_K$ has finite order s if its rate of approximation $\rho_m^{1/m}(f, K) \to 0$, as $m \to \infty$, is faster than $m^{-1/s}$, $s < \infty$.

We note that the following formula provides a way to calculate the exact order of approximation:

$$t = \varlimsup_{m \to \infty} \frac{m \ln m}{-\ln \rho_m(f, K)}. \tag{3}$$

Moreover, the word "order" is justified here. Indeed, if we consider $e_m^{1/m}(f, K)$ instead of the rate of approximation $\rho_m^{1/m}(f, K)$, then we will have an entire function $\tilde{f}$ in $\mathbf{C}^n$ of order t (see [19]).

3 Proofs of the Main Results

Proof of Theorem 1

Step 1. By the assumption of finite order $t < \infty$ and by the definition of rate of approximation $\rho_m = \rho_m (f, K)$, for any number $s > t$, there is sequence of rational functions

$$r_m (z) = \frac{p_m (z)}{q_m (z)}, \quad m = 1, 2, \ldots,$$

such that

$$\| f - r_m \|_K^{1/m} = \rho_m^{1/m} \le \frac{1}{m^{1/s}}, \quad m \ge m'.$$

Now we consider a subsequence $m_k = k^l$, where l is a fixed integer, $l \ge 8s$. We represent the function f in the form of a functional series

$$f (z) = r_{m_1} (z) + \sum_{k=1}^{\infty} \left[r_{m_{k+1}} (z) - r_{m_k} (z) \right], \quad z \in K. \tag{4}$$

Since the sequence of partial sums of the series $S_{k-1} (z) = r_{m_k} (z)$ and $\| f - r_{m_k} \|_K^{1/m_k} \le \frac{1}{m_k^{1/s}}$, $m_k \ge m'$, the series (4) converges uniformly rapidly to f on K.

Step 2. Next we prove convergence of the series (4) in capacity on $\mathbb{C}^n$. For this we estimate the difference of $r_{m_{k+1}} (z) - r_{m_k} (z)$, $m_k \ge m'$:

$$\left\| r_{m_{k+1}} (z) - r_{m_k} (z) \right\|_K \le \left\| r_{m_{k+1}} (z) - f (z) \right\|_K +$$
$$+ \left\| r_{m_k} (z) - f (z) \right\|_K \le \left\{ \frac{1}{[m_{k+1}]^{m_{k+1}/s}} + \frac{1}{[m_k]^{m_k/s}} \right\} \le \frac{2}{[m_k]^{m_k/s}}. \tag{5}$$

Multiplying the numerator and denominator of $r_m (z) = \frac{p_m(z)}{q_m(z)}$ by a constant we may assume that the maximum modulus of the coefficients of $q_m (z)$ is equal to 1. Then the numerator of

$$r_{m_{k+1}} (z) - r_{m_k} (z) = \frac{p_{m_{k+1}} (z) q_{m_k} (z) - p_{m_k} (z) q_{m_{k+1}} (z)}{q_{m_k} (z) q_{m_{k+1}} (z)}$$

is a polynomial of degree $d_k = m_k + m_{k+1} = k^l + (k + 1)^l$, and

$$\left\| p_{m_{k+1}} (z) q_{m_k} (z) - p_{m_k} (z) q_{m_{k+1}} (z) \right\|_K$$
$$\le \left\| r_{m_{k+1}} (z) - r_{m_k} (z) \right\|_K \cdot \left\| q_{m_k} (z) q_{m_{k+1}} (z) \right\|_K. \tag{6}$$

Since the maximum modulus of the coefficients of polynomials $q_m(z)$ is equal to 1, then

$$|q_m(z)| = \left|\sum_{|J| \leq m} a_J z^J\right| \leq \sum_{|J| \leq m} |z^J| \leq \left(\sum_{j=0}^{m} r^j\right)^n \leq [(m+1)r^m]^n$$

for $|z_1| < r, \ldots, |z_n| < r$, $r > 2$, and for $K \subset \{|z_1| < r, \ldots, |z_n| < r\}$

$$\|q_m(z)\|_K \leq [(m+1)r^m]^n \leq r^{(m+1)n} \frac{(m+1)^n}{2^n} \leq [mr^{m+1}]^n.$$

From here and according to (5) and (6) we have

$$\left\|p_{m_{k+1}}(z)q_{m_k}(z) - p_{m_k}(z)q_{m_{k+1}}(z)\right\|_K \leq A_1 \frac{1}{[m_k]^{m_k/s}}, \quad m_k \geq m', \qquad (7)$$

where

$$A_1 = 2\left[m_{k+1}^2 r^{2(m_{k+1}+1)}\right]^n = 2r^{2n}\left[m_{k+1}r^{m_{k+1}}\right]^{2n} =$$

$$= 2r^{2n}m_{k+1}^{2n}\left[r^{2n}\right]^{m_{k+1}} = 2r^{2n}(k+1)^{2nl}r^{2n(k+1)^l}.$$

Now we use the Bernstein-Walsh inequality. That is, for any polynomial $p(z)$ we have an inequality

$$|p(z)| \leq \|p(z)\|_K \cdot [\Phi(z, K)]^{\deg p}, \quad z \in \mathbf{C}^n,$$

where $\Phi(z, K) = \exp V^*(z, K)$, $V^*(z, K)$ is the extremal Green function of the compact set $K \subset \mathbf{C}^n$. Since K is not a pluripolar set, the function $\Phi(z, K) = \exp V^*(z, K)$ is locally bounded from above in $\mathbf{C}^n$. From this and (7) it follows that

$$\left|p_{m_{k+1}}(z)q_{m_k}(z) - p_{m_k}(z)q_{m_{k+1}}(z)\right| \leq A_1 \frac{1}{[m_k]^{m_k/s}} \cdot [\Phi(z, K)]^{d_k}, \quad z \in \mathbf{C}^n.$$

Consequently, we get the following estimate:

$$\left|r_{m_{k+1}}(z) - r_{m_k}(z)\right| \leq A_1 \frac{1}{[m_k]^{m_k/s}} \cdot \frac{[\Phi(z, K)]^{d_k}}{\left|q_{m_k}(z)q_{m_{k+1}}(z)\right|}, \quad z \in \mathbf{C}^n, \quad m_k \geq m'. \quad (8)$$

Step 3. Next, we use the so called τ–capacity (see, for example [22]). Let E be an arbitrary compact set, which belongs to a ball $B = B(0, R)$. It is easy to prove that for each m there is a polynomial T_m of degree $\deg T_m \leq m$, normalized with a condition $\|T_m\|_{\bar{B}} = 1$, for which the norm $\|T_m\|_E$ is minimal among all such polynomials.

Moreover, there exists a limit $\lim\limits_{m\to\infty} \|T_m\|_E^{1/m} = \tau(E)$, and $\|T_m\|_E^{1/m} \geq \tau(E)$ for all $m \geq 0$.

A value $\tau(E)$ is called τ-capacity of the set $E \subset B$. We note that if instead of normalization $\|T_m\|_{\bar{B}} = 1$, we considered the normalization with the highest modulus coefficient of T_m is equal to 1, then we obtain a capacity value that is equivalent to $\tau(E)$.

We have a following properties of $\tau(E)$.

(1) $\tau(E) = \exp\left\{ -\max\limits_{z\in\bar{B}} V^*(z, E) \right\} = \frac{1}{\max\limits_{z\in\bar{B}} \Phi(z, E)}$.

(2) $\tau-$ capacity is close to zero if and only if the condenser capacity $C(E) = C(E, B)$ is close to zero. More precisely, for any $\varepsilon > 0$ there is a $\delta(\varepsilon) > 0$ such that $C(E) < \varepsilon$ whenever $\tau(E) < \delta(\varepsilon)$ and, conversely, $\tau(E) < \varepsilon$ whenever $C(E) < \delta(\varepsilon)$. From here it follows that $\tau(E) = 0$ is equivalent to the fact that E is pluripolar set.

(3) *The value $\tau(E)$ is a convex set-function, i.e., for any compact sets K_1, $K_2 \subset B$ the following inequality holds:*

$$\tau\left(K_1 \bigcup K_2\right) + \tau\left(K_1 \bigcap K_2\right) \leq \tau(K_1) + \tau(K_2).$$

In particular, $\tau(E)$ is $\sigma-$ subadditive.

The second property can be proved using the first property, and the third property can be proved like one proves convexity of logarithmic capacity on the plane (see [19, p. 213]).

Step 4. We let

$$Q_k = \left\{ z \in \mathbf{C}^n : \ |q_{m_k}(z)|^{1/m_k} < \frac{1}{k^2} \right\}.$$

We consider a bounded domain in $\mathbf{C}^n$, say $B = B(0, R)$, $R > 1$. If we denote $Q_{k,B} = Q_k \bigcap B$, then the inner $\tau-$ capacity of this set satisfies $\tau\left(Q_{k,B}\right) \leq \frac{A_2}{k^2}$, where A_2 is a constant depending only on R. From $\sigma-$ subadditivity of $\tau-$ capacity we have $\tau\left(\bigcup_{k\geq N} Q_{k,B}\right) \to 0$, $N \to \infty$. Consequently, the set $E = \bigcap_{N=1}^{\infty} \bigcup_{k\geq N} Q_k$ is pluripolar.

If a point $z^0 \notin \bigcup_{k\geq N} Q_k$, then at this point

$$|q_{m_k}(z^0)|^{1/m_k} \geq \frac{1}{k^2}, \quad k \geq N,$$

and therefore,

$$|q_{m_k}(z^0) q_{m_{k+1}}(z^0)|^{1/d_k} \geq \left[\left(\frac{1}{k^{2m_k}}\right)\left(\frac{1}{(k+1)^{2m_{k+1}}}\right)\right]^{\frac{1}{d_k}} \geq$$

$$\geq \left[\frac{1}{(k+1)^2}\right]^{\frac{m_k+m_{k+1}}{m_k+m_{k+1}}} = \frac{1}{(k+1)^2}.$$

According to (8)

$$\left|r_{m_{k+1}}\left(z^0\right) - r_{m_k}\left(z^0\right)\right| \leq A_1 \left[\frac{1}{k^l}\right]^{\frac{k^l}{s}} \cdot \left[\Phi\left(z^0, K\right)\right]^{d_k} (k+1)^{2d_k} =$$

$$= A_1 k^{[2d_k - \frac{l}{s}k^l]} \cdot \left[\Phi\left(z^0, K\right)\right]^{d_k} \left(1 + \frac{1}{k}\right)^{2d_k} \leq$$

$$\leq A_1 k^{[2d_k - \frac{l}{s}k^l]} \cdot \left[4\Phi\left(z^0, K\right)\right]^{d_k}, \quad k \geq N. \tag{9}$$

Since $d_k = k^l + (k+1)^l$ and $\frac{l}{s} \geq 8$,

$$2d_k - \frac{l}{s}k^l = \left\{2\left[1 + \left(1 + \frac{1}{k}\right)^l\right] - \frac{l}{s}\right\} k^l \leq -3k^l$$

and, consequently, for a large enough $k \geq k_0$

$$\left|r_{m_{k+1}}\left(z^0\right) - r_{m_k}\left(z^0\right)\right| \leq A_1 k^{-3k^l} \left[4\Phi\left(z^0, K\right)\right]^{d_k}, \quad k \geq \max\{k_0, N\}, \quad m_k \geq m'. \tag{10}$$

We note that

$$\left|r_{m_{k+1}}\left(z^0\right) - r_{m_k}\left(z^0\right)\right|^{1/d_k} \leq A_1^{1/d_k} k^{-3\frac{k^l}{d_k}} \left[4\Phi\left(z^0, K\right)\right], \quad k \geq \max\{k_0, N\}, \quad m_k \geq m'.$$

Since, as $k \to \infty$

$$\frac{k^l}{d_k} = \frac{k^l}{k^l + (k+1)^l} \to \frac{1}{2},$$

and

$$\lim_{k \to \infty} A_1^{1/d_k} = \lim_{k \to \infty} \left\{2r^{2n}(k+1)^{2nl} r^{2n(k+1)^l}\right\}^{\frac{1}{k^l+(k+1)^l}} = \lim_{k \to \infty} \left[r^{2n}\right]^{\frac{(k+1)^l}{k^l+(k+1)^l}} = r^n,$$

then for a large enough $k \geq k\left(z^0\right)$ we have the following inequality

$$\left|r_{m_{k+1}}\left(z^0\right) - r_{m_k}\left(z^0\right)\right|^{1/d_k} < \frac{4r^n \Phi\left(z^0, K\right)}{k}.$$

As can be seen from the proof, for $z \in \bar{B} \setminus \bigcup_{j \geq N} Q_j$, where $B = B(0, R)$, this inequality is satisfied uniformly, i.e., there is an integer $k_0(R, N) \in \mathbf{N}$ such that

$$\left| r_{m_{k+1}}(z) - r_{m_k}(z) \right|^{1/d_k} < \frac{A_3(R)}{k}, \quad k \geq k_0(R, N), \ z \in \bar{B} \setminus \bigcup_{j \geq N} Q_j, \ A_3(R) - const.$$

(11)

Step 5. Since the set $Z = \bigcup_{k=1}^{\infty} \{z \in \mathbf{C}^n : q_k(z) = 0\}$ is pluripolar, there is a plurisubharmonic function $v \in psh(\mathbf{C}^n)$ such that $v \not\equiv -\infty$, $v|_Z = -\infty$. We put $Z_N = \{z \in \mathbf{C}^n : v(z) < 1/N\}$ and

$$S_N = \left\{ \bigcup_{k \geq N} Q_k \right\} \bigcup Z_N, \quad F_N = \mathbf{C}^n \setminus S_N, \quad S = \bigcap_{N=1}^{\infty} S_N, \quad F = \mathbf{C}^n \setminus S = \bigcup_{N=1}^{\infty} F_N.$$

From the $\sigma-$ subadditivity of the capacity,

$$\tau \left(S_N \bigcap B \right) \leq \sum_{k \geq N} \tau \left(Q_k \bigcap B \right) + \tau \left(Z_N \bigcap B \right) \to 0.$$

as $N \to \infty$. Hence S is a pluripolar set. Outside of S the series (4) converges rapidly and its sum defines a function $\tilde{f}(z)$ in $\mathbf{C}^n \setminus S$.

According to (11), for fixed $N \in \mathbf{N}$ there is $k_0(R, N) \in \mathbf{N}$ such that we have a uniform estimation

$$\left| r_{m_k}(z) - \tilde{f}(z) \right| = \left| \sum_{t \geq k} \left(r_{m_{t+1}}(z) - r_{m_t}(z) \right) \right| \leq \sum_{t \geq k} \left| r_{m_{t+1}}(z) - r_{m_t}(z) \right| \leq$$

$$\leq \sum_{t \geq k} \left(\frac{A_3(R)}{t} \right)^{d_t} = A_4 \left(\frac{A_3(R)}{k} \right)^{d_k}, \quad k \geq k_0(R, N), \ A_4 - const$$

(12)

on a compact set $\bar{B} \bigcap F_N$. Hence

$$\left| r_{m_k}(z) - \tilde{f}(z) \right|^{1/m_k} \leq A_4^{1/m_k} \left(\frac{A_3(R)}{k} \right)^2 \leq \frac{A_5(R)}{k^2},$$

(13)

where $z \in \bar{B} \bigcap F_N$, $k \geq k_0(R, N)$, $A_5(R) - const.$

Step 6. The above constructed function $\tilde{f}(z)$ is fine-analytic in $\mathbf{C}^n$. Indeed, the set $\mathbf{C}^n \setminus \bigcup_N F_N = S$ is pluripolar. For arbitrary ball $B \subset \mathbf{C}^n$ and for arbitrary $N \in \mathbf{N}$ the set $B \bigcap F_N$ is compact, and according to (13) we have

$$\lim_{k \to \infty} \left\| r_{m_k}(z) - \tilde{f}(z) \right\|_{F_N}^{1/m_k} = 0.$$

It means that $\tilde{f}|_{F_N} \in R(F_N)$ and $\tilde{f}(z)$ is fine-analytic in $\mathbf{C}^n$. *The proof of Theorem 1 is complete.*

Proof of Theorem 2

We fix a radius $R > 0$, and a whole number $l \geq 8s$. As in Theorem 1 we consider a subsequence $m_k = k^l$. Then as it was shown in the proof of the Theorem 1 the sum of the series (2)

$$\tilde{f}(z) = r_{m_1}(z) + \sum_{k=1}^{\infty} \left[r_{m_{k+1}}(z) - r_{m_k}(z) \right]$$

is a fine-analytic function in $\mathbf{C}^n$. Moreover, according to (13) there is a Borel-type F_σ set $F = \bigcup_{N=1}^{\infty} F_N$ such that $F_N \subset F_{N+1}$, $\mathbf{C}^n \backslash F$ is pluripolar and for a fixed $N \in \mathbf{N}$ on the set $F_{N,R} = B(0, R) \bigcap F_N$ we have a uniform estimate $\left| r_{m_k}(z) - \tilde{f}(z) \right|^{1/m_k} \leq \frac{C_1}{k^2}$, $k \geq k_0$, $C_1 - const.$

We estimate $\left| r_m(z) - \tilde{f}(z) \right|$ for arbitrary fixed $m \in \mathbf{N}$, with $m_k \leq m < m_{k+1}$. For $z \in F_{N,R}$, $k \geq k_0$

$$\left| r_m(z) - \tilde{f}(z) \right| \leq \left| r_{m_k}(z) - \tilde{f}(z) \right| + \left| r_m(z) - r_{m_k}(z) \right| \leq$$

$$\leq \left(\frac{C_1}{k^2} \right)^{m_k} + \left| r_m(z) - r_{m_k}(z) \right|, \tag{14}$$

where

$$r_m(z) - r_{m_k}(z) = \frac{p_m(z) q_{m_k}(z) - p_{m_k}(z) q_m(z)}{q_m(z) q_{m_k}(z)}.$$

Hence the following polynomials of degree $d_k = m_k + m$ admit the estimate

$$\left\| p_m(z) q_{m_k}(z) - p_{m_k}(z) q_m(z) \right\|_K \leq \left\| r_m(z) - r_{m_k}(z) \right\|_K \cdot \left\| q_m(z) q_{m_k}(z) \right\|_K.$$

Since $\left\| r_m(z) - r_{m_k}(z) \right\|_K \leq 2 [m_k]^{-m_k/s}$, $\| q_m(z) \|_K \leq \left[m r^{m+1} \right]^n$ and $\left\| q_{m_k}(z) \right\|_K \leq \left[m_k r^{m_k+1} \right]^n$ (see proof of Theorem 1), we have

$$\left\| p_m(z) q_{m_k}(z) - p_{m_k}(z) q_m(z) \right\|_K \leq C_2 \frac{1}{[m_k]^{m_k/s}}, \quad m_k \geq m', \tag{15}$$

where $C_2 = 2 r^{2n} m_{k+1}^{2n} r^{2nm_{k+1}} = 2 r^{2n} (k+1)^{2nl} r^{2n(k+1)^l}$.

Using the Bernstein-Walsh inequality we get

$$\left| p_m(z) q_{m_k}(z) - p_{m_k}(z) q_m(z) \right| \leq C_2 \frac{1}{[m_k]^{m_k/s}} \cdot [\Phi(z, K)]^{d_k}, \quad z \in \mathbf{C}^n,$$

and from this we obtain

$$\left|r_m(z) - r_{m_k}(z)\right| \leq C_2 \frac{1}{[m_k]^{m_k/s}} \cdot \frac{[\Phi(z, K)]^{d_k}}{\left|q_{m_k}(z)\, q_m(z)\right|}, \quad z \in \mathbf{C}^n, \ m_k \geq m'.$$

If we remove two open sets from $\mathbf{C}^n$

$$Q_m = \left\{ z \in \mathbf{C} : \ |q_m(z)|^{1/m} < \frac{1}{k^2} \right\}, \quad Q_{m_k} = \left\{ z \in \mathbf{C} : \ |q_{m_k}(z)|^{1/m_k} < \frac{1}{k^2} \right\},$$

then, for large enough $k \geq k_0$ we have

$$\left|r_m(z) - r_{m_k}(z)\right|^{1/d_k} \leq C_2^{1/d_k} k^{-3\frac{k^l}{d_k}} \cdot [4\Phi(z, K)], \quad z \in B(0, R) / \left(Q_m \bigcup Q_{m_k} \right),$$

where

$$C_2^{1/d_k} \to r^n, \quad \frac{k^l}{d_k} \to \frac{1}{2},$$

as $k \to \infty$(see Step 4 of the proof of Theorem 1). Consequently,

$$\left|r_m(z) - r_{m_k}(z)\right|^{1/d_k} \leq \frac{C_3}{k}, \quad k \geq k_0, \ z \in B(0, R) / \left(Q_m \bigcup Q_{m_k} \right), \qquad (16)$$

where C_3 is a constant. From this and according to (14) for $k \geq k_0, \ m_k \leq m < m_{k+1}$ we have

$$\left|r_m(z) - \tilde{f}(z)\right| \leq \left(\frac{C_1}{k^2}\right)^{m_k} + \left(\frac{C_3}{k}\right)^{d_k} \leq \left(\frac{C_4}{k^2}\right)^{m_k}, \quad z \in F_{N,R}/\left(Q_m \bigcup Q_{m_k} \right), \tag{17}$$

where C_4 is a constant.

We note that the capacities of these intersections will be estimated as follows:

$$\tau\left(B(0, R) \bigcap Q_m \right) \leq \frac{C_5}{k^2}, \quad \tau\left(B(0, R) \bigcap Q_{m_k} \right) \leq \frac{C_5}{k^2}, \qquad (18)$$

where C_5 is a constant. Thus,

$$\tau\left(B(0, R) \bigcap \left(Q_m \bigcup Q_{m_k} \right) \right) \leq \frac{2C_5}{k^2}$$

then the capacity of the set

$$B(0, R) \setminus \left[F_{N,R} \setminus \left(Q_m \bigcup Q_{m_k} \right) \right] = [B(0, R) \setminus F_{N,R}] \bigcup \left[B(0, R) \bigcap \left(Q_m \bigcup Q_{m_k} \right) \right]$$

does not exceed the value

$$\tau\left(B\left(0,R\right)\setminus F_{N,R}\right) + \tau\left(B\left(0,R\right)\bigcap\left(Q_m\bigcup Q_{m_k}\right)\right) \le \delta\left(N\right) + \frac{2C_5}{k^2},$$

where $\delta\left(N\right) \to 0$ as $N \to \infty$. From this we obtain

$$\lim_{m\to\infty} \tau\left\{z \in B\left(0,R\right): \left|r_m\left(z\right) - \tilde{f}\left(z\right)\right|^{1/m} > \varepsilon\right\} \le \delta\left(N\right)$$

for any $\varepsilon > 0$. The value $\delta\left(N\right)$ can be made as small as we want which implies that $\{r_m\left(z\right)\}$ rapidly converges to $\tilde{f}\left(z\right)$ in capacity. *The proof of Theorem 2 is complete.*

Corollary 1 *If a function $f\left(z\right) \in R_K$ is of finite order, then it defines a unique fine-analytic function $\tilde{f}\left(z\right)$ in $\mathbf{C}^n$. In other words, if the sequences of rational functions $\left\{r_m^1\left(z\right)\right\}$ and $\left\{r_m^2\left(z\right)\right\}$ converge on K to the function $f\left(z\right)$ with speed of finite order, then the fine analytic functions $\tilde{f}_1\left(z\right)$, $\tilde{f}_2\left(z\right)$ defined by them coincide almost everywhere with respect to the capacity in $\mathbf{C}^n$.*

Indeed, if we take a sequence with alternating sequence members

$$\left\{r_m^1\left(z\right)\right\}, \left\{r_m^2\left(z\right)\right\}: \{R_m\left(z\right)\} = \left\{r_1^1\left(z\right), r_1^2\left(z\right), r_2^1\left(z\right), r_2^2\left(z\right), r_3^1\left(z\right), r_3^2\left(z\right), \ldots\right\},$$

then the sequence $\{R_m\left(z\right)\}$ also rapidly converges to $f\left(z\right)$, with speed of the finite order, on K. Consequently, according to Theorem 1 it defines fine-analytic function $\tilde{F}\left(z\right)$ in the entire space $\mathbf{C}^n$. It is easy to see that $\tilde{F}$ almost everywhere coincides in capacity with both $\tilde{f}_1$ and $\tilde{f}_2$, as required.

Proof of Theorem 3

We assume that the function $f \in R_K$ is of finite order and $\{r_m\left(z\right)\}$ is a sequence of rational functions such that $\|f - r_m\|_K^{1/m} \le \frac{1}{m^{1/s}}$, $m \ge m'$, $s < \infty$. Then according to Theorem 1 (see inequality (13)), there is a subsequence $\{m_k\} \subset \mathbf{N}$ and there is a sequence of closed sets $F_N \subset \mathbf{C}^n$, $F_N \subset F_{N+1}$: $\mathbf{C}^n\setminus \bigcup_{N=1}^\infty F_N$ is pluripolar and for any ball $B = B\left(0,R\right)$ and for any number $N \in \mathbf{N}$ we have

$$\left|r_{m_k}\left(z\right) - \tilde{f}\left(z\right)\right|^{1/m_k} \le \frac{A_5\left(R\right)}{k^2}, \quad z \in \bar{B}\bigcap F_N, \quad k \ge k_0\left(R,N\right). \tag{19}$$

If $r_m^*\left(z\right)$, $\deg r_m^* \le m$, is a sequence of rational functions which rapidly converges to the function $f\left(z\right)$ on the compact set K (it is not required that the convergence of $r_m^*\left(z\right)$ is of finite order), then

$$\left|r_{m_k}^*\left(z\right) - \tilde{f}\left(z\right)\right| \le \left|r_{m_k}\left(z\right) - \tilde{f}\left(z\right)\right| + \left|r_{m_k}\left(z\right) - r_{m_k}^*\left(z\right)\right| \le$$

$$\le \left(\frac{A_5\left(R\right)}{k^2}\right)^{m_k} + \left|r_{m_k}\left(z\right) - r_{m_k}^*\left(z\right)\right|, \quad z \in \bar{B}\bigcap F_N, \quad k \ge k_0\left(R,N\right). \tag{20}$$

Now we use rapidly convergence of the sequences $\{r_m(z)\}$ and $r_{m_k}^*(z)$ to the same function $f(z)$ in the compact set K

$$\|f - r_m\|_K^{1/m} \leq \frac{1}{m^{1/s}}, \quad m \geq m', \quad \|f - r_m^*\|_K^{1/m} \leq \sigma_m,$$

where $s < \infty$, $\sigma_m \to 0$ as $m \to \infty$. Accordingly,

$$\left\|r_{m_k}(z) - r_{m_k}^*(z)\right\|_K \leq \left\|r_{m_k}(z) - f(z)\right\|_K + \left\|r_{m_k}^*(z) - f(z)\right\|_K \leq \frac{1}{m_k^{m_k/s}} + \sigma_{m_k}^{m_k},$$

where $m_k > m'$. Consequently, if

$$r_{m_k}(z) - r_{m_k}^*(z) = \frac{p_{m_k}(z)\, q_{m_k}^*(z) - p_{m_k}^*(z)\, q_{m_k}(z)}{q_{m_k}(z)\, q_{m_k}^*(z)},$$

then a polynomial of degree $2m_k$ admits an estimate (we assume a priori that the maximal modulus of the coefficients of polynomials $q_{m_k}(z)$, $q_{m_k}^*(z)$ are equal to 1 and $K \subset \{|z_1| < r, |z_2| < r, \ldots, |z_n| < r\}$)

$$\left\| p_{m_k}(z)\, q_{m_k}^*(z) - p_{m_k}^*(z)\, q_{m_k}(z) \right\|_K \leq$$

$$\leq \left(\frac{1}{m_k^{m_k/s}} + \sigma_{m_k}^{m_k}\right) \left\| q_{m_k}(z)\, q_{m_k}^*(z) \right\|_K \leq r^{2n} m_k^{2n} r^{2nm_k} \delta_{m_k}^{m_k}, \quad m_k > m', \quad (21)$$

where $\delta_m = \max\left\{\frac{1}{m^{1/s}}, \sigma_m\right\}$.

Now we use the Bernstein-Walsh inequality as in the proof of Theorem 1, to obtain

$$\left| p_{m_k}(z)\, q_{m_k}^*(z) - p_{m_k}^*(z)\, q_{m_k}(z) \right| \leq r^{2n} m_k^{2n} r^{2nm_k} \delta_{m_k}^{m_k} \cdot [\Phi(z, K)]^{2m_k},$$

for $z \in \mathbf{C}^n$, $m_k > m'$. Hence

$$\left| r_{m_k}(z) - r_{m_k}^*(z) \right| \leq r^{2n} m_k^{2n} r^{2nm_k} \delta_{m_k}^{m_k} \cdot \frac{[\Phi(z, K)]^{2m_k}}{\left| q_{m_k}(z)\, q_{m_k}^*(z) \right|}, \quad z \in \mathbf{C}^n, \ m_k \geq m'.$$

If we fix a closed ball $\bar{B} = \bar{B}(0, R)$ and we remove two open sets

$$Q_{m_k} = \left\{ \left| q_{m_k}(z) \right|^{1/m_k} < \delta_{m_k}^{1/4} \right\}, \quad Q_{m_k}^* = \left\{ \left| q_{m_k}^*(z) \right|^{1/m_k} < \delta_{m_k}^{1/4} \right\}$$

from $\bar{B}$, then for a large enough $k \geq k_0$ we obtain

$$\left| r_{m_k}(z) - r_{m_k}^*(z) \right|^{1/m_k} \leq A_6 \delta_{m_k}^{1/2}, \quad z \in \bar{B}/\left(Q_{m_k} \bigcup Q_{m_k}^*\right), \quad A_6 - \text{const}.$$

According to (20) we have

$$\left| r_{m_k}^*(z) - \tilde{f}(z) \right| \le \left(\frac{A_5(R)}{k^2} \right)^{m_k} + \left(A_6 \delta_{m_k}^{1/2} \right)^{m_k} \le A_7 \gamma_k^{m_k},$$

$$z \in \bar{B} \bigcap F_N / \left(Q_{m_k} \bigcup Q_{m_k}^* \right), \quad k \ge k_0, \quad A_7 - const,$$

where $\gamma_k = \max \left\{ \frac{1}{k^2}, \delta_{m_k} \right\} \to 0$ as $k \to \infty$. This inequality shows that the sequence $r_{m_k}^*(z)$ converges rapidly to $\tilde{f}(z)$ on $\bar{B} \bigcap F_N / \left(Q_{m_k} \bigcup Q_{m_k}^* \right)$.

Since $\tau \left(\left(\bar{B} \bigcap Q_{m_k} \right) \bigcup \left(\bar{B} \bigcap Q_{m_k}^* \right) \right) \le 2 A_8 \delta_{m_k}^{1/4} \to 0$, as $k \to \infty$ and A_8 is a constant, we can prove that the sequence $r_m^*(z)$ converges rapidly in capacity to the function $\tilde{f}$ as in Theorem 2. *The proof of Theorem 3 is complete.*

Acknowledgements I would like to express my warm thanks to the referee of this paper for numerous corrections, in particular for the wonderful Remark 1.

References

1. Bedford, E., Taylor, B.A.: Fine topology, Šilov boundary and $(dd^c)^n$. J. Funct. Anal. **72**(2), 225–251 (1987)
2. Brelot, M.: On Topologies and Boundaries in Potential Theory. Lecture Notes in Mathematics, vol. 175. Springer, Berlin (1971)
3. Doob, J.L.: Classical Potential Theory and its Probabilistic Counterpart. Springer, Berlin (1984)
4. Edigarian, A., Wiegerinck, J.: The pluripolar hull of the graph of a holomorphic function with polar singularities. Indiana Univ. Math. J. **52**, 1663–1680 (2003)
5. Edigarian, A., Wiegerinck, J.: Determination of the pluripolar hull of graphs of certain holomorphic functions. Ann. Inst. Fourier (Genoble) **54**, 2085–2104 (2004)
6. Edigarian, A., El Marzguioui, S., Wiegerinck, J.: The image of a finely holomorphic map is pluripolar. Ann. Pol. Math. **97**, 137–149 (2010)
7. Edlund, T.: Complete pluripolar curves and graphs. Ann. Pol. Math. **84**, 75–86 (2004)
8. Edlund, T., Jöricke, B.: The pluripolar hull of a graph and fine analytic continuation. Ark. Mat. **44**, 39–60 (2006)
9. El Kadiri, M., Fuglede, B., Wiegerinck, J.: Plurisubharmonic and holomorphic functions relative to the plurifine topology. J. Math. Anal. Appl. **381**(2), 706–723 (2011)
10. Fuglede, B.: Fine topology and finely holomorphic functions. In: Proceedings of the 18th Scandinavian Congress of Mathematicians Aarhus 1980, Birkhäuser, pp. 22–38 (1981)
11. Fuglede, B.: Finely holomorphic functions. A survey. Rev. Roum. Math. Pures Appl. **33**, 283–295 (1988)
12. Gonchar, A.A.: A local condition for single-valuedness of analytic functions. Math. Sbornic **89**(131), 148–164 (1972). English Transl. Math. USSR Sbornic **18** (1972)
13. Gonchar, A.A.: On the convergence of Pade approximants. Math. Sbornic **92**(134), 152–164 (1973). English Transl. Math. USSR Sbornic **21** (1973)
14. Gonchar, A.A.: A local condition for single-valuedness of analytic functions of several variables. Math. Sbornic **93**(135), 296–313 (1974). English Transl. Math. USSR Sbornic **22** (1974)
15. Imomkulov, S.A.: On holomorphic continuation of functions defined on a pencil of boundary of complex lines. Izv. RAN. Ser. Mat. **69**(2), 125–144 (2005)
16. Landkof, N.S.: Foundations of Modern Potential theory, Nauka, Moscow, 1966, English Transl. Springer, Berlin (1972)

17. Levenberg, N., Martin, G., Poletsky, E.A.: Analytic disks and pluripolar sets. Indiana Univ. Math. J. **41**, 515–532 (1992)
18. Levenberg, N., Poletsky, E.A.: Pluripolar hulls. Mich. Math. J. **46**, 151–162 (1999)
19. Levin, B.Y.: Distribution of zeros of entire functions, GITTL, Moscow, 1956, English Transl. American Mathematical Society, Providence (1964)
20. El Marzguioui, S., Wiegerinck, J.: Continuity properties of finely plurisubharmonic functions. Indiana Univ. Math J. **59**(5), 1793–1800 (2010)
21. Sadullaev, A.S.: Plurisubharmonic measures and capacities on complex manifolds. Russ. Math. Surv. **36**(4), 61–119 (1981)
22. Sadullaev, A.S.: Rational approximation and pluripolar sets. Math. USSR Sbornic **47**(1), 91–113 (1984)
23. Sadullaev, A.S.: A criterion for rapid rational approximation in $\mathbf{C}^n$. Math. USSR Sbornic **53**(1), 271–281 (1986)
24. Sadullaev, A.S.: Plurisubharmonic functions. Several Complex Variables II. Encyclopaedia of Mathematical Sciences, pp. 59–106. Springer, Berlin (1994)
25. Sadullaev, A.S.: Pluripotential Theory: Applications. Palmarium Akademic Publishing, Germany (2012). (in Russian)
26. Sadullaev, A.S., Chirka, E.M.: On continuation of functions with polar singularities. Math. USSR Sbornic **60**(2), 377–384 (1988)
27. Sadullaev, A.S., Imomkulov, S.A.: Continuation of holomorphic and pluriharmonic functions with thin singularities on parallel sections. Proc. Steklov Inst. Math. **253**(2), 144–159 (2006)
28. Sadullaev, A.S., Ibragimov, Z.S.: Class *R* of Gonchar and fine-analytic functions. Russ. Math. Surv. (accepted)
29. Wiegerinck, J.: Plurifine potential theory. Ann. Pol. Math. **106**, 275–292 (2012)

On a Generalised Samarskii-Ionkin Type Problem for the Poisson Equation

Aishabibi A. Dukenbayeva, Makhmud A. Sadybekov
and Nurgissa A. Yessirkegenov

Abstract In this paper, we consider a generalised form of the Samarskii-Ionkin type boundary value problem for the Poisson equation in the disk and show its well-posedness. The possibility of separation of variables is justified. We obtain an explicit form of the Green function for this problem and an integral representation of the solution.

Keywords Poisson equation · Periodic boundary conditions · Samarskii-Ionkin type boundary value problem · Green function

1 Introduction

The Dirichlet and Neumann boundary value problems play an important role in the theory of harmonic functions and have been widely studied in literatures. Another important problem, called periodic boundary value problem, arises when one considers the problem in a one-dimensional case or in a multidimensional parallelepiped. For the first time, in [8, 9], a new class of the boundary value problems for the Poisson equation in a multidimensional ball $\Omega \subset R^n$, $(k = 1, 2)$ was introduced:

The problem P_k. *Find a solution of the Poisson equation*

$$-\Delta u = f(x), \ x \in \Omega$$

satisfying the following periodic boundary conditions

A. A. Dukenbayeva · M. A. Sadybekov (✉)
Institute of Mathematics and Mathematical Modeling, Almaty, Kazakhstan
e-mail: sadybekov@math.kz

A. A. Dukenbayeva
e-mail: dukenbayeva@math.kz

N. A. Yessirkegenov
Department of Mathematics, Imperial College London, London, UK
e-mail: n.yessirkegenov15@imperial.ac.uk

© Springer Nature Switzerland AG 2018

Z. Ibragimov et al. (eds.), *Algebra, Complex Analysis, and Pluripotential Theory*,
Springer Proceedings in Mathematics & Statistics 264,
https://doi.org/10.1007/978-3-030-01144-4_17

$$u(x) - (-1)^k u(x^*) = \tau(x), \quad x \in \partial\Omega_+,$$

$$\frac{\partial u}{\partial r}(x) + (-1)^k \frac{\partial u}{\partial r}(x^*) = v(x), \quad x \in \partial\Omega_+.$$

Here as usual $\partial\Omega_+$ is a part of the sphere $\partial\Omega$, for which $x_1 \geq 0$; each point $x = (x_1, x_2, \ldots, x_n) \in \Omega$ is matched by its "opposite" point $x^* = (-x_1, \alpha_2 x_2, \ldots, \alpha_n x_n) \in \Omega$, where the indices $\alpha_j \in \{-1, 1\}$, $j = 2, \ldots, n$. Clearly, if $x \in \partial\Omega_+$, then $x^* \in \partial\Omega_-$.

These problems are analogous to the classical periodic boundary value problems. In [8], the well-posedness of these problems are investigated. They also showed the existence and uniqueness of the solution of the problem P_1, while the solution of the problem P_2 is unique up to a constant term and exists if the necessary condition of the well-posedness is held. The uniqueness and existence are shown by using the extremum principle and Green's function, respectively. In [9], they considered the problem P_k in the two dimensional case and showed the possibility of using the method of separation of variables. Moreover, in this case, the self-adjointness of these problems and its spectral properties are studied.

We also refer to Ammari and Kang [1, Chs. 2, 8], and Kapanadze, Mishuris, and Pesetskaya [4, 5] for a large variety of applications of the periodic boundary value problems, and refer to [6, 7] for the problems in periodic domains.

If we turn to the non-classical problems, then one of the most popular problems is the Samarskii-Ionkin problem, arisen in connection with the study of the processes occurring in the plasma in the 70s of the last century by physicists (see e.g. [2, 3]). In [10], an analog of the Samarskii-Ionkin type boundary value problem for the Poisson equation in a disk is considered. The authors proved the well-posedness of this problem and constructed its Green's function in the explicit form. Then, in [11], the spectral properties of this problem were studied. They constructed the eigenfunctions of this problem and proved its completeness.

We also note that in [12, 13], there were investigated problems generalizing the periodic problem P_k.

In this paper we continue the research from [10, 11] devoted to the well-posedness and spectral properties of the Samarskii-Ionkin type boundary value problem for the Poisson equation. Namely, we consider a generalised form of the Samarskii-Ionkin type boundary value problem for the Poisson equation in the disk and investigate its well-posedness.

2 Formulation of the Problem

Let $\Omega = \{z = (x, y) = x + iy \in C : |z| < 1\}$ be the unit disk, $r = |z|$ and $\varphi = \arctan(y/x)$. Consider the following problem $(S - I)_\alpha$.

Problem $(S - I)_\alpha$. *Find a solution of the Poisson equation*

$$- \Delta u = f(z), \quad |z| < 1 \tag{1}$$

satisfying the following boundary conditions

$$u(1, \varphi) - \alpha u(1, 2\pi - \varphi) = \tau(\varphi), \quad 0 \le \varphi \le \pi, \tag{2}$$

$$\frac{\partial u}{\partial r}(1, \varphi) - \frac{\partial u}{\partial r}(1, 2\pi - \varphi) = \nu(\varphi), \quad 0 \le \varphi \le \pi \tag{3}$$

or

$$\frac{\partial u}{\partial r}(1, \varphi) + \frac{\partial u}{\partial r}(1, 2\pi - \varphi) = \nu(\varphi), \quad 0 \le \varphi \le \pi, \tag{4}$$

where $\alpha \in \mathbb{R}$, $f(z) \in C^{\gamma}(\overline{\Omega})$, $\tau(\varphi) \in C^{1+\gamma}[0, \pi]$ and $\nu(\varphi) \in C^{\gamma}[0, \pi], 0 < \gamma < 1$.

It is clear that a necessary condition for the existence of the solution of the problem (1)–(3) in the class $C^1(\overline{\Omega})$ is given by the following condition:

$$\nu(0) = \nu(\pi) = 0. \tag{5}$$

The antiperiodic boundary value problem (1)–(3) for $\alpha = -1$ and the periodic boundary value problem (1)–(2), (4) for $\alpha = 1$ were studied in [8, 9]. When $\alpha = 0$, these problems were investigated in [10, 11].

We say that the problem (1)–(3) is a non-noetherian problem, if the homogeneous problem (1)–(3) has infinite number of independent solutions.

3 On the Uniqueness of the Solution of the Problem $(S - I)_\alpha$

In this section, we discuss the uniqueness of the solution of the problem $(S - I)_\alpha$ and application of the extremum principle.

Theorem 1 (a) Let $\alpha \ne 1$. Then the solution of the problem (1)–(3) is unique in the class of continuous functions in $\overline{\Omega}$.

(b) Let $\alpha = 1$. Then the problem (1)–(3) is a non-noetherian problem. In particular, the homogeneous problem (1)–(3) has infinite number of independent solutions in the following form

$$u_n(z) = r^n a_n \cos n\varphi, \quad n = 0, 1, \ldots, \quad 0 \le \varphi \le 2\pi, \quad 0 < r \le 1$$

for arbitrary complex numbers a_n, where $r = |z|$ and $\varphi = \arctan(y/x)$.

Proof (a) Suppose that there are two functions $u_1(r, \varphi)$ and $u_2(r, \varphi)$ satisfying the conditions of the problem (1)–(3). We show that the function $u(r, \varphi) = u_1(r, \varphi) - u_2(r, \varphi)$ is equal to zero. It is obvious that the function $u(r, \varphi)$ is harmonic and satisfies the homogeneous conditions (2) and (3):

$$u(1, \varphi) - \alpha u(1, 2\pi - \varphi) = 0, \ 0 \le \varphi \le \pi, \tag{6}$$

$$\frac{\partial u}{\partial r}(1, \varphi) - \frac{\partial u}{\partial r}(1, 2\pi - \varphi) = 0, \ 0 \le \varphi \le \pi. \tag{7}$$

Let us denote $\vartheta(r, \varphi) = u(r, \varphi) - u(r, 2\pi - \varphi)$ for $0 \le \varphi \le 2\pi$. Then it follows that $\Delta\vartheta(r, \varphi) = 0$ and $\frac{\partial\vartheta}{\partial r}(1, \varphi) = 0, 0 \le \varphi \le 2\pi$. Hence, $\vartheta = Const$. Since we have $\vartheta(r, \varphi) = -\vartheta(r, 2\pi - \varphi), 0 \le \varphi \le 2\pi$ from the representation of $\vartheta(r, \varphi)$, one has $u(r, \varphi) = u(r, 2\pi - \varphi)$ for $0 \le \varphi \le 2\pi$. From this and (6), we deduce that

$$(1 - \alpha)u(1, \varphi) = 0, \ \text{and} \ (1 - \alpha)u(1, 2\pi - \varphi) = 0, \ 0 \le \varphi \le \pi.$$

Since we have $\alpha \ne 1$, then $u(1, \varphi) = 0, 0 \le \varphi \le 2\pi$. Taking into account that $u(r, \varphi)$ is a harmonic function, we obtain $u(r, \varphi) = 0$.

(b) Now we consider the case $\alpha = 1$. Let us implement the following auxiliary functions

$$c(r, \varphi) = \frac{1}{2}(u(r, \varphi) + u(r, 2\pi - \varphi)),$$

$$s(r, \varphi) = \frac{1}{2}(u(r, \varphi) - u(r, 2\pi - \varphi)) \tag{8}$$

for $0 \le \varphi \le 2\pi$ and $0 < r \le 1$. It is easy to see that $u(r, \varphi) = c(r, \varphi) + s(r, \varphi)$. Putting this in the homogeneous boundary conditions (2) and (3), one obtains that $s(r, \varphi) = 0$ and $c(r, \varphi)$ has the following form

$$c(z) = c_n(z) = r^n a_n \cos n\varphi, \ n = 0, 1, \dots, \ 0 \le \varphi \le 2\pi, \ 0 < r \le 1$$

for arbitrary complex numbers a_n.

By a similar argument as the one of the proof of Part b of Theorem 1, we obtain the following result:

Theorem 2 *Let $\alpha = -1$. Then the problem* (1), (2) *and* (4) *is a non-noetherian problem. In particular, the homogeneous problem* (1), (2) *and* (4) *has infinite number of independent solutions in the following form*

$$u_n(z) = r^n b_n \sin n\varphi, \ n = 1, 2, \dots, \ 0 \le \varphi \le 2\pi, \ 0 < r \le 1,$$

for arbitrary complex numbers b_n, where $r = |z|$ and $\varphi = \arctan(y/x)$.

4 On the Existence of a Solution of the Problem $(S - I)_\alpha$

Now we analyse the existence of a solution of the problem $(S - I)_\alpha$ and justify the possibility of using the method of separation of variables for solving the formulated problems. Let us first consider the case $f = 0$. In polar coordinates the homogeneous equation (1) has the form

$$r\frac{\partial}{\partial r}\left(r\frac{\partial}{\partial r}u\right) + \frac{\partial^2 u}{\partial \varphi^2} = 0. \tag{9}$$

We seek the regular solutions of the Eq. (9) in the form

$$u(r, \varphi) = \frac{a_0}{2} + \sum_{n=1}^{+\infty} r^n(a_n \cos n\varphi + b_n \sin n\varphi), \tag{10}$$

where $\{a_n\}_{n=0}^{\infty}$ and $\{b_n\}_{n=1}^{\infty}$ are the Fourier coefficients that will be defined later. Noting the representation $u(r, \varphi) = c(r, \varphi) + s(r, \varphi)$ and by a direct calculation, we obtain from the problem (1)–(3) the Dirichlet problem

$$\Delta c(z) = 0, \ z \in \Omega; \ c(1, \varphi) = \frac{\widetilde{\tau}(\varphi)}{1 - \alpha}, \ 0 \le \varphi \le 2\pi, \tag{11}$$

for the finction $c(z)$, and the Neumann problem

$$\Delta s(z) = 0, \ z \in \Omega; \ \frac{\partial s}{\partial r}(1, \varphi) = \frac{\widetilde{v}(\varphi)}{2}, \ 0 \le \varphi \le 2\pi, \tag{12}$$

for the function $s(z)$, where

$$\widetilde{\tau}(\varphi) = \begin{cases} \tau(\varphi) - (1 + \alpha)s(1, \varphi), \ 0 \le \varphi \le \pi \\ \tau(2\pi - \varphi) - (1 + \alpha)s(1, 2\pi - \varphi), \ \pi \le \varphi \le 2\pi, \end{cases} \tag{13}$$

and

$$\widetilde{v}(\varphi) = \begin{cases} v(\varphi), \ 0 \le \varphi \le \pi \\ -v(2\pi - \varphi), \ \pi \le \varphi \le 2\pi. \end{cases} \tag{14}$$

As in the classical case, one can find the solutions of (11) and (12) in the series form, and taking into account $u(r, \varphi) = c(r, \varphi) + s(r, \varphi)$ and (10), we compute the coefficients in (10):

$$a_n = \frac{2}{\pi}\left(\frac{\tau(\varphi) - (1 + \alpha)\sum_{k=1}^{+\infty} b_k \sin k\varphi}{1 - \alpha}, \cos n\varphi\right)_{L_2(0,\pi)},$$

$$b_n = \frac{1}{\pi n}(\nu(\varphi), \sin n\varphi)_{L_2(0,\pi)}, n = 0, 1, 2, \dots.$$

Thus, one obtains the formal solution of the problem (9), (2), and (3).

From condition (5) we have that $\nu(\pi) = 0$. Therefore, from (14) we find that $\widetilde{\nu}(\pi) = 0$. Hence, by virtue of $\nu(\varphi) \in C^\gamma[0, \pi]$, we have $\widetilde{\nu}(\varphi) \in C^\gamma[0, 2\pi]$. From (5) we also have that $\nu(0) = 0$. Therefore $\widetilde{\nu}(0) = \widetilde{\nu}(2\pi) = 0$. Consequently, $\widetilde{\nu}(\varphi) \in C^\gamma(\partial\Omega)$.

Hence we also have that the solution of the Neumann problem (12) belongs to the class $s(r, \varphi) \in C^{1+\gamma}(\overline{\Omega})$. Therefore, $s(1, \varphi) \in C^{1+\gamma}(\partial\Omega)$. Accordingly, since $\tau(\varphi) \in C^{1+\gamma}[0, \pi]$, from (13) we find that $\widetilde{\tau}(\varphi) \in C^{1+\gamma}[0, \pi] \cap C^{1+\gamma}[\pi, 2\pi]$ and $\widetilde{\tau}(\pi - 0) = \widetilde{\tau}(\pi + 0)$. Hence we have $\widetilde{\tau}(\varphi) \in C^\gamma[0, 2\pi]$. Since $\widetilde{\tau}(0) = \widetilde{\tau}(2\pi)$, we finally obtain that $\widetilde{\tau}(\varphi) \in C^\gamma(\partial\Omega)$.

Thus we have shown that $\widetilde{\nu}(\varphi), \widetilde{\tau}(\varphi) \in C^\gamma(\partial\Omega)$. Therefore, the justification of the uniform convergence of the obtained series (up to the boundary) and their term by term differentiation are conducted in the same way as for the classical Dirichlet and Neumann problems, since the solution $u(z)$ is the sum of the solutions of the Dirichlet (11) and Neuman (12) problems.

Summing up the obtained series as in [17, p.314], we obtain an explicit form of the solution

$$u(r, \varphi) = \frac{1}{2\pi(1 - \alpha)}$$

$$\times \int_0^\pi \tau(\varphi_1)\left(\frac{1 - r^2}{1 - 2r\cos(\varphi_1 + \varphi) + r^2} + \frac{1 - r^2}{1 - 2r\cos(\varphi_1 - \varphi) + r^2}\right)d\varphi_1$$

$$+\frac{1}{2\pi}\int_0^\pi \nu(\varphi_1)\ln\sqrt{\frac{1 - 2r\cos(\varphi_1 + \varphi) + r^2}{1 - 2r\cos(\varphi_1 - \varphi) + r^2}}d\varphi_1$$

$$-\frac{1 + \alpha}{4\pi^2(1 - \alpha)}$$

$$\times \int_0^\pi \nu(\varphi_2)\int_0^\pi \left(\frac{1 - r^2}{1 - 2r\cos(\varphi_1 + \varphi) + r^2} + \frac{1 - r^2}{1 - 2r\cos(\varphi_1 - \varphi) + r^2}\right)$$

$$\times \ln\sqrt{\frac{1 - \cos(\varphi_2 + \varphi_1)}{1 - \cos(\varphi_2 - \varphi_1)}}d\varphi_1 d\varphi_2. \tag{15}$$

Thus, the following theorem is proven.

Theorem 3 *Let* $\alpha \neq 1$, $f(z) = 0$, $\tau(\varphi) \in C^{1+\gamma}[0, \pi]$, $\nu(\varphi) \in C^\gamma[0, \pi]$, $0 < \gamma < 1$, *and let the condition (5) hold. Then there exists a unique solution of (1)–(3), and it is represented in the form (15).*

Remark 1 We note that when $v(\varphi) = 0$, the solution of the homogeneous problem (1)–(3) has the following property

$$u(r, \varphi) = u(r, 2\pi - \varphi).$$

5 On the Green's Function of the Problem $(S - I)_\alpha$

Various construction methods of the Green's function of the Dirichlet problem exist. The Green's function was constructed in the explicit form for many types of the domain. But for the Neumann problem in the multidimensional spaces the construction of the Green's function was an open problem. In our previous works we built the Green's functions of the classical Neumann problem of arbitrary dimension: in the unit ball [14] and for the exterior of the unit ball [15, 16].

In this section, we construct the Green's function of the investigated problem and give an integral representation of the solution. We introduce the auxiliary functions

$$\vartheta(z) = \frac{1}{2}(u(z) + u(\bar{z})), \ \omega(z) = \frac{1}{2}(u(z) - u(\bar{z})), \ \bar{z} = (x, -y).$$

It is obvious that $u(z) = \vartheta(z) + \omega(z)$. By a direct calculation we find the problems for these functions: the function ϑ is a solution of the Dirichlet problem

$$- \Delta\vartheta(z) = f_+(z), \ z \in \Omega; \ \vartheta(1, \varphi) = \frac{\widetilde{\tau}(\varphi)}{1 - \alpha}, \ 0 \le \varphi \le 2\pi, \qquad (16)$$

where $\alpha \ne 1$ and the function $\omega(z)$ is a solution of the Neumann problem

$$- \Delta\omega(z) = f_-(z), \ z \in \Omega; \ \frac{\partial\omega}{\partial r}(1, \varphi) = \frac{\widetilde{v}(\varphi)}{2}, \ 0 \le \varphi \le 2\pi. \qquad (17)$$

Here $f_+(z) = (f(z) + f(\bar{z}))/2, \ f_-(z) = (f(z) - f(\bar{z}))/2,$

$$\widetilde{\tau}(\varphi) = \begin{cases} \tau(\varphi) - (1 + \alpha)\omega(1, \varphi), \ 0 \le \varphi \le \pi, \\ \tau(2\pi - \varphi) - (1 + \alpha)\omega(1, 2\pi - \varphi), \ \pi \le \varphi \le 2\pi, \end{cases}$$

and

$$\widetilde{v}(\varphi) = \begin{cases} v(\varphi), \ 0 \le \varphi \le \pi, \\ -v(2\pi - \varphi), \ \pi \le \varphi \le 2\pi. \end{cases}$$

The Dirichlet problem (16) has a unique solution. It is represented as follows

$$\vartheta(z) = \int\int_\Omega G_D(z, \varsigma)f_+(\varsigma)d\varsigma - \frac{1}{1 - \alpha} \int_0^{2\pi} \frac{\partial G_D(z, \varsigma)}{\partial n_\varsigma}\Big|_{|\varsigma|=1}\widetilde{\tau}(\varphi_1)d\varphi_1, \quad (18)$$

where $G_D(z, \zeta)$ is a Green's function of the Dirichlet problem. It is easy to see that the function $\vartheta(z)$ has the symmetric property $\vartheta(z) = \vartheta(\bar{z})$.

Since

$$\int\int_\Omega f_-(\zeta)d\zeta = 0 \quad \text{and} \quad \int_0^{2\pi} \tilde{v}(\varphi)d\varphi = 0$$

are held, we can use the criterion of the solution existence of the Neumann problem (17). Its solution is unique up to an arbitrary constant and can be represented in the following form

$$\omega(z) = \int\int_\Omega G_N(z, \zeta) f_-(\zeta)d\zeta + \frac{1}{2}\int_0^{2\pi} G_N(z, \zeta)\big|_{|\zeta|=1}\tilde{v}(\varphi_1)d\varphi_1 + C_1, \quad (19)$$

where $G_N(z, \zeta)$ is the Green's function of the Neumann problem. We note that the function $\omega(z)$ has the symmetric property $\omega(z) = -\omega(\bar{z})$ if and only if $C_1 = 0$. Thus, we further assume that this condition is fulfilled.

The Green's functions of the Dirichlet and Neumann problems are well known. In the unit disk they have the form

$$G_D(z, \zeta) = \frac{1}{2\pi}\left(-\ln|z - \zeta| + \ln\left|z|\zeta| - \frac{\zeta}{|\zeta|}\right|\right),$$

$$G_N(z, \zeta) = \frac{1}{2\pi}\left(-\ln|z - \zeta| - \ln\left|z|\zeta| - \frac{\zeta}{|\zeta|}\right|\right) + C.$$

By substituting them in (18) and (19), after elementary transformations, we obtain a representation of the solution (1)–(3), and the constant C from the representation of the Green's function of the Neumann problem disappears. The smoothness of the solution of the problem follows from the smoothness of the solutions of the corresponding Dirichlet (16) and Neumann problems (17). Thus, the following theorem is proven.

Theorem 4 *Let $\alpha \neq 1$, $f(z) \in C^\gamma(\overline{\Omega})$, $\tau(\varphi) \in C^{1+\gamma}[0, \pi]$; $v(\varphi) \in C^\gamma[o, \pi]$, $0 < \gamma < 1$, and let the condition (5) hold. Then there exists a unique solution of (1)–(3), and it is represented in the form:*

$$u(z) = \int\int_\Omega G(z, z_2) f(z_2)dz_2 - 2\int_0^\pi \frac{\partial G(z, z_2)}{\partial n_{z_2}}\Big|_{|z_2|=1}\tau(\varphi_2)d\varphi_2$$

$$+ \int_0^\pi G(z, z_2)\big|_{|z_2|=1}v(\varphi_2)d\varphi_2.$$

Here $G(z, z_2)$ is the Green's function of the problem (1)–(3), which has the form

$$G(z, z_2) =$$

$$= \frac{G_D(z, z_2) + G_D(z, |z_2|, 2\pi - \varphi_2) + G_N(z, z_2) - G_N(z, |z_2|, 2\pi - \varphi_2)}{2}$$

$$+ \frac{1 + \alpha}{2(1 - \alpha)} \left\{ \int_0^\pi (G_N(z_1, z_2) - G_N(z_1, |z_2|, 2\pi - \varphi_2)) \right.$$

$$\left. \times \left(\frac{\partial G_D(z, z_1)}{\partial n_{z_1}} + \frac{\partial G_D(z, |z_1|, 2\pi - \varphi_1)}{\partial n_{z_1}} \right) \Big|_{|z_1|=1} d\varphi_1 \right\}.$$

Acknowledgements The authors were supported in parts by the MES RK grant AP051333271 as well as by the MES RK target grant BR05236656.

References

1. Ammari, H., Kang, H.: Polarization and Moment Tensors. Applied Mathematical Sciences, vol. 162. Springer, New York (2007)
2. Ionkin, N.I.: Solution of a boundary-value problem in heat conduction theory with a nonclassical boundary condition. Differentsial'nye Uravneniya [Differ. Equ.] **13**(2), 294–304 (1977)
3. Ionkin, N.I., Moiseev, E.I.: A problem for a heat equation with two-point boundary conditions. Differentsial'nye Uravneniya [Differ. Equ.] **15**(7), 1284–1295 (1979)
4. Kapanadze, D., Mishuris, G., Pesetskaya, E.: Exact solution of a nonlinear heat conduction problem in a doubly periodic 2D composite material. Arch. Mech. (Arch. Mech. Stos.) **67**, 157–178 (2015)
5. Kapanadze, D., Mishuris, G., Pesetskaya, E.: Improved algorithm for analytical solution of the heat conduction problem in doubly periodic 2D composite materials. Complex Var. Elliptic Equ. **60**, 1–23 (2015)
6. Lanza de Cristoforis, M., Musolino, P.: A singularly perturbed nonlinear Robin problem in a periodically perforated domain: a functional analytic approach. Complex Var. Elliptic Equ. **58**, 511–536 (2013)
7. Musolino, P.: A singularly perturbed Dirichlet problem for the Laplace operator in a periodically perforated domain: a functional analytic approach. Math. Methods Appl. Sci. **35**, 334–349 (2012)
8. Sadybekov, M.A., Turmetov, B.Kh.: On analogues of periodic boundary value problems for the Laplace operator in a ball. Eurasian Math. J. **3**(1), 143–146 (2012)
9. Sadybekov, M.A., Turmetov, B.Kh.: On an analog of periodic boundary value problems for the Poisson equation in the disk. Differ. Equ. **50**(2), 268–273 (2014)
10. Sadybekov, M.A., Torebek, B.T., Yessirkegenov, N.A.: On an analog of Samarskii-Ionkin type boundary value problem for the Poisson equation in the disk. AIP Conf. Proc. **1676**, 020035 (2015)
11. Sadybekov, M.A., Yessirkegenov, N.A.: Spectral properties of a Laplace operator with Samarskii-Ionkin type boundary conditions in a disk. AIP Conf. Proc. **1759**, 020139 (2016)
12. Sadybekov, M.A., Turmetov, B.Kh., Torebek, B.T.: Solvability of nonlocal boundary-value problems for the Laplace equation in the ball. Electron. J. Differ. Equ. **2014**, 1–14 (2014)
13. Sadybekov, M.A., Yessirkegenov, N.A.: On a generalised Samarskii-Ionkin type problem for the Poisson equation. Kazakh Math. J. **17**(1), 115–116 (2017)
14. Sadybekov, M.A., Torebek, B.T., Turmetov, B.Kh.: Representation of Green's function of the Neumann problem for a multi-dimensional ball. Complex Var. Elliptic Equ. **61**(1), 104–123 (2016)

15. Sadybekov, M.A., Torebek, B.T., Turmetov, B.Kh.: On an explicit form of the Green function of the third boundary value problem for the Poisson equation in a circle. AIP Conf. Proc. **1611**, 255–260 (2014)
16. Sadybekov, M.A., Torebek, B.T., Turmetov, B.Kh.: Representation of the Green's function of the exterior Neumann problem for the Laplace operator. Sib. Math. J. **58**(1), 153–158 (2016)
17. Tihonov, A.N., Samarskii, A.A.: Equations of Mathematical Physics, Courier Corporation (2013)

Ergodicity Properties of p-Adic (2, 1)-Rational Dynamical Systems with Unique Fixed Point

Iskandar A. Sattarov

Abstract We consider a family of (2, 1)-rational functions given on the set of p-adic field Q_p. Each such function has a unique fixed point. We study ergodicity properties of the dynamical systems generated by (2, 1)-rational functions. For each such function we describe all possible invariant spheres. We characterize ergodicity of each p-adic dynamical system with respect to Haar measure reduced on each invariant sphere. In particular, we found an invariant spheres on which the dynamical system is ergodic and on all other invariant spheres the dynamical systems are not ergodic.

Keywords p-adic numbers · Rational function · Dynamical system · Ergodic

1 Introduction

In this paper we will give some results concerning discrete dynamical systems defined over the p-adic field Q_p. In [2] the behavior of a p-adic dynamical system $f(x) = x^n$ in the fields of p-adic numbers Q_p and complex p-adic numbers C_p was investigated. Some ergodic properties of that dynamical system have been considered in [3].

In [1, 5, 7–9] the trajectories of some rational p-adic dynamical systems in the complex p-adic field C_p were studied. In particular, it is proved in [5] that such kind of dynamical system is not ergodic on a unit sphere with respect to the Haar measure.

In this paper for a class of (2, 1)-rational functions we study the ergodicity properties of the dynamical systems on the spheres of p-adic numbers Q_p. For each such function we describe all possible invariant spheres. We characterize ergodicity of each 2-adic dynamical system with respect to the Haar measure restricted on each invariant sphere. In particular, we found an invariant sphere on which the dynamical system is ergodic and on all other invariant spheres the dynamical systems are not ergodic.

I. A. Sattarov (✉)
Institute of mathematics, 81, Mirzo Ulug'bek str., 100170 Tashkent, Uzbekistan
e-mail: sattarovi-a@yandex.ru

© Springer Nature Switzerland AG 2018
Z. Ibragimov et al. (eds.), *Algebra, Complex Analysis, and Pluripotential Theory*,
Springer Proceedings in Mathematics & Statistics 264,
https://doi.org/10.1007/978-3-030-01144-4_18

1.1 p-Adic Numbers

Let Q be the field of rational numbers. The greatest common divisor of the positive integers n and m is denotes by (n, m). Let p be a fixed prime number. Every rational number $x \neq 0$ can be represented in the form $x = p^{\gamma(x)} \frac{n}{m}$, where $\gamma(x), n \in Z, m$ is a positive integer, $(p, n) = 1$, $(p, m) = 1$.

The p-adic norm of x is given by

$$|x|_p = \begin{cases} p^{-\gamma(x)}, & \text{for } x \neq 0, \\ 0, & \text{for } x = 0. \end{cases}$$

It has the following properties:

(1) $|x|_p \geq 0$ and $|x|_p = 0$ if and only if $x = 0$,
(2) $|xy|_p = |x|_p |y|_p$,
(3) the strong triangle inequality holds

$$|x + y|_p \leq \max\{|x|_p, |y|_p\},$$

(3.1) if $|x|_p \neq |y|_p$ then $|x + y|_p = \max\{|x|_p, |y|_p\}$,
(3.2) if $|x|_p = |y|_p$ then $|x + y|_p \leq |x|_p$.
Thus $|x|_p$ is a non-Archimedean norm.

The completion of Q with respect to the p-adic norm defines the p-adic field which is denoted by Q_p.

For any $a \in Q_p$ and $r > 0$ denote

$$U_r(a) = \{x \in Q_p : |x - a|_p \leq r\}, \quad V_r(a) = \{x \in Q_p : |x - a|_p < r\},$$

$$S_r(a) = \{x \in Q_p : |x - a|_p = r\}.$$

A function $f : U_r(a) \to Q_p$ is said to be *analytic* if it can be represented by

$$f(x) = \sum_{n=0}^{\infty} f_n (x - a)^n, \quad f_n \in Q_p,$$

which converges uniformly on the ball $U_r(a)$.

1.2 Dynamical Systems in Q_p

In this section we recall some known facts concerning dynamical systems (f, U) in Q_p, where $f : x \in U \to f(x) \in U$ is an analytic function and $U = U_r(a)$ or Q_p.

Now let $f : U \to U$ be an analytic function. Denote $x_n = f^n(x_0)$, where $x_0 \in U$ and $f^n(x) = \underbrace{f \circ \cdots \circ f}_{n}(x)$.

Let us first recall some the standard terminology of the theory of dynamical systems (see for example [6]). If $f(x_0) = x_0$ then x_0 is called a *fixed point*. A fixed point x_0 is called an *attractor* if there exists a neighborhood $V(x_0)$ of x_0 such that for all points $y \in V(x_0)$ it holds that $\lim_{n \to \infty} y_n = x_0$. If x_0 is an attractor then its *basin of attraction* is

$$A(x_0) = \{y \in Q_p : y_n \to x_0, \ n \to \infty\}.$$

A fixed point x_0 is called *a repeller* if there exists a neighborhood $V(x_0)$ of x_0 such that $|f(x) - x_0|_p > |x - x_0|_p$ for $x \in V(x_0)$, $x \neq x_0$.

A set V is called invariant for f, if $f(V) \subset V$.

Let x_0 be a fixed point of a function f. The ball $V_r(x_0)$ (contained in U) is said to be a *Siegel disk* if each sphere $S_\rho(x_0)$, $\rho < r$ is an invariant sphere for f. The union of all Siegel disks with the center at x_0 is said to *a maximum Siegel disk* and is denoted by $SI(x_0)$.

Let x_0 be a fixed point of an analytic function f. Put

$$\lambda = \frac{d}{dx} f(x_0).$$

The point x_0 is *attractive* if $0 \leq |\lambda|_p < 1$, *indifferent* if $|\lambda|_p = 1$, and *repelling* if $|\lambda|_p > 1$.

2 Ergodicity of (2, 1)-Rational p-Adic Dynamical Systems

A function is called an (n, m)-rational function if and only if it can be written in the form $f(x) = \frac{P_n(x)}{T_m(x)}$, where $P_n(x)$ and $T_m(x)$ are polynomial functions (without a common factor) with degree n and m respectively ($T_m(x)$ is a non zero polynomial).

In this paper we consider the ergodicity properties of the dynamical system associated with the (2, 1)-rational function $f : Q_p \to Q_p$ defined by

$$f(x) = \frac{x^2 + ax + b}{x + c}, \quad a, b, c \in Q_p, \ a \neq c, \ c^2 - ac + b \neq 0 \tag{1}$$

where $x \neq \hat{x} := -c$.

Note that $f(x)$ has the unique fixed point $x_0 = \frac{b}{c-a}$.

For any $x \in Q_p$, $x \neq \hat{x}$, by simple calculations we get

$$|f(x) - x_0|_p = |x - x_0|_p \cdot \frac{|(x - x_0) + (x_0 + a)|_p}{|(x - x_0) + (x_0 + c)|_p}. \tag{2}$$

Denote
$$\mathscr{P} = \{x \in Q_p : \exists n \in N \cup \{0\}, \ f^n(x) = \hat{x}\},$$

$$\alpha = |x_0 + a|_p \quad \text{and} \quad \beta = |x_0 + c|_p.$$

Consider the following functions (see [1]):
For $0 \leq \alpha < \beta$ define the function $\varphi_{\alpha,\beta} : [0, +\infty) \to [0, +\infty)$ by

$$\varphi_{\alpha,\beta}(r) = \begin{cases} \frac{\alpha}{\beta} r, & \text{if } r < \alpha, \\ \alpha^*, & \text{if } r = \alpha, \\ \frac{r^2}{\beta}, & \text{if } \alpha < r < \beta, \\ \beta^*, & \text{if } r = \beta, \\ r, & \text{if } r > \beta, \end{cases}$$

where α^* and β^* are some given numbers with $\alpha^* \leq \frac{\alpha^2}{\beta}$, $\beta^* \geq \beta$.
For $0 \leq \beta < \alpha$ define the function $\phi_{\alpha,\beta} : [0, +\infty) \to [0, +\infty)$ by

$$\phi_{\alpha,\beta}(r) = \begin{cases} \frac{\alpha}{\beta} r, & \text{if } r < \beta, \\ \beta', & \text{if } r = \beta, \\ \alpha, & \text{if } \beta < r < \alpha, \\ \alpha', & \text{if } r = \alpha, \\ r, & \text{if } r > \alpha, \end{cases}$$

where α' and β' some positive numbers with $\alpha' \leq \alpha \leq \beta'$.
For $\alpha \geq 0$ we define the function $\psi_\alpha : [0, +\infty) \to [0, +\infty)$ by

$$\psi_\alpha(r) = \begin{cases} r, & \text{if } r \neq \alpha, \\ \hat{\alpha}, & \text{if } r = \alpha, \end{cases}$$

where $\hat{\alpha}$ is a given number.
Using the formula (2) we easily get the following:

Lemma 1 *If $x \in S_r(x_0)$, then the following formula holds*

$$|f^n(x) - x_0|_p = \begin{cases} \varphi^n_{\alpha,\beta}(r), & \text{if } \alpha < \beta, \\ \phi^n_{\alpha,\beta}(r), & \text{if } \alpha > \beta, \quad n \geq 1. \\ \psi^n_\alpha(r), & \text{if } \alpha = \beta. \end{cases}$$

Thus the p-adic dynamical system $(f, Q_p \setminus \mathscr{P})$ is related to the real dynamical systems generated by $\varphi_{\alpha,\beta}$, $\phi_{\alpha,\beta}$ and ψ_α.

Theorem 1 ([1]) *The p-adic dynamical system generated by f has the following properties:*

1. *If $\alpha < \beta$, then $A(x_0) = V_\beta(x_0)$ and the spheres $S_r(x_0)$ are invariant with respect to f for all $r > \beta$.*

2. If $\alpha = \beta$, then $SI(x_0) = V_\beta(x_0)$ and the spheres $S_r(x_0)$ are invariant with respect to f for all $r > \beta$.
3. If $\alpha > \beta$, then the inequality $|f(x) - x_0|_p > |x - x_0|_p$ satisfies for $x \in V_\alpha(x_0)$, $x \neq x_0$ and the spheres $S_r(x_0)$ are invariant with respect to f for all $r > \alpha$.
4. $f(S_r(x_0)) \not\subset S_r(x_0)$ for any $r \in \{\alpha, \beta\}$.
5. 5.1. If $\alpha \leq \beta$, then $\mathcal{P} \subset S_\beta(x_0)$.
   5.2. If $\alpha > \beta$, then $\mathcal{P} \subset U_\alpha(x_0)$.

We define the following sets

$$I_1 = \{r : r > \max\{\alpha, \beta\}\} \text{ if } \alpha \neq \beta;$$

$$I_2 = \{r : r \neq \beta\} \text{ if } \alpha = \beta;$$

and we denote $I = I_1$ if $\alpha \neq \beta$ and $I = I_2$ if $\alpha = \beta$.

Using the Theorem 1 we get the following

Corollary 1 *The sphere $S_r(x_0)$ is invariant for f if and only if $r \in I$.*

In this paper we are interested to study ergodicity properties of the dynamical system on the invariant sphere.

Lemma 2 *For every closed ball $U_\rho(s) \subset S_r(x_0)$, $r \in I$ the following equality holds*

$$f(U_\rho(s)) = U_\rho(f(s)).$$

Proof From inclusion $U_\rho(s) \subset S_r(x_0)$ we get $|s - x_0|_p = r$.
Let $x \in U_\rho(s)$, i.e. $|x - s|_p \leq \rho$, then

$$|f(x) - f(s)|_p = |x - s|_p \cdot \frac{|(s - x_0)(x - x_0) + (x_0 + c)[(x - x_0) + (s - x_0)] + (x_0 + c)(x_0 + a)|_p}{|[(x - x_0) + (x_0 + c)][(s - x_0) + (x_0 + c)]|_p}. \quad (3)$$

We have $|x - x_0|_p = r$, because $x \in U_\rho(s) \subset S_r(x_0)$. Consequently,

$$|f(x) - f(s)|_p = |x - s|_p \cdot \frac{\max\{r^2, \beta r, \alpha\beta\}}{(\max\{r, \beta\})^2}.$$

If $r \in I_1$, then $\max\{r^2, \beta r, \alpha\beta\} = r^2$ and $\max\{r, \beta\} = r$. Using these equalities by (3) we get $|f(x) - f(s)|_p = |x - s|_p \leq \rho$.
If $\alpha = \beta$, then $r \in I_2$. Consequently, $r < \beta$ or $r > \beta$.
If $r < \beta$, then $\max\{r^2, \beta r, \alpha\beta\} = \beta^2$ and $\max\{r, \beta\} = \beta$. Then we get $|f(x) - f(s)|_p = |x - s|_p \leq \rho$.
If $r > \beta$, then $\max\{r^2, \beta r, \alpha\beta\} = r^2$ and $\max\{r, \beta\} = r$. Consequently $|f(x) - f(s)|_p = |x - s|_p \leq \rho$. This completes the proof.

Recall that $S_r(x_0)$ is invariant with respect to f iff $r \in I$.

Lemma 3 *If $x \in S_r(x_0)$, where $r \in I$, then*

$$|f(x) - x|_p = \begin{cases} \frac{r|a-c|_p}{\beta}, & \text{if } 0 < r < \beta = \alpha \\ |a - c|_p, & \text{if } r > \beta = \alpha \\ \max\{\alpha, \beta\}, & \text{if } r \in I_1. \end{cases}$$

Proof Since, if $\alpha \neq \beta$, then $|a - c|_p = |(x_0 + a) - (x_0 + c)|_p = \max\{\alpha, \beta\}$.
Thus

$$|f(x) - x|_p = \left| \frac{(a - c)(x - x_0)}{(x - x_0) + (x_0 + c)} \right|_p = \begin{cases} \frac{|a-c|_p r}{\beta}, & \text{if } 0 < r < \beta = \alpha \\ |a - c|_p, & \text{if } r > \beta = \alpha \\ \max\{\alpha, \beta\}, & \text{if } r \in I_1. \end{cases}$$

The proof is completed.

By Lemma 3 we see that $|f(x) - x|_p$ depends on r, but does not depend on $x \in S_r(x_0)$ itself, therefore, we define $\rho(r) = |f(x) - x|_p$, if $x \in S_r(x_0)$. Then the following theorem holds as Theorem 11 in [8].

Theorem 2 *If $s \in S_r(x_0)$, $r \in I$ then*

1. For any $n \geq 1$ the following equality holds

$$|f^{n+1}(s) - f^n(s)|_p = \rho(r). \tag{4}$$

2. $f(U_{\rho(r)}(s)) = U_{\rho(r)}(s)$.
3. If for some $\theta > 0$ the ball $U_\theta(s) \subset S_r(x_0)$ is invariant for f, then

$$\theta \geq \rho(r).$$

For each $r \in I$ consider a measurable space $(S_r(x_0), \mathcal{B})$, here $\mathcal{B}$ is the algebra generated by closed subsets of $S_r(x_0)$. Every element of $\mathcal{B}$ is a union of some balls $U_\rho(s) \subset S_r(x_0)$.

Let $\bar{\mu}$ be the Haar measure such that $\bar{\mu}(U_\rho(s)) = \rho$.

Note that $S_r(x_0) = U_r(x_0) \setminus U_{\frac{r}{p}}(x_0)$. So, we have $\bar{\mu}(S_r(x_0)) = r(1 - \frac{1}{p})$.

We consider the normalized Haar measure on $S_r(x_0)$:

$$\mu(U_\rho(s)) = \frac{\bar{\mu}(U_\rho(s))}{\bar{\mu}(S_r(x_0))} = \frac{p\rho}{(p-1)r}.$$

By Lemma 2 we conclude that f preserves the measure μ, i.e.

$$\mu(f(U_\rho(s))) = \mu(U_\rho(s)).$$

Consider the dynamical system (X, T, μ), where $T : X \to X$ is a measure preserving transformation, and μ is a measure. We say that the dynamical system is *ergodic* if for every invariant set V we have either $\mu(V) = 0$ or $\mu(V) = 1$ (see [10]).

2.1 Case $p \geq 3$

Theorem 3 *Let $p \geq 3$ and μ be the normalized Haar measure. Then the dynamical system $(S_r(x_0), f, \mu)$ is not ergodic if one of the following conditions holds:*

1. $r \in I_1$.
2. $r \in I_2$, $|a - c|_p < \beta$.
3. $r \in I_2$, $|a - c|_p = \beta$ and $r > \beta$.

Proof Under the conditions of the theorem, by Lemma 3, it follows that $\rho(r) < r$, i.e. $p\rho(r) \leq r$. Now by Theorem 2, we have that $\rho(r)$ is the minimal radius of the invariant ball in $S_r(x_0)$. Then

$$0 < \mu(U_{\rho(r)}(s)) \leq \frac{1}{p-1},$$

and $(S_r(x_0), f, \mu)$ is not ergodic. The proof is completed.

2.2 Case $p = 2$

Note that $Z_2 = \{x \in Q_2 : |x|_2 \leq 1\}$. So we have $1 + 2Z_2 = S_1(0)$. The following theorem gives a criterion of the rational functions which send $S_1(0)$ to itself:

Theorem 4 ([4]) *Let $f, g : 1 + 2Z_2 \rightarrow 1 + 2Z_2$ be polynomials whose coefficients are 2-adic integers. Set $f(x) = \sum_i a_i x^i$, $g(x) = \sum_i b_i x^i$, and*

$$A_1 = \sum_{i \, odd} a_i, \quad A_2 = \sum_{i \, even} a_i, \quad B_1 = \sum_{i \, odd} b_i, \quad B_2 = \sum_{i \, even} b_i.$$

The rational function $R = \frac{f}{g}$ is ergodic if and only if one of the following situations occurs:

1. $A_1 = 1(mod \ 4)$, $A_2 = 2(mod \ 4)$, $B_1 = 0(mod \ 4)$, and $B_2 = 1(mod \ 4)$.
2. $A_1 = 3(mod \ 4)$, $A_2 = 2(mod \ 4)$, $B_1 = 0(mod \ 4)$, and $B_2 = 3(mod \ 4)$.
3. $A_1 = 1(mod \ 4)$, $A_2 = 0(mod \ 4)$, $B_1 = 2(mod \ 4)$, and $B_2 = 1(mod \ 4)$.
4. $A_1 = 3(mod \ 4)$, $A_2 = 0(mod \ 4)$, $B_1 = 2(mod \ 4)$, and $B_2 = 3(mod \ 4)$.
5. *One of the previous cases with f and g interchanged.*

But, in this paper we will study ergodicity of the dynamical system $(S_r(x_0), f, \mu)$ for any $r \in I$. For this purpose we can not use Theorem 4 directly, because the radius of the sphere is arbitrary and the center is not 0.

Let $r = p^l$ and a function $f : S_{p^l}(x_0) \rightarrow S_{p^l}(x_0)$ be given. Denote $f \circ g = f(g(t))$.

Consider $x = g(t) = p^{-l}t + x_0$, $t = g^{-1}(x) = p^l(x - x_0)$ then it is easy to see that $f \circ g : S_1(0) \rightarrow S_{p^l}(x_0)$. Consequently, $g^{-1} \circ f \circ g : S_1(0) \rightarrow S_1(0)$.

Let $\mathcal{B}$ (resp. $\mathcal{B}_1$) be the algebra generated by closed subsets of $S_{p^l}(x_0)$ (resp. $S_1(0)$), and μ (resp. μ_1) be normalized Haar measure on $\mathcal{B}$ (resp. $\mathcal{B}_1$).

Theorem 5 ([8]) *The dynamical system $(S_{p^l}(x_0), f, \mu)$ is ergodic if and only if $(S_1(0), g^{-1} \circ f \circ g, \mu_1)$ is ergodic.*

Remark 1 We note that Theorem 4 gives a criterion of ergodicity for rational functions defined on the sphere with fixed radius ($=1$). Our Theorem 5 allows us to use Theorem 4 for the spheres with an arbitrary radius.

Now using the above mentioned results for $f(x) = \frac{x^2+ax+b}{x+c}$, when $p = 2$ and $f : S_r(x_0) \to S_r(x_0)$ we prove the following theorem.

Theorem 6 *If $p = 2$, then the dynamical system $(S_r(x_0), f, \mu)$ is ergodic iff $\alpha \neq \beta$ and $r = 2 \max\{\alpha, \beta\}$.*

Proof Let $r = 2^l, \alpha = 2^q$ and $\beta = 2^m$. Since $\alpha = |x_0 + a|_2$ and $\beta = |x_0 + c|_2$, then we have $x_0 + a \in 2^{-q}(1 + 2Z_2)$ and $x_0 + c \in 2^{-m}(1 + 2Z_2)$.

In $f : S_{2^l}(x_0) \to S_{2^l}(x_0)$ we change x by $x = g(t) = 2^{-l}t + x_0$. We note that $x \in S_{2^l}(x_0)$, then $|x - x_0|_2 = 2^l|t|_2 = 2^l$, $|t|_2 = 1$ and the function $g^{-1}(f(g(t)))$: $S_1(0) \to S_1(0)$ has the following form

$$g^{-1}(f(g(t))) = \frac{t^2 + 2^l(x_0 + a)t}{t + 2^l(x_0 + c)}. \tag{5}$$

For the numerator of (5) we have

$$|t^2|_2 = |t|_2 = 1, \quad |2^l(x_0 + a)t|_2 = 2^{q-l} \quad \text{and} \quad |2^l(x_0 + c)|_2 = 2^{m-l}.$$

If $r \in I_1$, then $l > q$ and $l > m$. Consequently,

$$t^2 + 2^l(x_0 + a)t =: \gamma_1(t), \quad \text{is such that} \quad \gamma_1 : 1 + 2Z_2 \to 1 + 2Z_2$$

and

$$t + 2^l(x_0 + c) =: \gamma_2(t) \quad \text{is such that} \quad \gamma_2 : 1 + 2Z_2 \to 1 + 2Z_2.$$

Let $r \in I_2$, i.e. $q = m$. Then $l > m$ or $l < m$. If $l > m$, then

$$\gamma_1, \gamma_2 : 1 + 2Z_2 \to 1 + 2Z_2.$$

If $l < m$, then we can write (5) the following form

$$g^{-1}(f(g(t))) = \frac{\frac{t^2}{2^l(x_0+a)} + t}{\frac{t}{2^l(x_0+a)} + \frac{x_0+c}{x_0+a}}. \tag{6}$$

For the numerator of (6) we have

$$|t|_2 = \left| \frac{x_0 + c}{x_0 + a} \right|_2 = 1, \text{ and } \left| \frac{t^2}{2^l(x_0 + a)} \right|_2 = \left| \frac{t}{2^l(x_0 + a)} \right|_2 = 2^{l-m} \leq \frac{1}{2}.$$

Consequently,

$$\frac{t^2}{2^l(x_0 + a)} + t =: \delta_1(t), \text{ is such that } \delta_1 : 1 + 2Z_2 \to 1 + 2Z_2$$

and

$$\frac{t}{2^l(x_0 + a)} + \frac{x_0 + c}{x_0 + a} =: \delta_2(t) \text{ is such that } \delta_2 : 1 + 2Z_2 \to 1 + 2Z_2.$$

Hence the function (5) satisfies all conditions of Theorem 4, therefore using this theorem we have

$$A_1 = 1, \quad A_2 = 2^l(x_0 + c), \quad B_1 = 2^l(x_0 + a) \text{ and } B_2 = 1.$$

Moreover,

$$A_1 = 1 (\mathrm{mod}\, 4), \quad A_2 \in 2^{l-m}(1 + 2Z_2), \quad B_1 \in 2^{l-q}(1 + 2Z_2) \text{ and } B_2 = 1 (\mathrm{mod}\, 4).$$

By this relations and Theorem 4 we get

$$A_2 = 0 (\mathrm{mod}\, 4) \text{ and } B_1 = 2 (\mathrm{mod}\, 4), \text{ iff } l - m \geq 2 \text{ and } l - q = 1$$

or

$$A_2 = 2 (\mathrm{mod}\, 4) \text{ and } B_1 = 0 (\mathrm{mod}\, 4), \text{ iff } l - m = 1 \text{ and } l - q \geq 2.$$

Therefore we conclude that the dynamical system $(S_1(0), g^{-1} \circ f \circ g, \mu_1)$ is ergodic if and only if $q > m$ and $l = q + 1$ or $q < m$ and $l = m + 1$, i.e. $\alpha > \beta$ and $r = 2\alpha$ or $\alpha < \beta$ and $r = 2\beta$. Consequently, by Theorem 5, $(S_r(x_0), f, \mu)$ is ergodic iff $\alpha \neq \beta$ and $r = 2 \max\{\alpha, \beta\}$.

Acknowledgements The author expresses his deep gratitude to U. Rozikov for setting up the problem and for the useful suggestions. He also thanks both referees for helpful comments. In particular, a suggestion of a referee was helpful to simplify the proof of Theorem 3.

References

1. Albeverio, S., Rozikov, U.A., Sattarov, I.A.: p-adic (2,1)-rational dynamical systems. J. Math. Anal. Appl. **398**(2), 553–566 (2013)
2. Albeverio, S., Khrennikov, A., Tirozzi, B., De Smedt, S.: p-adic dynamical systems. Theor. Math. Phys. **114**, 276–287 (1998)

3. Gundlach, V.M., Khrennikov, A., Lindahl, K.O.: On ergodic behavior of p-adic dynamical systems. Infin. Dimens. Anal. Quantum Probab. Relat. Top. **4**, 569–577 (2001)
4. Memić, N.: Characterization of ergodic rational functions on the set 2-adic units. Int. J. Number Theory **13**, 1119–1128 (2017)
5. Mukhamedov, F.M., Rozikov, U.A.: On rational p-adic dynamical systems. Methods Funct. Anal. Topol. **10**(2), 21–31 (2004)
6. Peitgen, H.-O., Jungers, H., Saupe, D.: Chaos Fractals. Springer, Heidelberg (1992)
7. Rozikov, U.A., Sattarov, I.A.: On a non-linear p-adic dynamical system. p-adic numbers, ultrametric. Anal. Appl. **6**(1), 53–64 (2014)
8. Rozikov, U.A., Sattarov, I.A.: p-adic dynamical systems of (2,2)-rational functions with unique fixed point. Chaos, Solitons Fractals **105**, 260–270 (2017)
9. Sattarov, I.A.: p-adic (3,2)-rational dynamical systems. p-Adic Numbers, Ultrametric. Anal. Appl. **7**(1), 39–55 (2015)
10. Walters, P.: An Introduction to Ergodic Theory. Springer, Berlin (1982)

Printed in the United States
By Bookmasters